Swiftwater Rescue

By

Slim Ray

A Manual for the Rescue Professional

To Alex Khanimarian,
without whose help and support this
book could not have been written.

Published by CFS Press, 8 Pelham Road, Asheville, NC 28803.

Book design by Purple Planet Studio

Art Direction: Cynthia Potter and Carlos Steward

Sketches by Slim Ray and Jan Miles

Photographs by Slim Ray, except where otherwise noted

Cover photograph by Rich Shveyda: Los Angeles County Firefighter Eric Featherston is lowered via aerial ladder to rescue a man whose truck was swept away during a flood. The truck's rear window was broken to access the man, who was rescued without injury.

Library of Congress Cataloging-in-Publication Data

Ray, Slim
    Swiftwater rescue: a manual for the rescue professional / Slim Ray.
    p. cm.
    Includes bibliographical references and index.

SECOND EDITION

ISBN 978-0-9882435-0-7

LCCN: 96-84128

    1. Aquatic sports--Safety measures. 2. Rescue work--Handbooks, manuals, etc. 3. Search and rescue operations--Handbooks, manuals, etc. 4. Floods.   I. Title.

GV770.6.R39 1996   797
                           QB196-40045

Printed in U.S.A.

Table of Contents

Introduction to the Second Edition

It seems unbelievable that fifteen years have passed since the first edition of this book rolled off the presses. As the first comprehensive book on swiftwater rescue it soon gained international acceptance and has been used as both a guide and textbook for fire, EMS, and rescue services. While the pace of innovation has slowed somewhat, new equipment and techniques continue to evolve and will be covered in this edition.

One of the developments in swiftwater rescue has been the increased use of helicopters in aquatic rescue in general and swiftwater rescue in particular. Part of this has been due to the greater number of helicopters available, and some to better agency cooperation. An example is the use of National Guard aviation resources here in North Carolina for flood rescue, in large part because of the partnership between the Charlotte Fire Department and the NC National Guard (other notable helo aquatic programs include that of Los Angeles County and Travis County STAR Flight in Texas). The availability of trained crews and specialized equipment has led to a modification of the dictum that helicopters are last means to be employed – depending on the situation they may be the first rescue tool of choice.

Thanks to the publicity surrounding large flood events like Fran and Floyd in North Carolina, Katrina on the Gulf Coast, and Sandy in the Northeast, awareness of the need for flood and swiftwater rescue is higher than ever. Nevertheless more needs to be done, and the recent recession has cut training and equipment budgets to the bone.

The big hurricanes mentioned above have led to another notable development, that of animal rescue in floods and swiftwater. The plight of animals in New Orleans during Katrina stirred national sympathy and led to congressional legislation. Since it's an issue that every agency needs to plan for, I have included a chapter on animal rescue in this edition.

Controversy continues in the rope rescue community about the effects of shock loading and the role of "equalizing" anchors. While disfavored for technical rope work, equalizing anchors still have a place for river rigging and remain in this edition.

In early February 2007 the swiftwater community lost a pioneer and the man who was probably the best instructor ever when Jim Segerstrom died suddenly after a massive stroke. Jim, who had as much to do with the development of swiftwater rescue as a discipline as anyone, is greatly missed by all who knew and loved him.

When reading this book, keep in mind that swiftwater rescue cuts across many different disciplines, including some very technical ones like high-angle rope rescue and scuba. It is impossible to cover all these subjects in detail, and in some, therefore, I have limited the discussion in most cases to ways in which they differ from conventional practice. Do not assume, therefore, that all the answers are in this book. You *cannot* learn swiftwater rescue from a book or a video, no matter how good. There is no substitute for professional training *and* experience. It is your responsibility to get professional, hands-on training in these areas *before* attempting to use them in an actual rescue. All techniques in this book should be thoroughly practiced in a safe environment before they are attempted under field conditions. Even so, there remains a risk of death or serious injury when using the techniques described in this book. *This risk must be accepted by the reader/rescuer.* **Neither the author nor publisher assumes any responsibility or liability for death or personal injury resulting from the use or misuse of information contained in this book.**

Many products are either shown or mentioned in this book. However, this is not a catalogue and appearance herein does not constitute an endorsement of any product, or of its fitness for a particular purpose. Anyone contemplating either buying or using any piece of equipment is urged to contact the manufacturer regarding its limitations and proper use.

Swiftwater rescue is a dynamic field. New techniques, equipment, and materials will become available that are not discussed in this book. It is the responsibility of the reader to remain abreast of new developments in the field.

I'd like to express special thanks to these people for their help and support (and apologize for any I leave out): Dennis Kerrigan; Jim Traub; Barry Miller;

Ken Phillips at Grand Canyon National Park; Horst Fürsattel; Alex Khanimarian; Chris Hawkesworth of Wild-Water; Steve Hudson and Beth Elliott at PMI; Gary Seidel; Don Tayenaka and Larry Collins of Los Angeles County Fire Department; Jim Segerstrom, Barry Edwards, Gaile Lane, Mike & Judy Turnbull of Rescue 3/Rescue Source; Charlie Walbridge; Deborah McMillan and Jeff Yeager, and the American Canoe Association; the Alpiner Kayak-Club; the British Canoe Union; Graham Wardle and Ray Rowe; Nancy Rigg; Mickey Gallagher of Los Angeles County Fire Department Lifeguard Division; Landis Arnold of Wildwasser; Gordon Grant; Tim Rogers; Kirk and Katie Mauthner; Trina Hudson; Bill Perry; Eric Berge; Paul McMinn; Emily King; Les Bechdel; Marshall Parks of the San Diego Lifeguards; Mike Fischesser; Anne Dickison; Tomas and Rebecca Gimenez; Carin Smith; Holly Wallace; Glenn Newell of Rescue Technology; Rex Myers of EMS options; Paul Ratcliffe of Sierra Shutterbug; John Weinel; Paul Ratcliff and Sierra Shutterbug; Jim Frank of CMC Rescue; Gordon and Wendy Dalton; Jeff Redding; Bryan & Michelle Stewart; Casey Ping; Bill Wortell and everyone at Del Norte County Sheriff's SAR; Peter Linde at King County Sheriff's Dept.; Jayson Mellom; John Weaver; and NRS.

Finally, although all or part of the manuscript has been reviewed by many acknowledged experts, the responsibility for any errors or omissions is mine alone.

Slim Ray
Asheville, NC
Spring 2013

Prologue
An Accident on the San Marcos River

by Tom Goynes

Under ordinary conditions the San Marcos River is a gentle little stream fed by large springs in the central Texas college town of San Marcos. Other than downed trees and occasional riffles, the most common obstacles are man-made: low-water bridges and dams, some of which are standing and some of which have fallen. Conditions on Friday, January 3, 1992, however, were anything but ordinary. December had been one of the wettest on record, and the river was running about five feet above normal.

Bruce R. was a trained kayaker who had taken lessons both in Texas and in North Carolina. He had been involved with marathon canoe racing over the years and was known as a strong boater. He owned his own kayak and, as a member of the Austin Fire Department, had taken some swiftwater training. He had made arrangements with friends to paddle on the Guadalupe River on Saturday. The Guadalupe, which flows just to the west of the San Marcos, is a far more technical whitewater river, and floodwater was being released from Canyon Dam at a rate of over 5,000 cfs. Saturday would be the day that many frustrated Texas boaters would finally get to see some big water. The years of drought since 1987 were over and just about every river in Texas was flowing great. Saturday would be the day for the big water, but Bruce had Friday off, so he decided to warm up on the more forgiving San Marcos.

Some have said that Bruce's biggest mistake was paddling floodwater alone, but many would disagree. The latter would argue that the San Marcos is tame enough to allow an experienced kayaker like Bruce a safe trip. Few, however, would argue that Bruce should have ever attempted to shoot Martindale Dam during such high water while alone.

Martindale Dam is a turn-of-the-century structure, created to power a cotton gin. The dam is approximately 12 feet tall and has a sloping (about a 60° angle), rather than vertical drop. At the bottom of the drop there is a concrete ledge extending about 10 feet downstream of the dam. In other words, the water doesn't fall vertically into a deep pool to cause what is known as a hydraulic, or reversal. In higher water than what the river was carrying that Friday, a hydraulic forms because there is then enough surface water above the ledge to allow the backflow of water and the dreaded undertow at the base of the dam. But that Friday there was no hydraulic; what there was, however, was a whole lot of water going down a twelve-foot drop, being turned horizontally by a ledge, and then plowing into the slower water downstream. What there was a huge stopper, or standing wave.

No one knows what went through Bruce's mind as he approached the dam. Certainly he was apprehensive; an eyewitness said he scouted the dam from his kayak and then went back upstream a bit to get up a full head of steam as he made his run. Considering his actions later, one must wonder if Bruce knew the difference between a stopper and a hydraulic. Did he believe that he had to blast over the dam to punch a reversal or to punch a wave? While both situations require a similar approach, self-rescue from each condition, if the run is unsuccessful, is different.

When Bruce hit the wave the force knocked his paddle out of his hands and spun his kayak sideways to the river and into the trough of the big wave. It was four p.m.; for the next two hours Bruce would be hand surfing on a very well-defined wave.

Meanwhile, the eyewitness put in a call to the local fire department, proceeded to retrieve Bruce's paddle, and attempted to get it to him. Due to the current and the fact that Bruce was too far away, these attempts failed.

The Martindale Fire Department is a small, volunteer organization with all of the usual problems associated with such groups. Funding and training is very limited and equipment is scarce. Nonetheless, it consists of a dedicated force of people willing to put their lives on the line for others. I have been a member for over a decade, and have worked under several different chiefs. Since I make my living through canoeing (I own a campground and am an outfitter for the San Marcos and other Texas rivers) past chiefs have relied on me for various river rescues and body searches. More than once I have been called to rescue people from floodwater in the middle of the night.

All of the past chiefs have been aware of my white-water ability, and I assumed that the present chief was no exception. What I had not fully considered was that the drought that had plagued boaters had also squelched the need for river rescues. Basically, the new chief was not aware of the whitewater skill level within his department and available in the community through several canoeing businesses on the river.

The fire department arrived within minutes and began surveying the situation (unfortunately, I was out of town at the time the call first came in and would not be in the area until an hour later). An incident commander—that is, a person selected to be in charge of the particular rescue—was chosen and a plan formu-lated. It consisted of getting a rope across the river upstream of the dam, with the help of some canoers who came paddling up, and then lowering the rope down the dam to Bruce, who could then be pulled to the side, and out of the "hydraulic" that had him trapped.

One of the things I have learned from this inci-dent is that most swiftwater rescue classes, especially those designed for fire departments, don't spend a lot of time differentiating between the different types of dams and the currents they produce. The average firefighter has probably seen footage of dams in flood stage and, although he can't explain exactly how it happens, he knows that such a "drowning machine" will suck in motorboats, rafts, trees, and just about anything else that comes too close. To an average firefighter, a dam is a thing to be avoided, a place of strange undertows and killing currents. Perhaps that is the real reason that Bruce didn't do the thing that most kayakers would have done in his situation—abandon ship. Had he turned over and come out of his boat, or even if he had just popped his sprayskirt and let the kayak fill up with water, the downstream current just below the surface of the wave would have yanked him down the river. It seems possible that Bruce, with his fire department training, believed that he was in such a drowning machine and that the saf-est thing to do was to wait for a rescue to happen.

When the rope was lowered over the dam it got stuck on a post sticking out of the top of the structure near the center of the river. Thus, it was impossible to actually pull the rope to the right or left, but Bruce was able to grab the rope and pull himself toward the right side of the dam. He got close enough to the side that he was flipped by a curling wave that existed where the side eddy entered the trough of the big wave. Unfortunately, he was able to stay in his kayak, and either the current rolled him up or he did a very quick hand roll. This was about the time that I arrived on the scene.

I heard a communication explaining that there was a kayaker stuck in the hydraulic below Martindale Dam. I quickly loaded my fourteen-foot raft, 200 feet of floating rescue rope, several throw bags, paddles, and PFDs (personal flotation devices), and headed for Martindale. I just happened to pull out onto the highway in front of a couple of local kayakers who were returning from a day of boating on a nearby river and they, realizing that I normally don't drive quite so fast or recklessly, decided that something was up and followed me to the scene.

Since I had been specially paged, I assumed that once I arrived on the scene I would be allowed to do whatever needed to be done. In retrospect, I should have found the incident commander, reported in, and explained my plan to him. The plan was fairly simple: I would use the available paddlers, including the two expert kayakers who had followed me. With my long rescue rope tied to the stern (just in case we got too close to the wave ourselves and needed to be pulled back), we would paddle up from downstream of the dam, getting close enough to Bruce to hit him with a throw bag and pull him over the wave and down-stream. By this time it was well after five o'clock and Bruce had been hand surfing for over an hour. Many firefighters and other emergency personnel were on the scene, so it was easy to get a crew of volunteers together to handle the end of my raft rescue rope.

As we approached the dam I immediately heard people on the dam telling us to get back, to get out of the river. I paid these warnings little atten-tion because, after years of river rescue work, I have become accustomed to spectators yelling warnings at me, especially at dams. What I didn't realize was that the person yelling at me was the incident commander. At that time, we were approaching Bruce from the left side of the river where the current was the strongest. I thought the current was the reason we couldn't quite get in close enough for a throwbag; I found out later that our bank crew was instructed to pull us in to shore. In effect, we were paddling and they were pull-ing, and the D-ring was taking all the heat. In retro-spect, the incident commander was thinking "drown-ing machine," and all he knew was that he didn't want to lose a whole raftload of people to this killer hydraulic. I, meanwhile, decided to move to the river right, where we would have a strong upstream current in the form of an eddy that would take us right to the wave and in easy range of Bruce.

We ferried the raft to the right side, deputized a whole new bank crew, and made another attempt. This time we got fairly close to Bruce before I heard my fire chief telling me to back off because a heli-copter was on the way. As the bank crew pulled us back, I looked up at the chief in disbelief and repeated the word, this time in the form of a question, "Helicopter?"

I got out of the raft and ran to the chief to ask him why a helicopter was being used for this situation. He explained that he didn't want to lose a raftload of four men trying to save this one kayaker. I then explained that the situation was simply not that dangerous, but that to bring in a helicopter, especially at dusk, would be. Furthermore, there were well over a hundred spectators around, and considering the power lines just downstream of the dam, the trees on the river right, and the buildings on the river left, a helicopter should be a last resort. I also explained that a kayaker who has been hand surfing for two hours is exhausted and certainly would not be able to hold on to a rope. I was thinking specifically about an incident several years ago when a helicopter dropped a girl to her death who had been in a bus that had been washed into the Guadalupe River. The chief then gave me permission to attempt the rescue.

We made one last run toward Bruce, confident that this time we would be allowed to save him. However, what I
didn't realize was that the chief was not the incident commander of this rescue; it was the assistant chief on the other side of the river, and once again, he called us off and once again our bank crew pulled us back.

It was at this time, about six p.m., in the fading daylight, that the Army MAST helicopter from San Antonio arrived. They landed to survey the situation from the ground and then approached from downstream and hovered just feet above the top of the dam. The man who was lowered from the helicopter was nineteen years old. It was his first rescue mission.

There are some situations where helicopters are excellent rescue vehicles, but this wasn't one of them. In cases where the victim is trapped on top of a car in floodwater or swimming in open seas, a helicopter is the obvious choice. In this case, the best rescue vehicle was a nonmotorized raft. The next best rescue strategy would have been to leave the kayaker alone and assume that eventually he would get tired and swim for it.

The young crewman was on a jungle penetrator, a rig that the kayaker was supposed to end up on as well. The plan was also to run a strap around Bruce to secure him further. But this was not an easy rescue; things started going wrong when the crewman was lowered all the way into the water. Immediately his legs were blasted downstream and he was at the mercy of the river and the winch operator. To compensate for the downstream pull he was pulled back upriver

and lifted, but this resulted in his collision with the dam itself and the swiftly falling water. At some point Bruce, who at 250 pounds outweighed the crewman by 75 pounds, was able to grab the crewman in a sort of bear hug. This resulted in the crewman losing his ability to use his arms either to signal to the helicopter or to secure Bruce to the jungle penetrator.

Meanwhile, the pilot was having his own problems. It seems that when you lower a crewman off one side of your aircraft you have to compensate for the resulting list. The crewman was on the right side, and the pilot had his controls as far to the left as possible. It was at this time that all 250 pounds of Bruce, and the weight of his kayak (he came out of his kayak only after he had the rescuer in his grip), was suddenly applied to the right side of the aircraft. The pilot's only choice was to go up and to go up quickly, otherwise he would slide sideways into the building on the left side of the dam. It is apparently Army policy to get the victim into the helicopter before the helicopter can land. Otherwise, the obvious thing to have done in this situation was to keep Bruce low and lower him as soon as possible to the ground. Nonetheless, the aircraft raised to a level of about fifty feet to clear the trees and then began hoisting Bruce and the crewman upward. The crewman stated that he told Bruce to hang on tight and not to try to get into the helicopter by himself. Whether Bruce heard this statement or not, or whether he was mentally coherent at his point (the cold temperature, his wet condition, and the downdraft of the aircraft might have made him hypothermic), no one knows. At any rate, at the last minute as the helicopter skid came by, Bruce grabbed for it and fell to the ground.

First responders on the scene could find no pulse. CPR was initiated and continued until the ambulance arrived and took Bruce to the hospital. The autopsy revealed several breaks to the spine, a ruptured spleen, and a torn lung, among other problems. It is believed that Bruce died almost instantly.

Towards a Philosophy of Swiftwater Rescue
Chapter 1

Fig. 1.0

Before getting into the specifics of swiftwater rescue, we should first establish a few basic principles—a rescue philosophy—to guide us.

Water rescue is a huge field. It includes swiftwater, lake rescue, surf and offshore rescue, ice rescue, underwater rescue, and much more. A very common mistake is to think that because you are expert in one of these areas that you are expert in them all. Swiftwater rescue, however, is a unique field, very unlike other water rescue specialty areas. What is swiftwater? The answer seems deceptively simple: water moving downhill. Yet, when water begins to move, it makes its own rules. The learning curve in swiftwater is steep and abrupt, and often there is no chance to retake the test. Many would-be rescuers have paid for this ignorance with their lives.

Our philosophy of swiftwater rescue, then, begins with knowledge—of the river, of techniques and equipment, and, finally, of ourselves. Rescue 3 International, a California-based rescue organization that was instrumental in developing many of the techniques in this book, teaches that there are four elements to any successful rescue: *training, practice, experience,* and *judgment.*

While you can learn much from books like this one, there is no substitute for hands-on *training* from knowledgeable instructors under realistic conditions. Still, this is only the first step. The novice must then *practice* those skills, both individual and team, so that their performance, even under the most adverse conditions, becomes automatic. By doing this he begins to get the *experience* he needs to make critical on-scene *judgments* during actual rescues. It is a long process with few shortcuts. Those skills must also be refreshed periodically or they will decay.

While there is certainly a place for standard operating procedures and common practices, we should avoid dogmatism in the way we do things. No two rescues or rivers are exactly alike, and the successful rescue is the one that works, not necessarily the one done by the book. Rescuers on-scene must have enough flexibility to do things differently when the situation warrants it. As the late Jim Segerstrom put it, "there are no 'always' and 'nevers' in swiftwater rescue."

Keep an open mind. Swiftwater rescue is a very dynamic and fast-changing field. Several subdisciplines, such as flood channel rescue, have literally come out of nowhere in the last few years. New techniques and

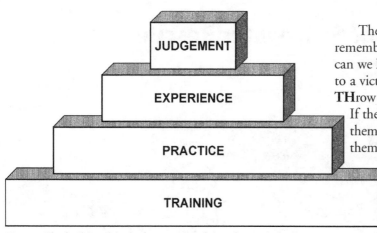

Fig. 1.1 *The building blocks for a successful swiftwater rescue.*

These methods and their priorities can be readily remembered by the mnemonic word **RETHROG**: can we **RE**ach something, like an oar or a tree limb, to a victim in the water? If not, is it possible to **TH**row them a life ring or a rope and haul them in? If they are too far away for this, can we **RO**w to them for a boat-based rescue? Finally, can we **G**o to them for a swimming contact rescue?

In these days of modern high technology we have added **HELO** to the list, which then becomes **REACH, THROW, ROW, GO, HELO**. Generally speaking, helicopters are a higher-risk option and should be used conservatively. Safe, effective rescues can be made by aircrews trained in swiftwater rescue, but more basic, low-tech approaches should be considered first.

Avoid common mistakes, such as:

- Stretching a line at right angles across the river to catch swimming victims. Unless the line is angled downstream, the pressure of the water will bow the rope downstream when the victim grabs the rope, holding him in the strongest part of the current.
- Rescuers tying themselves or victims to a rope. Since the force of the current usually forces anything tied like this underwater, this is an excellent recipe for drowning.

equipment are invented almost daily. Some of them may be just what you need for your local situation.

Emphasize training over equipment. In swiftwater, knowledge and experience are more important than gear. Yet, we often see organizations with garages full of shiny new gear and inadequate training. The equipment required for most rescues is minimal: personal protective gear for the rescuers, some throw bags, and perhaps an old paddle raft.

Value simplicity and speed. Most river rescues need not be complex, but time is almost always a factor.

Minimize risk where possible. Use the lowest-risk rescue option consistent with the situation's requirements.

Generally speaking, these swiftwater rescue situations are:

- **Shore-based rescues.** Any rescue during which the rescuers remain on shore is safer than one that puts them on or in the water. The most common shore-based rescues are those in which the rescuers reach or throw a rope to a victim.
- **Boat-assisted rescues.** By leaving the shore and entering a boat, the rescuers increase their risk level. However, they can minimize this risk by using the boat to assist rather than to perform the rescue. By this we mean using the boat for such tasks as ferrying equipment and patient evacuation rather than for actual rescue.
- **Boat-based rescues.** Here the rescuers use a boat (whether a rescue board or a 42-foot cutter) for the actual rescue. Boat use requires a much greater level of river skill by the rescuers, and being out on the river puts them in greater danger.
- **In-water contact rescue.** The rescuer may enter the water without a boat and contact the victim directly. This requires the highest level of skill from a rescuer and puts him in the most danger.

REACH

THROW

ROW

GO/TOW

HELO

Fig. 1.2 *The rescue sequence.*

Fig. 1.3 Do not tie yourself to a rope and enter moving water!

● Hanging a rope or a rescuer from a bridge to catch a swimming victim. Unless the rescuers are prepared to firmly grasp and quickly raise the victim, this situation will quickly degenerate into a tug of war with the river, which it will inevitably win.

When weighing rescue options, we want to use the lowest-risk option consistent with the situation. There are, however, other considerations, including:

● the time it takes to set up and operate a given system,
● available resources, and
● rescuer training and experience.

To properly determine the first factor, you must screen personnel for river experience and knowledge. A swimming rescue might be a high-risk option for a firefighter, but low- to medium-risk for a professional river guide. Similarly, a helicopter rescue by a crew trained in swiftwater rescue techniques might be a medium risk operation, but a boat-based rescue a high risk operation for a rescue squad without a lot of moving water expertise.

We also want to back up rescue systems in depth. Set up several systems downstream so that if one fails the next one may work. *At no time do we want a victim to get downstream of the last rescue system.* Backup applies to time as well. Resource and time-intensive systems, such as boat lowers, can be set up while faster and simpler systems are being tried. Alternative higher-risk systems can be set up at the same time.

The safety of the rescuers always comes first—before that

of the victim. No purpose is served by adding additional victims. Rescuers must quickly establish and remember rescue priorities:

● Rescue themselves—hence the importance of learning self-rescue as the first step.
● Rescue and ensure the safety of fellow team members.
● Rescue the victim.

Remember that "self-sacrifice in rescue services is traditional and commendable ... and a useless waste. Rescue instructors would rather appear as expert witnesses to testify why nothing was done, than as to why a rescuer was injured or died."

Compassion can kill. The urge to save lives and compassion for others brings many people into the field of search and rescue. However, the desire to save the victim at all costs, especially sympathetic ones like children, can be fatal to rescuers who ignore safety. This is particularly true when the operation is a body recovery rather than an actual rescue. Steve Fleming, a Colorado firefighter, relates a diving incident in which he almost became trapped in a vehicle during a search for the body of a 16-year-old boy under very hazardous conditions.

> **PRINCIPLE:** In any swiftwater rescue we want to get the victim away from the power of the river and out of the current as fast as we can. Avoid, if possible, any rescue technique that tries to hold the victim stationary in the river against the current. Try instead to use the river's power to move the victim into a safer position, i.e., toward the bank. You may also try to lift the victim out of the water, but this is much more difficult.

"Compassion," he says, "overruled my better judgment, and because of compassion I almost lost my own life trying to recover a body." The safety of the living must come first.

With these principles in mind, let's look at the environment in which these rescues take place—the river.

Fig. 1.4 Use the lowest technology solution consistent with the situation's requirements!

The River: An Introduction to Hydrology
Chapter 2

Fig. 2.0

Knowledge of the river and the habits of swiftwater are the basis upon which all rescues are built. There is tremendous power in any river. If you work against it, it can defeat or even kill you. The most basic principle of swiftwater rescue is to avoid or minimize this power or, even better, to make it work for you. The dynamics of rivers are relatively simple, and understanding swiftwater does not require an engineering degree. Any rescuer must have at least a basic awareness of the power and hazards of swiftwater.

The Force Of The Water

Current Velocity mph/kmh	On Legs lbf/N	On Body lbf/N	On Swamped Boat lbf/N
3/4.8	16.8/75	33.6/149	168/752
6/9.7	67.2/299	134/596	672/2989
9/14.5	151/672	302/1343	1512/6726
12/19.3	269/1196	538/2393	2688/11,957

Adapted from: RIVER RESCUE, Ohio Department of Natural Resources, 1980.

Characteristics of Swiftwater

All swiftwater has three common characteristics: it is *powerful*, it is *relentless*, and it is *predictable* to the experienced eye.

Swiftwater is powerful. As the speed of the current increases, so does the power. But it does not do so in a linear fashion, as you might expect. When the speed of the current is doubled, the force of the water against an object in the current is *quadrupled*; that is, the force of the current increases as the *square* of its speed, as shown in the above table.

How fast does the current flow on a typical whitewater river? Peter Reithmaier, an Austrian engineer and river safety expert, measured river speeds in the Alps. The highest velocity of water he found, even on steep, runoff-swollen Alpine rivers, was 17.5 km/hr (11 mph). Sections in which observers thought the river moved "very fast" were actually measured at 12 km/hr (7.5 mph).

As the river's speed and force increase, so does its carrying capacity. Rivers in flood can carry heavy objects like boulders and trucks, and can ravage entire districts. When the river's velocity and volume decrease, however, heavier objects begin to settle out, leaving the characteristic debris trail of a flood.

Swiftwater is relentless. Unlike an ocean wave, which breaks and then ebbs, river currents push against an object in the river continuously, without letup. Once in the current, you cannot expect any reprieve from the river's force.

Swiftwater is predictable. To the inexperienced eye, a river's currents and rapids seem random and chaotic. *They are not.* What happens is actually an orderly and predictable phenomenon *if* you know what you are looking at. Water acts exactly the same in a small brook as in a large river. It is simply a matter of scale.

Variables of Swiftwater

The nature of any river, swift or not, is determined by three things: the amount of water, how fast it moves, and what is in its bed and along its banks. Rivers, like people, have different personalities. Change any one of these variables and you change the river. This can sometimes happen rather abruptly.

The amount or *volume* of water in a river is usually expressed as the amount of water moving past a certain point during a given period of time. In the United States it is usually expressed in cubic feet per second, or cfs, while in most of the rest of the world it is expressed as cubic meters per second, or cms. A large river like the Niagara may flow at 100,000 cfs (2,800 cms), the Mississippi over a million at times. Water is also quite heavy. A cubic foot weighs 62.4 pounds (28.3 kg).

A river is a gravity-driven system. The steeper the *gradient*—that is, the amount of drop between two given points—the faster the water moves. Gradient is usually expressed as feet per mile or meters per kilometer. However, overall figures can be deceptive. A drop and pool river alternates between drops, or rapids, which may be very steep, and pools,

> To determine a river's volume, multiply the width by the depth times the speed of the current. For example, a channel ten feet deep and twenty feet wide moving at a velocity of five fps equals a volume of 1,000 cfs (10 x 20 x 5 = 1,000). Doubling the speed of the current doubles the volume. When a river is constricted, such as by narrow banks or a shallow section, the water must speed up to move the same amount of water through the reduced space. Conversely, when the size of the riverbed is increased, as in a wide section or a deep pool, the water flows more slowly.

which may be nearly flat. Other rivers give up their elevation in a more continuous fashion, a little at a time. Generally speaking, as the gradient and volume increase, so do the size and number of rapids.

Rivers do not have a uniform current speed, nor does all water flow downstream. There are two types of river currents: a *laminar* flow that accounts for most of the water going downstream, and a *helical* flow of the water along the banks. The friction of the water against the banks causes the water next to them to flow in a corkscrew motion downstream between the banks and the main current. It rises up to the surface next to the main current and flows toward the bank, then dives down along the bottom until it reaches the main current again. The main current, on the other hand, flows downstream in laminar fashion, with different layers of water moving at different speeds. Friction from the bottom and sides of the river slow the outer layers of water somewhat, so that the fastest water is usually just below the surface at the center of the river.

A river carries things in one of three ways: a *top load*, a *suspended load*, and a *bottom load*. A top load is anything that floats, whether it is a boat or a piece of debris. The suspended load is carried inside the river, neither floating nor touching the bottom. It can be fine like sand or it can be larger, heavier objects like boulders. The faster the water moves and the greater its volume, the greater the capacity for the suspended load. Loads too heavy to suspend move along the bottom, pushed by the current. As a river moves to flood stage it begins to pick up more of each kind of load; as it recedes, it will begin to drop the loads it carried.

Much of the character of a river derives from the nature of its bed and banks—the obstructions, constrictions, and bends. There is a complex interactive process between the river and its bed, and much of what happens on the surface of a river is determined by what lies beneath it. Geology plays an important part here, as does the hand of man. In extreme cases, like the Los Angeles River, the river is transformed into an almost entirely artificial flood channel. The natural obstructions in a river's bed, such as boulders and ledges, slow the water's speed somewhat. When these are removed, as they are in a flood channel, the water's speed rises dramatically. Water speed in flood channels has been measured at over 30 mph (48 km/hr).

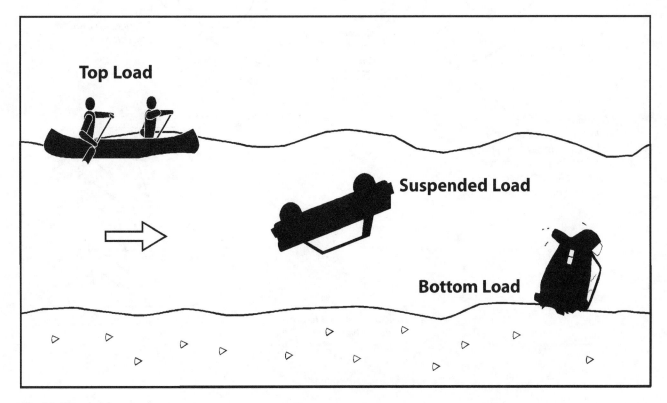

Fig. 2.1 River loads.

River Cross Section

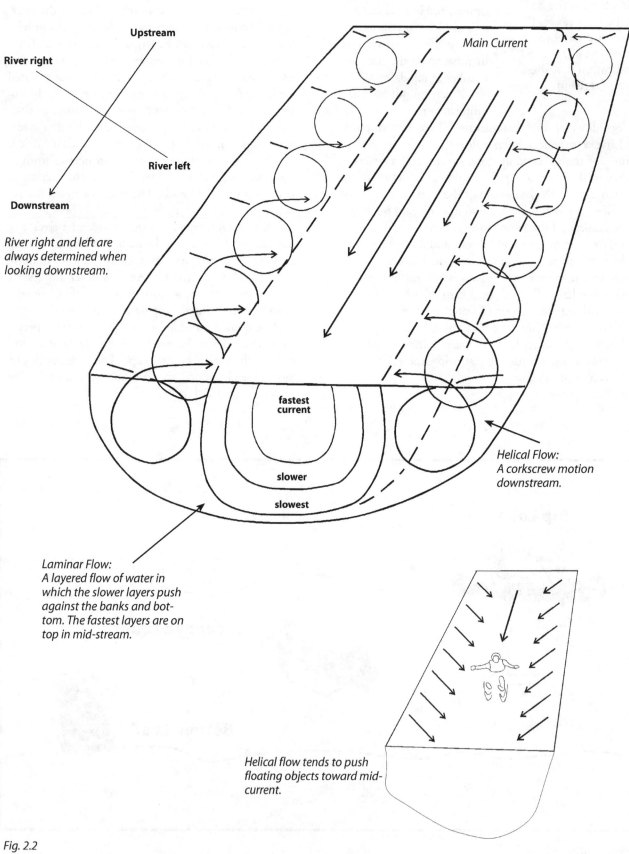

Upstream

River right

River left

Downstream

River right and left are always determined when looking downstream.

Main Current

*Helical Flow:
A corkscrew motion downstream.*

fastest current

slower

slowest

*Laminar Flow:
A layered flow of water in which the slower layers push against the banks and bottom. The fastest layers are on top in mid-stream.*

Helical flow tends to push floating objects toward mid-current.

Fig. 2.2

One example of an interactive process is where the river bends. The main current of the river does not flow straight downstream but instead meanders both side to side and up and down. Where the faster current pushes against the bank it begins to erode it. As the process continues, a bend forms. Since the current on the outside of the bend moves faster, it has both more power and a greater carrying capacity, causing it to erode the bank still more. The current on the inside of the bend, on the other hand, moves more slowly and so begins to drop sand and other debris on the inside of the bend. As the bend becomes steeper, the outside current begins to dive down along the bank, scouring soil away, then moving underneath the main current to drop it on the inside of the bend. This causes the characteristic river bend—deep with a swift current on the outside,

shallow and slow on the inside. The up and down meanders result in alternating deep, still pools and shallow, swift spots. On whitewater rivers the shallow spots are usually rapids. Water also "piles up" on the outside of river bends: that is, it is physically higher on the outside of a bend than on the inside.

As water speeds up, it gains energy. At some point, this energy must be given up. If the available energy exceeds the ability of a fluid to hold it, it begins to push the surface up into waves. River waves, unlike those in the ocean (which are mostly caused by wind), stay in one place, tied to the feature that caused them. If the height of a wave exceeds about one-seventh of its length, it will begin to break back upon itself. Really large waves may seem to explode rather than break.

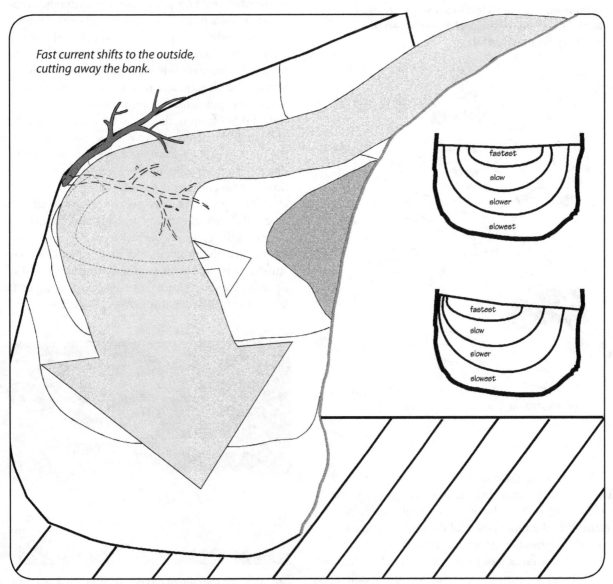

Fig. 2.3 The river's main current meanders up and down as well as side to side.

River waves have many causes, the most common of which are:
- A fast-moving current entering a slower section
- A sudden constriction or change in the river's cross section or gradient
- Water flowing over a submerged obstacle
- Current hitting a hard obstacle like a boulder or bank
- Converging currents

The wave troughs, or low spots in the river's surface, are as important as the peaks. A large wave breaking back upstream into a trough is quite capable of stopping and holding a floating object.

River orientation. A river has four directions: upstream, downstream, river right, and river left. While the first two are self-explanatory, right and left are not. To avoid confusion, directions right and left are always given looking downstream, and are referred to as river right and river left.

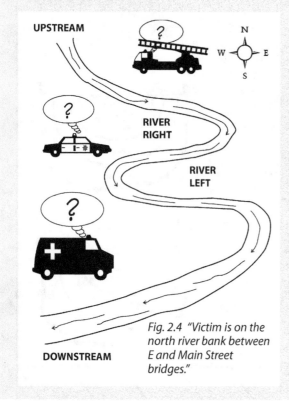

Fig. 2.4 "Victim is on the north river bank between E and Main Street bridges."

Not all water flows downstream. Looking at a flooded river this may seem an astonishing statement, but it is true. In fact, the more powerful the downstream flow, the more powerful the upstream flows become. The most common upstream flow is an eddy. As water flows past an obstacle like a boulder, several things happen. First, a reaction feature called a *pillow* forms as water piles up on the upstream side

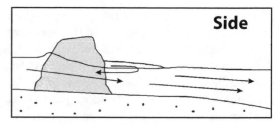

Fig. 2.5 An eddy is formed by an obstacle in the current.

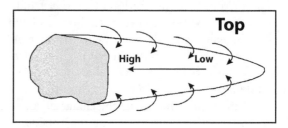

of the obstacle. This pillow forms a hydraulic cushion that tends to push floating objects away from it. Second, while water piles up higher than the river level on the upstream side of the obstacle, the water on the downstream side is lower, so water flows around the obstacle and then back toward it to fill in the low spot. This upstream current is called an *eddy*. Eddies form both behind midstream obstacles and along the banks. As we will see, eddies are one of the most important features of the river for boat handling and rescue. The line between the upstream and downstream currents is the *eddy line*.

On larger rivers there may be a substantial height differential between an eddy and the main current, called an *eddy fence* or *wall*. Big rivers may have extremely turbulent eddy lines several feet wide stuffed with whirlpools randomly forming and dissipating. Instead of being a haven, the surging eddies of a big river may be as much of a hazard as a rapid.

Fig. 2.6 When the current hits an obstacle, a pillow forms on the upstream side.

When water flows over a submerged obstacle, it gives rise to other upstream currents, *hydraulics* and *holes*. A hydraulic (reversal, pourover, keeper) forms when water flows over an obstacle near the surface, causing a recirculating current as the water dives down behind the obstacle and then flows back upstream. The water below a hydraulic is typically very aerated and presents a white, foamy appearance. The boiling, bubbly area downstream of a hydraulic where the water rises to the surface is called the *boil line*. The water upstream of the boil line (that is, the water flowing back toward the obstacle) is white and frothy, while the water downstream of the boil line has a darker, smoother appearance. The recirculating current can trap and hold an object for extended periods of time.

Fig. 2.7 *As the water level rises and flows over the top of the obstacle . . .*

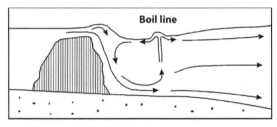

. . . it forms a hydraulic.

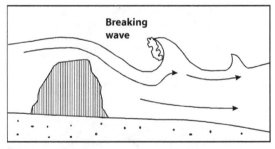

As the current deepens, it forms a breaking wave, or "hole."

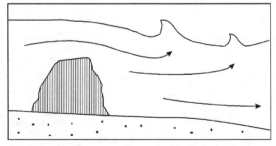

As the water deepens still more, the waves become smaller and stop breaking.

A hole (sousehole, stopper, breaking hole), on the other hand, is generally considered to be a standing wave that breaks back upstream. While also capable of holding floating objects, it is not as dangerous as the hydraulic. In practice, the two terms are often used interchangeably.

These features often happen in sequence as the water rises or falls. For instance, water flowing around a boulder causes an eddy to form behind it. If the water rises, the water begins to flow over the boulder, causing a hydraulic. As the water level rises even more, the hydraulic becomes a breaking wave and, finally, with the boulder submerged far below the surface, a smooth wave.

No discussion of the river would be complete without mention of one of the most durable legends of the river—the "undertow." This troll-like current, it is said, lurks somewhere beneath the surface, ready to drag the unwary swimmer under. The only real undertows are found off very steep ocean beaches, where waves piling up on shore seek an escape seaward (it is rarely fatal). By now it should be obvious that while there are dangerous currents on the river, even ones that can pull a swimmer underwater, they are not at all mysterious.

In the incident on the San Marcos River, recounted at the beginning of this book, confusion between the characteristics of a breaking wave and a hydraulic was a critical mistake both by the victim and the rescuers.

Hazards

So far our discussion of rivers has been somewhat academic. Now it's time to see what these river features and others can do to boats and people. Any river feature that poses a threat to life, limb, or navigation is generally called a *hazard*. River hazards also have specific names and causes.

Natural Hazards

Hydraulics. We have already seen that the recirculating current of a hydraulic can hold boats and swimmers for extended periods of time. Natural hydraulics generally come in two varieties—*smiling* and *frowning*—as seen from upstream. A smiling hydraulic has its outer edges curving downstream, so that the recirculating water (and anything caught in it) feeds out into the main current. In a frowning hydraulic, however, the outer edges curve upstream, feeding the water back into the center of the hydraulic. These are best avoided when possible, since escape from them may be very difficult.

Fig. 2.8 A typical low-head dam. Anything upstream of the boil line will be drawn back into the hydraulic.

BACKWASH

OUTWASH

BOIL LINE

Holes. A breaking hole is quite capable of holding or upsetting a boat, but since water is flowing through underneath it, it will not hold a swimmer. Once out of the boat and in the river, however, a swimmer may then be faced with a long, dangerous swim.

Strainers. Anything that allows water but not solid objects like boats and people to pass through is called a *strainer*. The most common strainers are fallen trees in the river, although fences, guard rails, and other objects effectively serve the same purpose during flood conditions. The force of the current will hold any solid object against the strainer.

Boulder/Debris Sieves. Boulders and debris piles can also function as strainers. A common phenomenon on some rivers is a *boulder sieve* caused by a side creek flash flooding into the main channel.

Cold Water. Water temperature has an important effect on survival. Even still water takes heat away from an unprotected body about twenty-five times as fast as air. If the body is stationary and the water moving, the effect is even more dramatic. A 5 mph (8 km/hr) current flowing past an unprotected stationary body may take up to 250 times the amount of heat away from the body as still air. While less than one percent of swimmers actually die from hypothermia, *hypothermia-induced debility* (HID) will quickly rob even the strongest swimmer of his strength, and is thus a causative factor in many drownings. "Cold" water means anything below body temperature, although obviously the colder the water, the more rapid its effect. Studies have shown that even a strong swimmer becomes exhausted after swimming for only twelve minutes in still 50°F (10°C) water. Cold water also has adverse physiological effects on people age 40 and over and those with heart conditions.

Fig. 2.9 Breaking Hole vs. Hydraulic

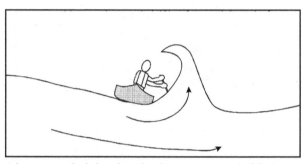

A large wave hole breaking back upstream can hold a buoyant object for an extended period, however . . .

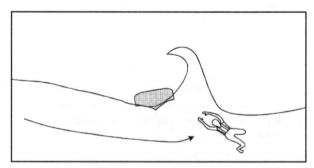

. . . a submerged object will quickly be swept downstream.

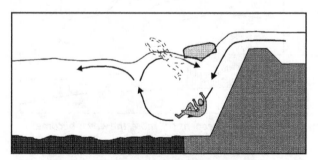

A hydraulic, however will hold both buoyant and submerged objects in its recirculating current.

Big Water (flush drowning). Swimming though big waves can lead to *flush drowning*, especially for unprepared or inexperienced persons. A person suddenly falling into cold water experiences an automatic *inhalation reflex* as the body suddenly tries to increase its metabolic rate. This reflex, as well as hyperventilation, can cause a swimmer to aspirate water when swimming through big waves, even though he is wearing a PFD. The effect is lessened by thermal protection for swimmers and training to breathe at the proper time in big waves.

Geomorphic Hazards. In addition to the ones already mentioned, all rivers contain natural geomorphic hazards. Some are worse than others in this respect. These include cracks or crevices in the river bed into which a leg, or even the entire body, can be wedged and held by the current. This is called *entrapment*. A very common peril is *foot entrapment*. This is usually caused when a swimmer attempts to stand up in moving water and gets a foot jammed into a crevice in the riverbed. The force of the current is usually enough to prevent self-rescue, and drowning often results. Another common danger is *undercut rocks*. These usually occur on geologically older rivers. Boats and people may be shoved into an undercut and held there by the current.

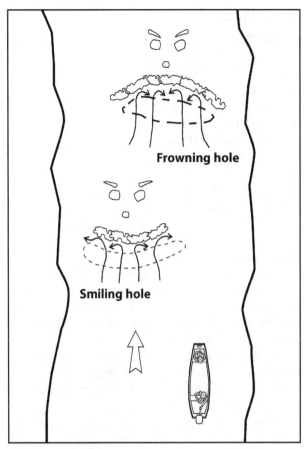

Fig. 2.10 Water flows out of the ends of a smiling hole, making escape easier. Water flows back into a frowning hole, making escape more difficult.

Fig. 2.11 A natural strainer – a tree down in the river's main current.

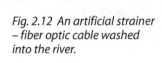

Fig. 2.12 An artificial strainer – fiber optic cable washed into the river.

Man-Made Hazards

In addition to the river hazards imposed by nature, we must also contend with those put there by man. These include:

Man-Made Debris. Old river structures like dams and mills are seldom removed. Usually, they are simply abandoned. The structures may wash away but leave reinforcing rods in the riverbed. These leave little "signature" and can be very hard to detect. Industrial waste is also frequently dumped into the river, as are old cars and appliances. One kayaker became entangled and drowned on a discarded conveyor belt.

Bridge Abutments. Unlike natural rocks and boulders, which usually cause an upstream pillow, bridge abutments usually have little in the way of a hydraulic cushion. Consequently, they are often the site of boat and raft pinnings, and frequently collect debris piles that become dangerous strainers.

Low-Head Dams. A low-head dam, or weir, is often referred to as a "drowning machine." These dams form hydraulics which, unlike natural ones, are regular features, all the way across the river, leaving no escape routes. Indeed, the sides of the dam are frequently blocked by concrete walls, and there may be debris piles at the base and exposed rebar. Some dams, like the infamous Brookmont Dam on the Potomac River, have been the site of numerous drownings. This is one area where man has unquestionably "improved" on nature—by making the danger greater.

Flood Channels. These are another improvement on nature. In order to reduce the danger of flooding in urban areas, engineers have devised concrete channels to move the water out of town as quickly as possible. Indeed, the water velocity may be double that of a natural river, and these channels are usually sprinkled liberally

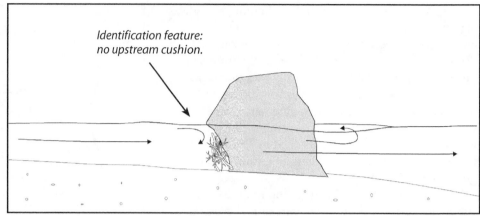

Identification feature: no upstream cushion.

Fig. 2.13 An undercut rock can trap boats and swimmers.

with other hazards like low-head dams and bridge abutments. Typically, they have few eddies and often go underground for part of their length.

Toxic Environments. After using the rivers of this country for sewers and dumping grounds for the last few hundred years, it should surprise no one that many are heavily polluted, both with pathogens like coliform bacteria and chemical pollutants like heavy metals, fertilizers, and pesticides. Both pose major health hazards to rescuers.

Drugs, Alcohol and Tobacco. Use of any or all of these can greatly increase the danger of drowning of both rescuers and victims. Alcohol use is involved in over half of the drownings in this country, and can be deadly even in small quantities when mixed with cold water. Tobacco acts as a vasoconstrictor and reduces lung capacity.

Fig. 2.14 Searching undercut rocks is difficult and risky. This one held a swimmer for over two weeks.

Rivers in Flood

Since rescuers are frequently called upon to make rescues during flood conditions, it is worthwhile to take a closer look at this subject. Most rivers have a normal water level, or rather a range of normal levels. While there may be low water during drought and high water after a storm, the river flows within its banks and is relatively predictable. Floods, however, are outside the usual range of river conditions. In flood, a river overtops its banks and begins to flow through the *flood plain*, or the low areas adjacent to the river. In the process, it becomes less predictable and more dangerous. Under flood conditions:

- The size and power of the river are both greatly increased, as is its carrying capacity. The river's swollen main current, instead of gently rounding a bend, strikes the bank and ricochets off.
- Almost all of the river hazards already mentioned, like low-head dams, become much more dangerous during a flood. There are often additional ones as well. Bridges, for example, may not have enough clearance for boats to pass under.
- Floodwaters are laden with debris, which can clog intakes and foul propellers on rescue boats. Trees and other large, heavy objects join the river's flow, often collecting against bridges to form strainers or natural dams, or forming "mobile strainers" in the current.
- Water flows through things on the flood plain like trees, fences, brush, and debris, which greatly adds to the danger of being "strained."
- As the river flows through "civilized" areas like streets, fields, and neighborhoods, the danger of contamination from pesticides, fecal matter, dead livestock, and chemicals greatly increases. Water treatment plants may be flooded and stop functioning altogether, dumping raw sewage into the water.
- Eddies and eddy lines become a danger. Eddies are wide and laced with whirlpools, eddy fences high and difficult to cross, and the eddies themselves are rapidly-moving whirlpools from which escape is difficult.

Some of the fallacies of flood control structures are also worth mentioning. Man often builds in the flood plain, then tries to keep the river out by building levees and dikes. This is quite often self-defeating. Recall that if we constrict the river, then add more water, the water must move faster, which increases its erosive power on the dikes, and causes it to rise higher, eventually to overtop the dikes. Urban development paves over areas that normally absorb floodwaters and increases runoff from storm drains and flood channels, which then become hazards in their own right. The best solution is to let the river occupy the flood plain when it needs to, but this is not always politically feasible.

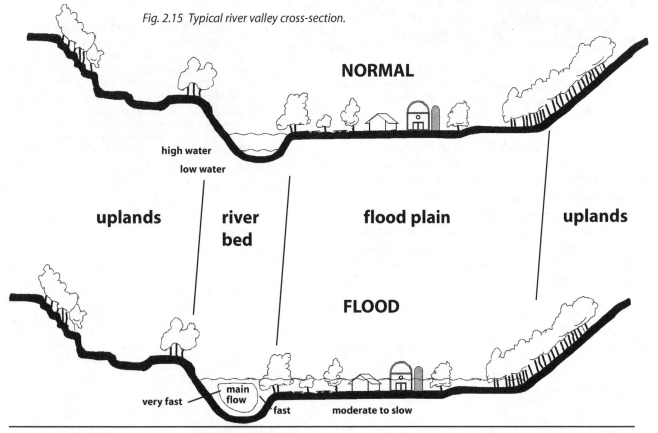

Fig. 2.15 Typical river valley cross-section.

Fig. 2.16 The Cuyama River gouged out a huge chunk of Highway 166 near Santa Maria, CA, during a 1998 flood. Two California Highway Patrolmen drove into it and were killed. Two other motorists and a sheriff's deputy looking for the pair narrowly escaped the same fate.

Dams are often built as flood control structures, but these have their limitations, too. In fact, some of the worst flooding has been due to the failure of these control structures. Dams and flood channels are also extremely dangerous for citizens and rescuers alike.

Finally, there is one other hazard of flooded rivers that should be mentioned. Rising waters often force snakes from their burrows near the river. Watch where you step!

Rapids rating system

Since all rivers are different and constantly changing, it's impossible to make exact comparisons, particularly of the difficulty of rapids. Whitewater paddlers, who have had a lot of experience in the field, have come up with an international rating system in which all rapids are rated on a scale from I (easy) to VI (extreme). While this system was designed with paddlers in mind, it is also useful for the rescuer.

International Scale of River Difficulty

This is the American version of a rating system used to compare river difficulty throughout the world. This system is not exact; rivers do not always fit easily into one category, and regional or individual interpretations may cause misunderstandings. It is no substitute for a guidebook or accurate firsthand descriptions of a run.

Paddlers attempting difficult runs in an unfamiliar area should act cautiously until they get a feel for the way the scale is interpreted locally. River difficulty may change each year due to fluctuations in water level, downed trees, geological disturbances, or bad weather. Stay alert for unexpected problems!

As river difficulty increases, the danger to swimming paddlers becomes more severe. As rapids become longer and more continuous, the challenge increases. There is a difference between running an occasional Class IV rapid and dealing with an entire river of this category. Allow an extra margin of safety between skills and river ratings when the water is cold or if the river itself is remote and inaccessible.

International Scale of River Difficulty

- **Class I:** Easy. Fast-moving water with riffles and small waves. Few obstructions, all obvious and easily missed with little training. Risk to swimmers is slight; self-rescue is easy.

- **Class II:** Novice. Straightforward rapids with wide, clear channels that are evident without scouting. Occasional maneuvering may be required, but rocks and medium-sized waves are easily missed by trained paddlers. Swimmers are seldom injured and group assistance, while helpful, is seldom needed. Rapids that are at the upper end of this difficulty range are designated "Class II+."

- **Class III:** Intermediate. Rapids with moderate, irregular waves that may be difficult to avoid and can swamp an open canoe. Complex maneuvers in fast current and good boat control are often required; large waves or strainers may be present but are easily avoided. Strong eddies and powerful current effects can be found, particularly on a large-volume river. Scouting is advisable for inexperienced parties. Injuries while swimming are rare; self-rescue is usually easy, but group assistance may be required to avoid long swims. Rapids that are at the lower or upper end of this difficulty range are designated "Class III-" or "Class III+" respectively.

- **Class IV:** Advanced. Intense, powerful, but predictable rapids requiring precise boat handling in turbulent water. Depending on the character of the river, it may feature large, unavoidable waves and holes or constricted passages demanding fast maneuvers under pressure. A fast, reliable eddy turn may be needed to initiate maneuvers, scout rapids, or rest. Rapids may require "must" moves above dangerous hazards. Scouting is necessary the first time down. Risk of injury to swimmers is moderate to high, and water conditions may make self-rescue difficult. Group assistance for rescue is often essential but requires practiced skills. A strong Eskimo roll is highly recommended. Rapids that are at the lower or upper end of this difficulty range are designated "Class IV-" or "Class IV+" respectively.

- **Class V:** Expert. Extremely long, obstructed, or very violent rapids that expose a paddler to added risk. Drops may contain large, unavoidable waves and holes, or steep, congested chutes with complex, demanding routes. Rapids may continue for long distances between pools, demanding a high level of fitness. What eddies exist may be small, turbulent, or difficult to reach. At the high end of the scale, several of these factors may be combined. Scouting is recommended but may be difficult. A very reliable Eskimo roll, proper equipment, extensive experience, and practiced rescue skills are essential. Because of the large range of difficulty that exists beyond Class IV, Class 5 is an open-ended, multiple-level scale designated by class 5.0, 5.1, 5.2, etc... each of these levels is an order of magnitude more difficult than the last. Example: increasing difficulty from Class 5.0 to Class 5.1 is a similar order of magnitude as increasing from Class IV to Class 5.0.

- **Class VI:** Extreme and Exploratory Rapids. One grade more difficult than Class V. These runs often exemplify the extremes of difficulty, unpredictability and danger. The consequences of errors are very severe and rescue may be impossible. For teams of experts only, at favorable water levels, after close personal inspection and taking all precautions. After a Class VI rapids has been run many times, its rating may be changed to an apppropriate Class 5.x rating.

The rating system works best when comparing rapids on the same river and not very well when comparing different types of rivers. For instance, it makes sense to say that a certain rapid is "Class III at low water and Class V at high water." However, it is not easy to compare a rapid on the Colorado River, which has huge waves but requires little technical maneuvering, to one on the Gauley in West Virginia, a much smaller river but one that is geomorphically dangerous and requires precise boat-handling skills. Nor is it easy to compare a river with continuous rapids with one that might have large drops separated by calm pools.

River Tactics and Basic Boat Handling

It is now time to apply the theories of hydrology to the practice of boat handling. While this is not a book about whitewater technique, swiftwater boat handling is an essential skill for rescuers. This means:

- Having a working knowledge of the river
- Developing self-rescue skills
- Developing boat-handling skills

River craft come in all shapes and sizes, from tiny stunt kayaks to giant motor rigs. They can be divided into two major categories: powered—meaning powered with a gasoline engine such as an outboard motor—and nonpowered—meaning muscle-powered by oar or paddle. In practical terms it means that a powered craft can be used at least to some extent for upstream travel, while nonpowered craft are mostly limited to downstream travel. Otherwise, the basics of boat handling are similar for any river craft.

Before looking at what effect the river has on a boat, let's look at a typical rapid. A series of rocks, perhaps washed down by a side creek, obstructs the river. The water above it backs up into a deep pool. Above each rock is a small reaction feature called an *upstream V*, which signals the presence of an obstruction, even if it is underwater. Where the water flows between the rocks it assumes a "V" shape pointing downstream, called a *downstream V*. This feature usually marks where the deepest current flows. Larger **V**s—that is, those that rise above the surface of the surrounding water—are called *tongues* or *chutes*. At the bottom of the chute is a series of standing waves called *tail waves*. On the river right side of the center tongue is a submerged rock, causing a hole. One of the chutes is blocked with a debris pile that forms a *strainer*. Behind each rock, and at places along the bank, are *eddies*. The rocks on the far right form a *boulder sieve*. Below this rapid, where the banks constrict the river's flow, is another series of tail waves. Here, the river picks up speed in the constriction, then gives up the energy in the form of waves when it hits the slower water below it.

TIP: If there is a break or irregularity in a series of standing waves, it is probably caused by an underwater obstacle like a rock.

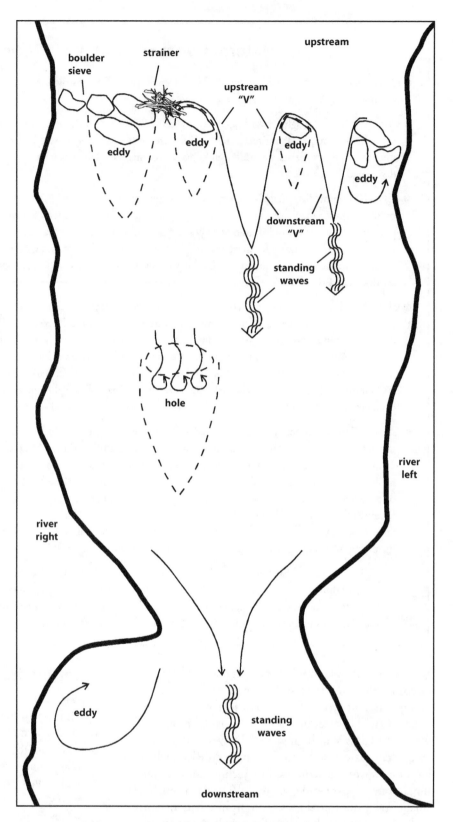

Fig. 2.17 Rapids are caused by one or more of the following:
1) Obstacles in the river bed
2) Change in gradient
3) Constriction

Fig. 2.18 The smooth, dark tongue of water between two large rocks just in front of this inflatable kayak marks the best route downstream. Expect a series of standing waves just below.

Route Finding

Let's look now at how we would get through a rapid like the one just described. The process of analyzing a rapid—usually just prior to running it—is called *scouting*. The most effective way to scout is to find a spot where you can see the whole rapid. Start from the bottom, where you'd like to end up, and work upstream. There will usually be several ways to run through each section of the rapid. The ultimate object is to connect each section into a coherent whole so that you can pass all the way down as easily and safely as possible. To do that, we will normally look for a route that will avoid hazards and put us as near as possible to where we'd like to go. In most rapids the deepest, smoothest water is usually marked by a downstream "V."

In our example rapid, we'd like to end up near the center of the river, below the tail waves. On the right side of the river (river right, as seen from upstream), the chutes are either too narrow to run or are blocked by debris. The center chute is wide and open, but there is a hole off to the right side that we want to avoid. Depending on the size of the tail waves, we may want to run to one side of them in order to avoid swamping. We could also run through the right-hand chute, but this would put us closer to the shore.

In the chapter on boat-based rescues we'll discuss more complex route finding, as well as ways to use other river features like eddies to our advantage.

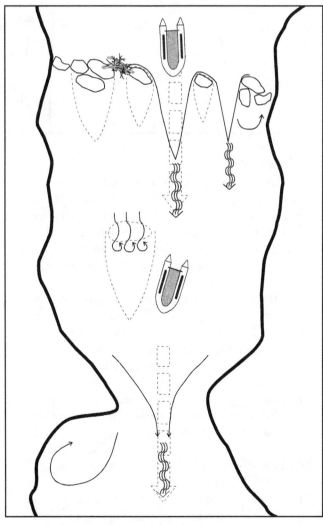

Fig 2.19 Route Finding: In the most basic terms, the route through a rapid follows the main current (marked by downstream "V"s) while avoiding hazards such as strainers and obstacles like holes.

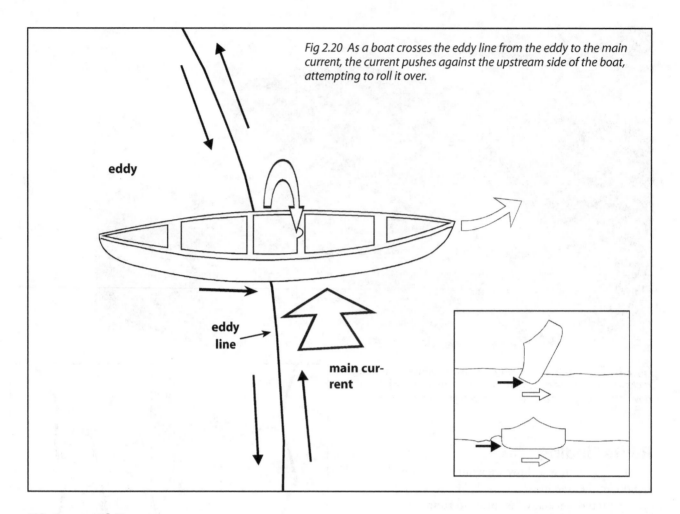

Fig 2.20 As a boat crosses the eddy line from the eddy to the main current, the current pushes against the upstream side of the boat, attempting to roll it over.

eddy

eddy line

main current

Water and Boats

So far we've learned that the speed of river currents can vary quite a lot. These differences, or current differentials, are important for boats. The two primary reasons that boats get into trouble on the river are current differentials and waves. Both current differentials and waves can capsize a boat, and waves can also swamp a boat.

As long as a boat moves at or near the same speed as the current, there are few problems. It is when a boat suddenly crosses from one area of the current to another that is moving at a different speed, or even in a completely different direction, that the fun begins. This happens most commonly when a boat enters or leaves an eddy. In an eddy, you will recall, the water is actually flowing *upstream*, completely in the opposite direction of the downstream current. Thus, when a boat enters or leaves an eddy, the capsizing force on the boat shifts rapidly from one side of the boat to the other. While some types of boat, like inflatable rafts, are so inherently stable as to be more or less immune from capsizing on an eddy line, most are not.

The solution is to anticipate this phenomenon and nullify its effect by either leaning the boat or shifting the weight within it. Smaller boats like canoes and

kayaks benefit more from leaning the boat into an *eddy turn* like a bicycle. As the boat crosses the eddy line, the paddler leans the boat so that the water flowing against the side of the boat tends to flow under it rather than piling up on the side and causing a capsize. With practice, both eddy turns (entering an eddy from the main current) and *peel-outs* (entering the main current from an eddy) can be done quickly and effectively.

In larger boats, a "high side" works better. Here, the boat crew shifts their weight away from the current pushing against the side of the boat. When the current pushes down on one side of the boat, it pushes the other side up. The *high side* must then be pushed down, usually by shifting the crew's weight over onto it, to restore the boat's equilibrium. If this is not done, and quickly, the boat will capsize. The same principle applies to a boat that is hit by a big wave or gets caught in a hole or hydraulic. An extreme form of high-siding is called *tube-punching*, where crew members literally throw themselves against the tubes to keep the boat from flipping.

If the boat hits a large breaking wave or a hole, the wave will break either directly or diagonally upstream, attempting to roll the boat over upstream.

Fig 2.21 High Siding

If a boat moving downstream hits a wave . . .

. . . it will roll the boat upstream

. . . unless the crew shifts their weight to the high side, or downstream side, of the boat.

If the wave manages to stop the boat in the current without capsizing it, this too will create a current differential as the upstream current pushes against the now-stationary boat. The remedy in either case is the same: the crew must aggressively shift their weight to the downstream side of the boat, *into the wave*. This sometimes requires an exercise of faith. When a huge wave towers up over the boat, the natural instinct is to lean away from it. This is exactly the wrong thing to do, as it will cause the boat to roll over upstream with truly amazing suddenness.

For the same reasons, boaters should also be prepared to lean into obstacles like rocks. Again, the idea is to keep the weight off the upstream side of the boat, letting the current flow under it and not against the side. As with a wave, the instinctive reaction is to lean away from the rock. As soon as the upstream gunwale dips underwater, however, the boat flips upstream and usually ends up pinned against the rock.

Large waves can also swamp a boat as they break over it, either sinking it or making it so heavy as to be unmanageable. Decked boats or self-bailing boats avoid this problem. Other types of boats must either avoid large waves or find a means of bailing quickly enough to prevent swamping.

Fig. 2.22 A classic "low side." The crew has failed to shift their weight toward the rock, resulting in a flip.

A boat caught in a hydraulic requires a special technique. A hydraulic, unlike a breaking hole, will not eject a capsized boat. Instead, both boat and crew may end up recirculating in the hydraulic. A boat that remains upright has a much better chance of escape or rescue from a hydraulic. If caught, the boaters must quickly move to the *downstream* side of the hydraulic, away from the water flowing down into it. Since boats in this situation often get spun around in the hole, the boaters must be prepared to move quickly within the boat to stay on the downstream side.

Fig. 2.23 High Siding

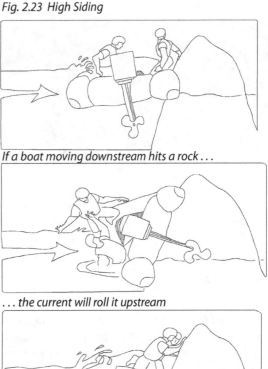

If a boat moving downstream hits a rock . . .

. . . the current will roll it upstream

. . . unless the crew shifts their weight against the rock.

Basic Boat Handling

In addition to traveling up and down stream, boats make three basic river maneuvers:

- Eddy turns
- Peel-outs
- Ferries

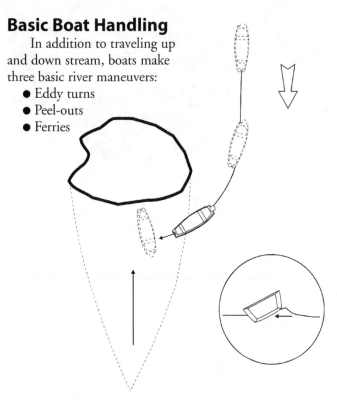

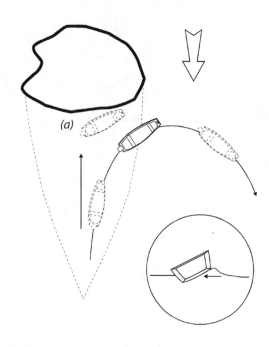

Fig 2.24 Eddy Turn: When entering an eddy from the main current, the boat must be leaned into the turn.

Fig 2.26 Peel Out: When crossing the eddy line, boat operators must lean downstream to avoid flipping. They should also avoid being pulled up against the rock. (a)

Fig. 2.25 An eddy turn.

Fig. 2.27 A peel-out into the current.

In an *eddy turn*, a boat in the river's main current goes into an eddy. Eddies often represent safe havens on a river where a boat can interrupt its downstream progress, even in the midst of a rapid, and stop. An eddy is strongest near the top (i.e., nearest the obstacle that caused it) and weakest at the bottom. The boundaries of an eddy, the eddy line or eddy fence, tend to push objects away from it. Therefore, any boat trying to enter an eddy must do so with some speed

in order to "punch" across the eddy line. Boaters must also be aware that the current *reverses direction* when crossing an eddy line and be prepared to either lean the boat or shift their weight accordingly.

In a *peel-out*, a boat in an eddy goes out into the main current. Many of the same factors apply as in an eddy turn: the need for speed when crossing the eddy line and the downstream lean or weight shift.

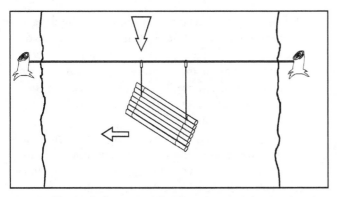

Fig 2.28 The basic ferry principle: The current pushes against the upstream side of the boat, moving it across the river.

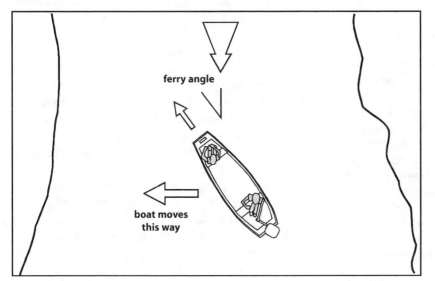

Fig. 2.29 Ferry: By applying power at an angle to the current, the boat moves across the current without moving downstream.

The *ferry* is one of the most useful river techniques for rescuers. The object is the same as it was in the days of ferry boatmen: to get across the river without going downstream. The ferry can also be used for crossing river currents from one eddy to another. The boat is set at an angle to the current and either held in position with a line or pushed forward with a motor or paddle. The resulting interaction of forces pushes the boat sideways across the river without losing its position.

A very common situation for rescuers is to find a victim stranded on a rock or car in the middle of the river. Even in very swift water, it's possible for a rescue boat to ferry across from an eddy along the bank to the eddy behind the victim, pick him up, and return the same way.

Swiftwater boat handling is mostly a combination of route finding and these three techniques.

Self-Rescue

A vital and primary skill for anyone who works on or near the river is the ability to perform a self-rescue. No purpose is served by making would-be rescuers into additional victims.

First, anyone who will be working on or near the river *must* have proper training in swiftwater self-rescue. Second, all rescuers should dress for the occasion. Dressing for the river is covered in more detail in the Chapter 3, but in general, rescuers should avoid wearing loose and heavy clothing, should wear a helmet and thermal protection such as a wet- or drysuit when appropriate, and *must* wear a life jacket, or PFD.

The *ten-foot rule* is a good one: all persons within ten feet of the water must wear a life jacket, as well as anyone whose location (such as at the top of a steep bank) puts them at risk of going into the water.

If you do go into the water, your immediate objective is *to get back to shore as soon as possible.* However, this may not be as easy as it sounds. The river's helical currents may push you away from shore and out towards the main current. If you find yourself in this situation: *Do not try to stand up or to stop yourself from going downstream by pushing against the river bottom with your feet.*

Trying to stand in the current invites foot entrapment. If your foot becomes wedged in a crack or crevice, the force of the current makes rescue difficult and drowning likely. If you are unable to get back to shore immediately, assume a *safe swimming position*. If the water is shallow and there is a danger of foot entrapment, the best swimming position is lying on your back, facing downstream with your feet up and

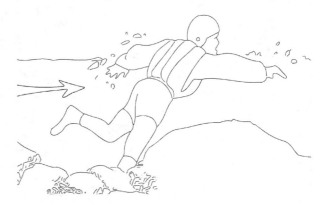

Fig. 2.30 Foot entrapment occurs when a swimmer's foot becomes wedged in a crack or crevice in the river bottom. The force of the current makes escape difficult.

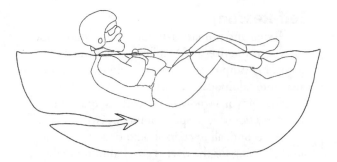

Fig 2.31 When swimming in swift water, face downstream and keep your feet near the surface. Fend off rocks with your feet. Do not try to stand until you have reached a safe eddy.

A Swimmer's Ferry

While swimmers have several disadvantages relative to boats, including reduced vision and slower speed, they can also use eddy turns, peel-outs, and ferries. A swimmer's ferry, for instance, can be very useful in self-rescue. To make an effective ferry, swim at about a 45° angle upstream to the current towards shore with either a backstroke or a crawl stroke. While swimmers will not, in most cases, be able to completely stall against the current, ferrying will slow their downstream progress and move them toward shore. Swimmers can also ferry away from hazards like strainers.

in front of you. This position minimizes the danger of entrapment, as well as allowing you to see downstream and to fend off obstacles with your feet. By backstroking at an angle against the current you can ferry towards the shore. In deeper water, where there is less danger of foot entrapment, rolling over and doing a crawl stroke works best.

As soon as possible after hitting the water, do a quick downstream survey. Try to determine if there are any hazards below and the location of any eddies, especially onshore eddies, that you might be able to get into. Work your way aggressively towards shore, avoid hazards, and aim for an eddy. *Do not attempt to stand until you are out of the main current and into shallow water.* This is sometimes known as the "safe eddy" rule.

Breathing. Since your mouth is close to the water, breathing in whitewater, especially in big waves, can be a problem. For starters, the inhalation reflex mentioned earlier may give a swimmer a mouthful of water immediately upon hitting the water. Turning your head to the side when going into waves will reduce the problem, as will experience. If you end up in a train of big waves, try to time your breathing so that you go for a breath after going through a wave, as you go down the back side of it. Proper thermal protection will reduce the body's reaction to being plunged into cold water.

Throw Ropes. If your team has properly set up the rescue site, they will have set up throw ropes to help swimmers get back to shore. If one lands near

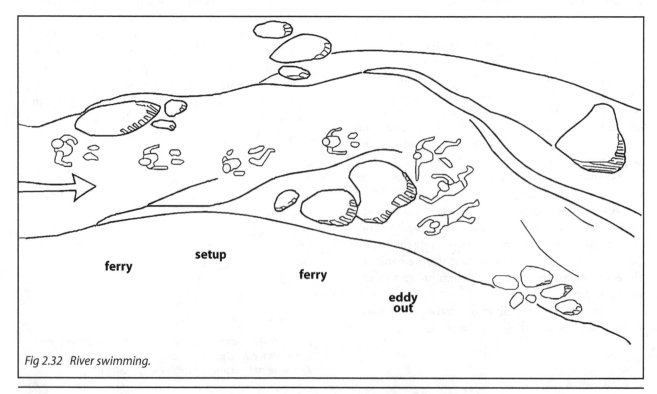

Fig 2.32 River swimming.

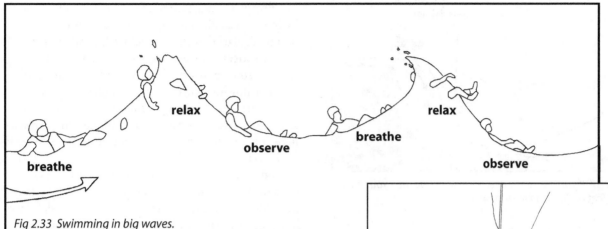

Fig 2.33 Swimming in big waves.

you, swim to it (without putting your feet down) and pull the rope to your chest. Roll over on your back and put the rope over the shoulder away from the bank you are headed toward. Face downstream with the rope over your shoulder, angling your body at about a 45° angle to the current. This body position will cause the water to flow around your back and past your head. Angling your body relative to the current will get you to shore more quickly and lessen the strain on the rescuers on shore. Do not grab the rope and face upstream. You will get a faceful of water and that may cause you to let go of the rope. In general, it

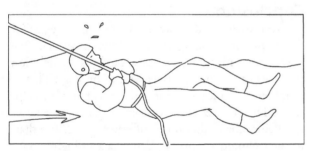

Fig 2.34 After grabbing a throw rope, roll over on your back and hold it against your chest.

is better to grab the rope and not the bag it is stuffed into, even though the bag is easier to hang on to. The reason this often isn't a good idea is because there may be more rope stuffed inside the bag, and you may go farther downstream than you intend to.

Also, *do not wrap or tie the rope around yourself or any part of your body.* If something goes wrong, or if you are pulled underwater, you need to be able to let go of the rope. If your hands are extremely cold and you are having trouble hanging on, however, you may clamp the rope under your armpits.

There are three special self-rescue situations that need to be considered separately.

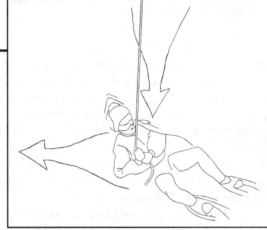

Fig 2.35 When holding a throw rope, angle your body toward shore. Hold the rope over the shoulder opposite shore.

Strainers. The best method of defense against strainers is to avoid them. If you see that you are headed for a strainer, swim aggressively away from it at a right angle to the current so that you will be carried past it. However, if you find yourself going into one anyway, you must quickly change your swimming position. Roll over and swim *forward* as fast as possible toward the strainer. Try to hit the strainer with some momentum and pull yourself up onto the strainer. The idea is, first, to avoid being swept under the strainer, and second, to get your body out of the water and away from the force of the current.

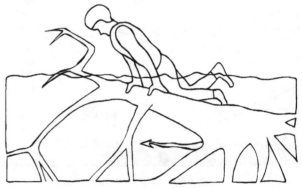

Fig 2.36 If you encounter a strainer, swim toward it head-first, then push yourself up and over it.

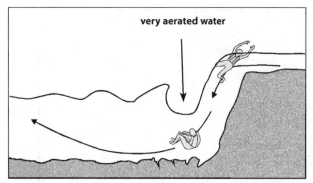

very aerated water

Fig 2.37 If you are swept over a vertical drop, pull your knees and feet up to avoid entrapment.

Vertical Drops. Even veteran whitewater paddlers have experienced foot entrapment or injury when swimming over a big drop in the river. The reason is simple: when a person in the safe swimming position (feet forward) goes over a vertical drop, he is suddenly pushed into a vertical position with his feet below him. The momentum from the fall then drives him under the surface and against the bottom. To guard against this, pull your knees up against your chest and "ball up" if going over a drop.

Holes, Hydraulics, and Low-Head Dams. In the accident on the San Marcos recounted at the beginning of this book, the kayaker was held by a breaking wave. Had he come out of his boat, he would have been carried downstream. Breaking holes and waves can bury a swimmer under tons of water, but he will swiftly be carried underneath them. Hydraulics, however, particularly those on low-head dams, present a much bigger problem. There, the recirculating current may trap and hold both boats and swimmers. A swimmer will find himself pulled to the face of the dam, pushed underwater, and then, upon coming to the surface downstream of the dam at the boil line, pulled back toward it again.

Escape from a hydraulic can be difficult. Natural hydraulics usually have open sides and often are not totally regular; that is, the backwash is often broken by downstream currents flowing through it. Here, the best strategy is usually either to try to find a jet of water flowing downstream, or to try to work your way to the sides of the hydraulic to catch the downstream current. This can be very difficult in a "frowning" hole in which the water tends to recirculate back to the center.

Low-head dams are more difficult still. Here, the ends of the dam are almost always blocked by retaining walls, and the dam has been deliberately designed as a regular feature with no breaks that might offer an escape route. In such cases, you may be able to escape by diving down under the backwash and catching the downstream-flowing current beneath it. Some hydraulics do not extend to the full depth of the river, while in others the recirculating current extends all the way to the bottom. Obviously, escape from a partial-depth hydraulic will be easier than from a full-depth one.

If escape fails, your options are limited. You may be able to hang on to a floating object in the hydraulic until help arrives. If you are able to work your way over to the side of the dam, others may be able to help you.

Conclusion

Knowledge of the ways of the river is the essential first step in the making of any swiftwater rescuer. One idea rescuers must give up at the outset is the delusion that they can "beat" the river with technology. They can, however, understand it and use its power to help them. The other essential concept is that swiftwater is different than other forms of water rescue, and that rescue techniques must be adapted accordingly.

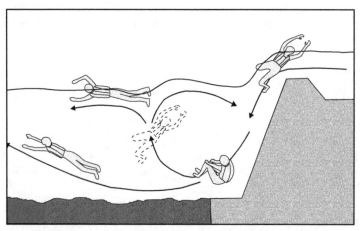

Fig 2.38 Escape from a low-head dam is difficult for a swimmer. Two possible methods are swimming downstream after surfacing at the boil line, or attempting to catch the water underneath.

Swiftwater Rescue

Equipment
Chapter 3

Fig 3.0 Gordon Grant , tethered with a rescue PFD, wades out to rescue a pinned kayaker on the Nantahala River.

After knowledge of the river and experience on it, the next most important component of river rescue is proper equipment. While essential, no piece of equipment can make up for lack of training and experience. Most river rescue equipment is simple and relatively inexpensive. After rescuers are provided with the basics, scarce agency dollars are often better spent on training.

The Basics—Individual Equipment

Individual equipment items either protect the rescuer or assist in rescues. For swiftwater rescue, personal protective equipment (PPE) items fall into three broad categories: flotation, thermal protection, and physical protection. All vary somewhat to meet on-site conditions. While there is much water rescue equipment on the market today, you should choose carefully. Many pieces of water rescue equipment look alike and have similar names, but may be designed for entirely different conditions. For example, gear designed for offshore rescue (e.g., life jackets and rescue harnesses) may be entirely unsuitable for swiftwater use.

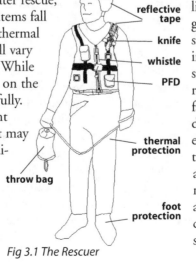

helmet

reflective tape

knife

whistle

PFD

thermal protection

throw bag

foot protection

Fig 3.1 The Rescuer

Personal Flotation Devices (life jackets, swim vests, flotation aids, buoyancy aids). The most basic piece of swiftwater rescue gear is the PFD. Its main purpose is to help keep a swimmer's mouth and nose above water, but the PFD serves many other purposes as well. It cushions the body from falls or blows—in or out of the water—adds effective insulation, and may be used to mount or store rescue gear.

Some agencies make a somewhat artificial distinction between a life jacket and a flotation or buoyancy aid. Under this definition, a life jacket is intended for extended emergency situations, such as those far from shore. Here, a victim is frequently in no immediate danger, but may face a lengthy stay in the water before being located and rescued. Therefore, life jackets are manufactured with more flotation and are designed to support an unconscious or exhausted person for extended periods of time. Since the victim is not expected to actively swim or to participate in his own rescue, bulk is not a major drawback, and there are often features like flotation collars intended to support the head. In swiftwater, however, it is often imperative

A Modern Swiftwater Rescue PFD

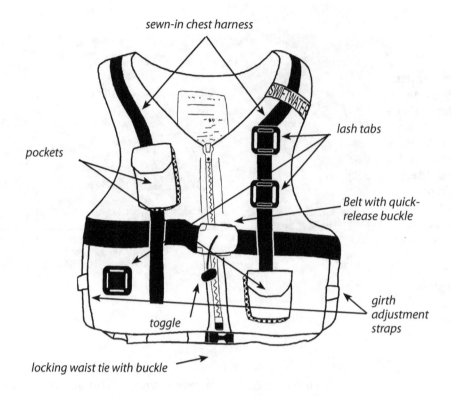

sewn-in chest harness

lash tabs

pockets

Belt with quick-release buckle

girth adjustment straps

toggle

locking waist tie with buckle

Fig 3.2

that a swimmer get out of the water as soon as possible, and that he be able to do it on his own. In that situation, a bulky offshore life jacket, while providing more flotation, may actually increase the danger to a river swimmer if it inhibits his swimming ability. Flotation or buoyancy aids (e.g. U.S. Coast Guard Type III) are intended for persons who are able to swim and help themselves, and where help (or shore) is close by. As such, they are designed not to interfere with or impede their activities. While swiftwater rescuers and whitewater paddlers normally use flotation aids, the terms life jacket, buoyancy aid, flotation aid, and PFD will be used more or less interchangeably in this book.

There are many types of PFDs available. While some are suitable for swiftwater use, many are not. For example, USCG-approved Type III water ski vests can be bought cheaply at many mass-market outlets, and while these work well for their intended purpose, they should not be used in moving water. The USCG approval process for PFDs is fairly generic, and does not test PFDs for suitability in swiftwater. In fact, the only in-water testing is done in a pool. The current regulatory process is cumbersome, time-consuming and expensive, and has done much to inhibit innovation in PFD design.

In addition, current regulations for wearing PFDs are confusing and inconsistent. In most cases, boating regulations require only that boaters carry an approved PFD on board, not that they wear it. Swimmers (who, one would think, are most in need of one) are not required to wear PFDs at all, nor are shore-based rescuers. There are, in fact, no national PFD regulations affecting all rescuers in the U.S. In almost all cases applicable to swiftwater rescue, either state boating regulations or federal agency regulations (e.g. Bureau of Land Management, National Park Service, U.S. Forest Service, etc.) will apply.

The bottom line is this: choose the best PFD for your application, not necessarily the one with the magic label on it.

What do we look for in a PFD for swiftwater rescue? Probably the most important single characteristic is fit. A PFD that is extremely bulky or that inhibits your movements, especially while in the water, should be rejected. The best swiftwater PFDs are vest-type jackets designed with a positive-locking waist tie and some sort of girth adjustment system so that it can be snugged down against the body. Most good designs have a cinching belt that fits under the rib cage to keep the jacket from riding up. It is vital that any PFD stay in position close to your body, even in turbulent water. If it rides up in your face, or entangles your arms, it is worse than useless. Often, the best choices are those PFDs designed for whitewater paddling or rafting, or specifically for river rescue. Do not hesitate to test any PFD *in moving water* before buying it.

The amount of flotation is another consideration. Do not assume that more is necessarily better. Most people have a nearly neutral buoyancy in the water. A wetsuit or drysuit can, by itself, add up to 10 pounds of flotation. The current mandated minimums are 15.5 lbs for the USCG and 6 kg for most European models. Most "rescue" vests have 18–24 lbs (8–11 kg) of flotation, and this is adequate for most people. If you expect to encounter really big water, or if you think you might have to rescue someone without a PFD, you may choose to go up to 28–30 lbs (13–14 kg). In no case, however, should the jacket interfere with your swimming ability. A PFD is not a boat, and no amount of flotation is going to keep your head above water at all times in really turbulent water.

Now let's consider specific PFD design features. In this country the Coast Guard requires that, in many types of PFDs, over half of a jacket's flotation be distributed in the front. The theory is that this will create a turning moment that will turn an exhausted or unconscious swimmer face up. While this may be true in calm water, *do not expect it to do so in swiftwater.* Flotation collars are another common design feature, and are required on some types of PFDs. The primary purpose of the collar is to protect the head (a helmet will do this much more effectively), as well as to give some additional support to the head of an exhausted or unconscious person. Both of these design features are derived from offshore rescue usage and cause problems in swiftwater. Concentrating flotation in the front of the jacket adds bulk where it may interfere with swimming and forces the designer to take it away from the back of the jacket, reducing impact protection. The protruding collar sometimes snags on throw ropes.

When evaluating a PFD for swiftwater use, consider its snag potential. Does it have anything on it that is likely to hang up or snag when used? Avoid PFDs with clip-type closures because these can clip into other things like lines when you least want them to. Horse-collar PFDs (USCG Type II) are totally unsuited for swiftwater, as are most inflatable PFDs.

One feature that is definitely worth looking for in a PFD intended for rescue use is pockets. Look for large, self-draining pockets with a positive locking closure. These can carry many useful personal items with less danger of snagging than if they were hung on the outside of the jacket.

Fig 3.3 Do not wear horsecollar PFDs in swiftwater!

Rescuers may want to consider making these modifications to their PFDs.

- Adding reflective tape to the shoulders. Some manufacturers have been adding these as an option.
- Attaching a knife (highly recommended, see the section on knives).
- Attaching a whistle.
- Installing crotch straps. This is a controversial option, with both proponents and detractors. Crotch straps *do* reduce the danger of the PFD coming off over the wearer's head in really turbulent water, especially for persons who are either very skinny (particularly children) or obese. Several recent PFD designs have the anchors for crotch straps built in, which allows the user the option of using them or not. Detractors point out that crotch straps increase the danger of snagging and make the PFD harder to get on and off.
- Attaching rescue gear such as carabiners, a Prusik loop, and perhaps a length of webbing to the PFD. This is also a controversial option. The best way to stow rescue gear is in pockets made for the purpose. Clipping carabiners to the outside of the PFD is certainly convenient, but there is a real risk that the carabiner will clip into something else inadvertently. If you choose to carry webbing or a Prusik loop on the outside of your jacket, be sure that it does not hang down and create a snagging hazard.

Chest Harness. A chest harness can be an effective piece of swiftwater rescue and safety gear. The applications will be discussed in more detail in later chapters, but suffice it to say for now that it can be used for shore safety, swimming rescues, and as a towing system.

A suitable chest harness for swiftwater use must:

- hold the maximum load without damage or slippage.
- release smoothly under maximum and minimum loads.
- have the attachment point for the tethering rope located so that the wearer's head stays above water in swift current (normally this is between the shoulder blades).
- have a release that can be operated with one hand in turbulent water.

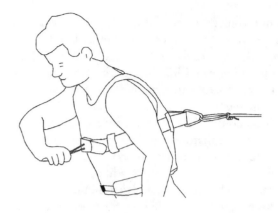

Fig 3.5 A quick-release buckle allows the wearer to release the tether when necessary. It is essential for swiftwater use.

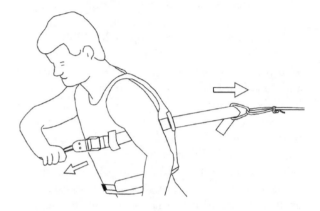

Fig 3.4 The chest harness in moving water.

A typical chest harness release system uses webbing threaded though a metal tri-glide retainer, then run into a plastic cam-lock buckle with a toggle on it. The metal tri-glide takes most of the force when the system is under pressure, allowing the weaker buckle to act merely as a retainer. The toggle allows the user to find the buckle in turbulent water. To release the system, the user pulls the toggle, opening the buckle and allowing the webbing to slip through the tri-glide and release. Most harnesses also include a D-ring cow's tail, or carabiner mounted on the back of the jacket to give a convenient point for clipping in.

An integral harness is designed as part of the life jacket. It cannot be removed, but is less likely to slip or snag, and is always instantly available when needed. The webbing, especially that run over the shoulders, adds substantial strength to the jacket. A separate chest harness may be worn over an existing life jacket. This avoids approval problems, can be adjusted for different people and life jackets, and can be switched easily to different PFDs. It is, however, more trouble to take on and off, can slip around more than an integral harness, and may snag. It may also be safely packed away when most needed. Combination harnesses—that is, harnesses having an attachment/release system designed for swiftwater use and a separate system for climbing

applications—have also been marketed. While ingenious, there have been accidents when inexperienced users attached themselves to the wrong system.

When choosing a chest harness, whether separate or built-in, be sure that it is specifically designed for swiftwater use. Some examples of chest harnesses that are *not* suitable for swiftwater use are: 1) those intended for tethering scuba divers, 2) offshore use, 3) climbing harnesses, 4) harnesses intended to be used by whitewater paddlers *only* for towing boats.

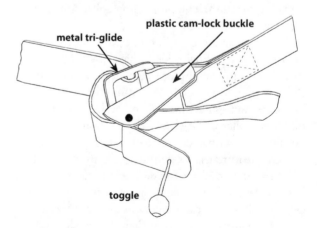

Fig 3.6 Detail of quick-release buckle.

These harnesses, while quite adequate for their intended purposes, are not suitable for swiftwater rescue use.

The primary accessory for the chest harness is the auxiliary tether or "cow's tail," usually a length of webbing attached to the back of the harness that allows the wearer to more easily clip himself into ropes, anchors, etc. A cow's tail is also an excellent way for a boater to tow a stray boat or an exhausted swimmer. Some cow's tails will extend under load, making towing easier. A cow's tail mounted on a life jacket increases the danger of snagging, so if you choose to attach one to your harness, make sure that it can be released at *both* ends.

Safety Tip: If you do not use a cow's tail and choose to clip a carabiner directly to the harness belt, be sure to use a locking model. There have been several instances of non-locking carabiners clipping into other parts of the jacket when abruptly pulled to the side, where they could not be released.

Fig 3.7 River knives come in all shapes and sizes.

Knife. Everyone who works in or around swiftwater should have a knife. This is particularly true when working with ropes and lines. If someone—victim or rescuer—gets entangled in a line, a quick cut will free them. With the right blade you may even be able to free someone entrapped in an inflatable raft or even a plastic kayak.

The ideal river knife should be compact and mounted so that it can be deployed one-handed with either hand. Some PFDs now come with a "lash tab" on which to mount a knife. The sheath should have a positive locking device to secure the knife. Rescuers should mount the knife so as to present as little snag hazard as possible. For general use, a sharp, thin, nonrusting, single-edged blade works best. For rescue work, however, a knife with a strong, thick blade suitable for prying may be better, particularly if rescuers need to break car windows during a vehicle rescue. The knife's handle should be substantial and made from nonslip material, since metal "skeleton" handles don't work very well in cold hands. While most people prefer a knife with a fixed blade, folding knives can also be used. Folding knives are more compact, and there are models that can be opened with one hand.

One situation rescuers may be faced with is having to cut a line that has wrapped around someone's arm or leg. Using a single-edged knife will reduce the chance of injury. Another solution is to use a safety knife, which was developed for use by emergency personnel to cut seat belts from auto accident victims without harming them. Although not as flexible for all-around use as a regular knife, they are definitely safer when working near a victim or around a loaded rope (e.g., high ropes) where the touch of a sharp blade may mean abrupt failure. Many river runners and rescuers also carry a small saw as part of their personal gear. For cutting through strainers and boats, a saw usually works better than a knife.

Clothing. Uncontrolled lowering of the body's core temperature, or hypothermia, is a major concern for rescuers. Almost all water that they can expect to be working in will be below body temperature, and may be contaminated as well. Moving water causes abrupt and extreme heat loss, and rescuers on shore are subject to wind-chill hypothermia. Obviously, some kind of thermal protection is needed.

For in-water protection, the most common form of thermal protection is a wetsuit or drysuit. *Wetsuits*, originally developed for diving, are form-fitting neoprene suits of varying thicknesses. They work by trapping a thin layer of water between the wearer's skin and the suit. The body warms this layer of water, thus providing an effective insulation layer. Although fit is critical for effective insulation, wetsuits need not be sealed and will work even when in poor condition. They also provide some cushioning if the wearer is swept into a rock, and compared to drysuits, they are

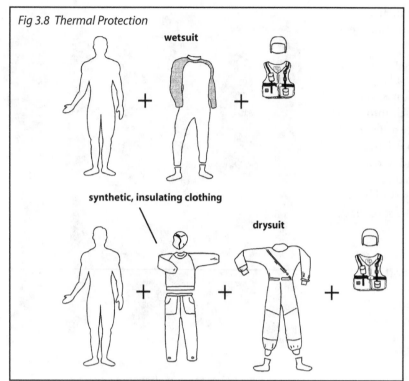

Fig 3.8 Thermal Protection

wetsuit

synthetic, insulating clothing

drysuit

clothes, in the event of a water rescue emergency. Drysuits are also better for rescues in toxic or polluted environments, since they do keep most of the water away from the rescuer's body. Drysuits, however, are more delicate and expensive than wetsuits. A torn gasket can reduce much of their effectiveness, nor do they provide much physical protection from rocks or other hazards.

There are also hybrids: suits made of neoprene like a wetsuit, but sealed like a drysuit and equipped with a windproof outer layer (many of these are designed for board sailing in cold conditions). These work very well in extreme conditions or for extended in-water use, but are too hot for most other work.

For warmer conditions, or for those rescuers who do not expect to actually get in the water, dressing is somewhat simpler. Since there is always the possibility of getting wet, rescuers should avoid cotton clothing when working near the river. Cotton soaks up water, dries slowly, and feels cold and clammy on the skin when wet. Synthetic piles or wool work much better as insulators when wet or damp. Use a layering system: several thin layers offer better and more flexible protection than a single thick one. For windy conditions, add an outer later of nylon. Coated nylon clothing gives protection against splashes but not immersion.

Use common sense when dressing for swiftwater rescue operations. While there is no sense in requiring everyone to wear a wetsuit on a warm day, it is equally silly to have people in cold weather operations standing on the bank in jeans and hunting boots. Because of the high rate of heat loss, however, in-water rescuers should always have appropriate thermal protection. Helicopter operations present yet another problem, which is discussed in more detail later.

We should also take a moment to dispel the myth that heavy clothing, waders, or a leaking drysuit will "drag you down." If you fill or soak anything with water and drop it in a river, it will assume an approximately neutral buoyancy—that is, it will neither float nor sink. A person with a PFD, even with soaked clothing or full waders, will

relatively inexpensive. Although recent wetsuit designs are much better in regard to fit, many are still bulky and restrictive. Unless carefully fitted, water can flow through them in swift current, greatly reducing their insulative value. Wetsuits work fine in the water, but not very well out of it, since there is then no insulating layer of water. In addition, unless the wetsuit has a windproof coating, users are very susceptible to wind chilling when out of the water. As with most other items, rescuers are advised to choose a wetsuit made for active swiftwater use as opposed to those intended for scuba use, since these tend to be somewhat restrictive. Look for a suit with full legs and extra padding on the lower extremities.

A *drysuit* is made from a waterproof synthetic fabric and, unlike the wetsuit, is sealed at the throat, wrists, and ankles with elastic gaskets (usually made of latex rubber). Entry and exit is usually through a waterproof zipper. Although not totally waterproof, the drysuit does generally seal water out and heat in. Instead of water, the user wears synthetic pile clothing as the insulating layer underneath. Drysuits work well in extreme conditions, although their heat-retaining properties may lead to overheating problems for those doing heavy physical labor out of the water. For rescue workers, an important advantage of a drysuit is that it may be quickly donned, even over street

Fig 3.9 Do not wear turnouts on swiftwater operations!

still float; however, these conditions will inhibit self-rescue. The weight and bulk of the absorbed or trapped water makes swimming difficult, and causes a "sea anchor" effect, making one tend to follow the current. These conditions, particularly for a poor swimmer not wearing a PFD, may lead to drowning. Any clothing a rescuer wears in or near the water should, therefore, fit snugly so as not to trap water or create drag when swimming.

> **TIP:** To gain traction on the river bottom, glue a thin piece of indoor-outdoor carpet or felt on the soles. Buy felt soles at shops catering to fly fishermen.

Use whitewater paddling clothing as a model for what to wear on the river. But whatever you wear, *make sure it is appropriate for river use*. Swiftwater rescue is not the primary job of most agency personnel like firefighters and park rangers. Their standard operating procedures, in fact, may specify that they wear clothing inappropriate for river work. Don't let this happen to you or your team—remember Mickey Gallagher's comment at the beginning of this chapter. Firefighters should not attempt river rescues in turn-outs any more than they would try to fight a structure fire in PFDs and wetsuits. NPS ranger Jim Traub also cautions that law enforcement personnel "don't need the 10 pounds or so that a duty belt adds. Plan to have a place to store weapons when you arrive on the scene and you won't be tempted to wear it on the river."

Foot Protection. Foot protection is often overlooked. River bottoms are usually slick, always cold, and very often strewn with debris, making some type of hard-sole boot or shoe designed for water use mandatory.

Helmets. Head protection is also mandatory for rescuers. Fortunately high-speed impacts are rare, so the choice of helmets is a wide one. Look for a helmet designed for water rescue that covers the forehead as well as the sides and back of the head. Avoid fire helmets, since they can cause neck injuries if the current catches them. Motorcycle helmets and flight helmets tend to be heavy and to restrict vision and hearing.

A large amount of heat is lost through the head, so it helps to get a helmet big enough so that a pile cap can be worn underneath.

Visibility. Rescues under conditions of poor visibility can be exceedingly difficult. One of the biggest problems is keeping track of the rescuers. Some solutions are:

● Reflective tape on the rescuers' helmets and on the shoulders of their PFDs.

water rescue helmet

climbing/rescue helmet

DO NOT wear fire service helmets on the river!

Fig 3.10

● Chemical light sticks either taped to the rescuer or kept in a mesh pocket on his PFD. Use different colors to identify different groups.
● A small, waterproof strobe clipped to the rescuer's arm or PFD. Strobes can be distracting, however, and are usually better used for things like getting someone's attention or for ground-to-air signals.

Swim Fins. A rescuer's speed in the water is greatly increased when using swim fins; however, using fins in moving current requires practice and the right fins. River fins (e.g. Shredder SAR, Churchill) are smaller than the fins normally used by scuba divers in calm water. Large fins are more likely to get caught by stray currents, and there has been at least one case where the larger fins were pressed against a rock and held there by the current. Smaller fins are also less clumsy when moving around on the river bank.

Communications. The simplest personal communication device (other than hand signals) is the whistle. Get a loud one.

Personal Rescue Gear. Some rescuers carry one or more carabiners and a pre-tied Prusik loop with them. If you choose to do this, carry them so that they won't snag anything when swimming.

Medical. Any modern rescuer is well-advised to carry and use personal protection when treating patients. A mask for CPR, even a simple one like the CPR MicroShield®, is a necessity, as are examination gloves. Carrying the mask and gloves on your person, as opposed to packed away in a medical kit, will ensure that you are prepared. It is yet another reason to have pockets on your PFD.

The Basics: Team Gear

Although many rescuers will have some of the more basic equipment in this section, like throw bags, in most cases this will be team equipment. This section is intended as an overview of what's available. Boats, helicopters, and technical rope gear will be discussed in more detail in the chapters dealing with their use.

Throw Bags. The most basic (and most often used) piece of gear for any swiftwater rescue team is the throw bag. It is simple, cheap, easy to use, and relatively safe for the rescuer. There are many variations, but the basic design is the same: a nylon sack with 25–70' (7.6–21 m) of rope stuffed inside, with a small chunk of foam included for flotation. The rope is usually a colored ⅜" (9.5 mm) braided polypropylene. Polypropylene, although weaker, is used rather than nylon because it floats, and its dynamic properties also reduce the shock loading on rescuers and victims alike. In practice, the rescuer throws the bag toward the victim, who then grabs it and is pulled to shore.

Throw bags come in a near-infinite variety, from small ones that can be clipped onto a PFD to "expedition" models with over 150' (46 m) of rope. Several manufacturers now make throw bags designed to be worn on the waist or even inside a PFD pocket. This makes swimming with a throw bag (normally a difficult thing to do) easy and safe.

Ropes. For the purposes of swiftwater rescue, ropes can be divided into three major categories:

- Floating ropes for water rescue. These are usually made of polypropylene and, with the exception of the Spectra™ core rope mentioned above, are not suitable for lifelines, technical rope applications, or mechanical haul systems. A coiled polypropylene rope can be effectively used in place of a throw bag.
- Dynamic ropes. These nylon ropes are usually intended for sport climbing and have a considerable dynamic stretch factor built into the rope in order to reduce the shock on a falling lead climber.
- Static or low-stretch ropes. These nylon ropes have a low stretch (typically 1–3% at working loads) and are suitable for the most demanding tasks of rescue work, including lifelines, technical rope applications, and mechanical haul systems.
- Rope Bags. Rope bags are the preferred way to store ropes prior to deployment. A bag protects the rope from UV deterioration and other environmental hazards, and reduces tangling and the danger of snagging. Rope bags are available with shoulder straps, which make transportation easy.

> **The poly rope used in most throw bags does not lend itself to mechanical haul systems. However, river runners, who do not have the luxury of trucks to haul their equipment, must make every piece of equipment do double duty. A new type of rope that is suitable for both throw bags and haul systems has a 6 mm (¼") Spectra™ core for strength and a braided polypropylene sheath for flotation, visibility, and abrasion protection. This results in a 3/8" (9.5 mm) low-stretch rope with a breaking strength of 4,500 lbf (20kN). While not perfect for mechanical haul systems (it has a low melting point and poor shock resistance), it is a significant improvement over polypropylene.**

> **TIP:** To avoid confusion, stencil the rope's length on the outside of each rope bag.

Fig. 3.12 Throw bags come in a variety of shapes and lengths. The bag on the lower right is made to be thrown like a football, the one above it to be carried about the waist.

Watercraft

While many rescues are made from shore or in the water without boats, there is no doubt that suitable watercraft can be not only useful but vital in many river rescues. These range in size all the way from rescue boards to Coast Guard cutters. We will concern ourselves here with the small end of things. As with most other things in swiftwater rescue, practice, experience, and judgment are the keys to successful rescues. Even marginal craft can often be used effectively by rescuers knowledgeable of the ways of the river, while the best boat in the world cannot make up for poor judgment and lack of experience.

In general, rescuers are better advised to choose watercraft that are light, simple and dependable than those that are technically sophisticated. Another highly desirable characteristic is a secondary propulsion capability—that it can be maneuvered with paddles or oars if the situation calls for it or if the motor fails. Consider your local conditions (and budget) when buying a boat. If only a small percentage of your calls are for swiftwater, you should be able to use the boat for other purposes as well. Narrow, rocky rivers call for small, shallow-draft boats, while bigger, deeper rivers dictate larger boats.

Canoes and Kayaks. Single-seat whitewater kayaks, in expert hands, have run some of the biggest and most difficult rapids on the planet. They can be a great asset to any rescue. However, the title of "expert" takes years to obtain, and this amount of training is beyond the means of most agencies. Expert kayakers are, however, sometimes available as volunteers in the local community. On the down side, whitewater kayaks have a very limited carrying capacity and are restricted in where they can travel by the muscle power of the paddler. Open canoes are generally not able to run as heavy a water as kayaks, but have a greater carrying capacity. Both are unforgiving in less than expert hands.

Rescue Boards. Originally marketed for surf use by Morey as the Boogie Board™, these small foam-core boards intended to be used with swim fins were quickly adapted for swiftwater rescue use by the California-based Rescue 3 company. Instructors found that although small, the boards added a substantial amount of flotation to a rescue swimmer and could be used for picking up a victim or as a rescue float. Although slow, they were very compact, easy to transport and store, and much easier to learn to use effectively than canoes or kayaks.

Several years later a larger board appeared, manufactured by Carlson. Designed especially for river rescue use, it was substantially larger than the Boogie Board™ and came with two sets of handles—one for

Fig. 3.13 The Boogie Board™ (left) and the larger Carlson Rescue Board (right).

the rescuer and one for the victim. Their disadvantages are that they are tiring to use and demand a high level of fitness, and because of their low position in the water give the rescuer limited visibility. Even larger boards have been introduced for casualty evacuation, and NRS now markets an inflatable version.

Inflatables. Inflatable watercraft represent an extraordinarily broad and varied category. At the small end are inflatable kayaks; at the big end are 45-foot semi-rigid-hull coastal rescue boats. Generally speaking, any inflatable depends for buoyancy to some extent on a flexible, air-filled chamber made of a tough rubberized fabric. While usually somewhat slower, inflatables are generally lighter, more stable and forgiving than rigid-hulled craft of similar capacity, and in most cases can be deflated for easier transportation and storage. Inflatables intended for swiftwater and off-shore use are designed with multiple air chambers, protecting them from the failure of any single chamber.

Rafts. By far the most common inflatable river craft is the raft. Rafting companies have sprung up all over the country and indeed the world and it often seems there is hardly a stream without them. Whitewater rafts share certain common features and are often very useful to the rescuer. Indeed, a four or six-person whitewater paddle raft often represents a simple, inexpensive solution to a wide variety of swiftwater rescue problems. Used ones can often be bought from outfitters at the season's end.

Fig. 3.14 An inflatable paddle raft.

Whitewater rafts usually range between 10–20' (3–6 m) long, with a width of about half their length. An inflated *tube* of heavy rubberized fabric, 15–20" (38–51 cm) diameter, runs around the outside of the boat, with 2–3 *thwarts* across it for rigidity. The ends of the raft have *rise* or *kick*, both to keep water out and to allow the boat to spin more quickly. The *floor* of the raft may be glued directly to the bottom of the tube, in which case the water must be bailed out manually, or an inflated floor can be attached by lacing or other means for a *self-bailing* raft. Water runs

out of a self-bailer between the floor and the tube or through scuppers as the inflated floor rises. Larger boats can carry six to ten passengers, and smaller ones three to four, although this depends on the difficulty of the water attempted. Rafts are propelled either by oars or with paddles. In paddle rafts, the crew sits on the outside tube with a paddle captain in back. In oar boats, on the other hand, the boatman uses a solid *rowing frame* of wood or metal to mount the 10–12' (3–4 m) oars. The oars are attached to the rowing frame either with oarlocks or by being clipped onto a thole pin ("pins and clips") attached to the frame. The frame is mounted either in the center or at the rear of the raft. Rafts with rear-mounted frames often use paddlers sitting in the front to add power and forward speed. Whitewater rafts are capable of running heavy water and carrying big loads, but require considerable skill and training to use effectively. On larger, deeper rivers in the western U.S., large motorized rafts (up to 39'/12 m) are often used.

Catarafts. A cataraft is the ultimate minimalist form of raft. It consists of two long tubes made of rubberized fabric held together with a metal frame. "Cats," like other rafts, can be powered by either paddles or oars, although some will take an auxiliary motor. In the right hands, catarafts are capable of running extremely heavy whitewater. Their long tubes and lack of a floor give little for the water to grab, making them extremely stable in rough water and able to punch through large hydraulics. For situations requiring entry in or near a hydraulic such as a low-head dam, the cataraft is the boat of choice. It can also be packed down into a very small package for travel into remote areas. On the down side, catarafts have a limited load capacity and, since they have no floor, are

Fig. 3.15 This motorized 23' (7 m) "Snout Rig" uses 36" (.9 m) inflatable pontoons joined with an aluminum deck/frame.

Swiftwater Rescue

Fig. 3.16 An oar-powered cataraft.

more difficult for rescuers to move around in. In addition, anything that is dropped is gone forever.

IRBs. IRB stands for Inflatable Rescue Boat (Australia, NZ) or Inshore Rescue Boat (U.S., Canada). These inflatable rafts are designed to be used with motors. They look much like their whitewater cousins, but instead of a tube going all the way around the boat, they have a wooden *transom* across the stern to accommodate the motor. Originally designed for surf rescue in Australia, the IRB has been adapted for river use as well. Most are 10–20' (3–6 m) long, with 15–20" (38–51 cm) tubes and multiple air chambers. Unlike whitewater rafts, IRBs seldom have thwarts or cross-tubes. Instead, they derive their rigidity from other design features such as the transom or from a rigid floor. Some also have handy features like a fuel bladder mounted in the

Fig. 3.17 An IRB used for surf rescue. Note the prop guard. The lifeguard on the left is wearing a waterproof radio pouch.

bow tube (which eliminates the need for a separate metal fuel can), foot cups for the crew, and self-bailing scuppers in the transom. Some IRBs, like whitewater rafts, have either rubberized fabric floors or inflatable floors to make them self-bailing. Others have rigid floors made of wood, fiberglass, or aluminum, which improves their speed. Some IRBs have inflatable keels or "speed tubes" on their hulls. While effective on open water they are unsuitable for shallow, rocky rivers.

Sometimes it's hard to tell exactly where an inflatable ends and a rigid-hull craft begins. There are rigid-hulled inflatables (RIBs or RHIBs) that have a rigid fiberglass boat hull with an inflatable sponson surrounding it. This combines the stability and unsinkability of an inflatable with the speed of a rigid-hulled boat. Some very large examples of this type of boat (over 45'/14 m) have been built for offshore use, complete with decks and cabins.

The advantages of IRBs are many:
- Ease of deployment. They are relatively light and easy to deploy, even without a trailer. If necessary they can be deflated and packed into remote locations, manhandled down steep river banks, or flown in by helicopter either inflated or deflated.
- Safety. They are stable and forgiving in swiftwater and capable of running fairly heavy whitewater or surf conditions. In the event of a mishap like a flip there are fewer hard surfaces to hit crewmen or victims. If the motor fails, most IRBs are light enough to be paddled.
- Training. Because of their stable and forgiving nature, they require less training and a lower skill level than most other types of rivercraft for successful mission accomplishment.

IRBs have disadvantages, too. Compared to rigid-hulled boats, they are expensive (typically costing three to four times as much), and relatively slow. Their rubberized skin, while quite tough, is subject to abrasion, tearing, and environmental deterioration. Rescuers must also allow time for inflation before use although rapid inflation kits can cut this down to a minute or two.

Rigid-Hulled Boats. For our purposes, a rigid-hulled boat is one that does not depend for stability or flotation to any extent on inflatable tubes or sponsons. Many different rigid-hulled craft are in use as rescue craft all over the world, ranging from lowly Jon boats to Coast Guard cutters. Most of the smaller ones that we will be concerned with here are converted recreational models. Boats adapted for rescue use generally have more flotation, additional lashing points, and reinforcing of critical points on the hull.

Fig. 3.18 This large rigid-hulled craft uses many of the same design features as IRBs.

Rigid-hulled boats are generally less expensive than comparable inflatables, have a very long service life, and are more readily available on the second-hand market. They are also, in most cases, quite a bit faster. However, they are less stable than rafts and IRBs, and cannot be deflated for transportation and storage. They are more bulky and slower to deploy, and most require a trailer and a launching ramp. In addition, there are more hard surfaces to hit in the event of a flip, and the higher freeboard of a rigid-hulled boat may cause difficulties in getting victims aboard.

Boats are made from many materials, the most popular being aluminum, fiberglass and, for larger vessels, steel. While there are a nearly infinite variety of hull shapes, few of these are designed for swiftwater. In general, the flatter a hull is on the bottom, the more *initial stability* it will have—that is, it will feel stable on flat water. More rounded hulls will feel less stable. In large waves, however, the effect is reversed. The flat bottom tends to follow the face of the wave, and may be lifted up to the point of capsize, whereas the rounded hull has less of a tendency to follow the wave's surface contour and so feels more stable in rough water. The part of the hull above water is called *freeboard*, and the transition between the bottom and sides of the hull is the *chine*. An abrupt curve is a hard chine, and a more gradual one a soft chine. A hard chine resists rolling, but when this effect is overcome, the transition is an abrupt one. For example, the typical Jon boat has a nearly flat bottom and a very hard chine. It is stable in calm water, but if tilted beyond a certain point it will capsize with little warning.

Much of the limitations of hull design can be overcome by proper training and by using techniques like weight shifts and boat leans. This is true even for craft like Jon boats. The Ohio Department of Natural Resources, for example, routinely uses modified Jon boats for swiftwater rescue training.

One exception to the statement that most rigid-hulled boats are not designed for whitewater is jet boats. These craft are propelled by a water jet pump. Since they lack a propeller, they have a very shallow draft. Although rather large, jet boats are extremely fast and have successfully made upstream runs of the Colorado River through the Grand Canyon. They are routinely used on whitewater rivers like the Salmon and Snake. Given proper operator training, jet boats can be a valuable rescue asset on many large-volume rivers. Jet drives are also available for more conventional craft.

Fig. 3.19 With proper operator training, jet boats are capable of handling whitewater rivers.

Personal Watercraft. Another type of watercraft that has recently been pressed into service for swiftwater rescue is the personal watercraft (e.g., Jet Ski™). Introduced in 1987, PWCs have become increasingly popular for recreation, rescue, and law enforcement use. They are very fast, quick to deploy, and are light enough to be launched and operated by one person, or even deployed from a helicopter. Since the personal watercraft lacks an exposed propeller, it is safe for both rescuer and victim to swim around while it is running. Their initial cost, upkeep, and maintenance are relatively low compared to conventional boats or IRBs. To reduce costs further, the Yamaha, Kawasaki, and Bombardier companies have instituted loaner programs for agencies.

On the down side, personal watercraft have a very limited carrying capacity, although most are able to tow a litter-type rescue device behind them. They require a thorough training program and, although they excel on moderate to easy water, they are not suited for heavy whitewater. In addition, the intake screen and pump are vulnerable to clogging in debris-laden flood waters. The Indiana River Rescue School has been a leader in developing applications and techniques for personal watercraft.

Hovercraft and Air Boats. Both hovercraft and air boats have also been used and promoted as river rescue craft. A *hovercraft* looks like an inverted bowl with a fan on top. The fan blows air into this plenum

Fig. 3.20 *A personal watercraft used for surf rescue. A rescue board is attached to the stern.*

chamber, which then escapes under a flexible skirt, lifting the hovercraft a few inches off the ground. Another fan mounted on the rear drives it forward. The primary advantage of the hovercraft is that its air cushion will float over terra firma, water, bog, muskeg, or nearly anything else, giving it unrivaled flexibility for mixed environments. It is also very fast. However, they are very expensive, carry a relatively small payload, and require a fair amount of training. Since the hovercraft does not touch the water but rather balances on a cushion of air, it is more difficult than with conventional craft to approach victims and get them on board. Since big waves will disrupt the air cushion, hovercraft are not suitable for any but the easiest whitewater (one agency completely trashed a hovercraft while attempting to run a Class III section of the Rio Grande). The fan blades of a hovercraft are also vulnerable to flood debris. Once disabled, a hovercraft, unlike a boat, cannot be propelled by a secondary system like paddles or oars.

Air boats were developed for traversing the swampy terrain of the Everglades when someone got the idea of mounting an aircraft engine on a flat-

Fig. 3.21 *Hovercraft (front) and air boat (rear) belonging to the Washington, D.C. Airport Authority.*

bottomed skiff. The resulting craft was extremely fast, had a very shallow draft, and could zip across open water and swampland with ease. Like a hovercraft, it can go places a conventional boat can't. Since an air boat has a conventional rigid boat hull of fiberglass or aluminum, it can stop dead in the water to pick up victims. Air boats have no exposed prop to endanger swimmers or to foul on lines or flood debris. They have been used in up to Class II whitewater.

Fig. 3.22 *An outboard jet drive (right) is paired with a conventional prop unit (left) that acts as a backup..*

A Word About Motors

Jet drive units are similar to props except that the motor drives an impeller mounted on the lower part of the engine that shoots a stream of water for propulsion. This is much safer for anyone in the water and allows for operation in very shallow water. However, jet drive units deliver less power than a prop, are somewhat heavier, are ineffective in reverse, and are more prone to clog with debris. They may be either outboard or inboard units.

The pump jet combines many of the advantages of both. Instead of an impeller the drive unit is an axial turbine similar to that on a jet engine. The pump jet gives the performance of a prop with the safety of a jet drive (since there are no exposed blades).

They are resistant to clogging, give excellent control at low speeds, and have more thrust per horsepower than props. Widely used by the military, they remain an expensive option.

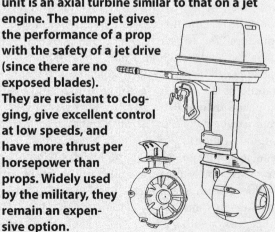

Fig. 3.23 *Pump jet*

However, their high center of gravity makes them unsuitable for heavy whitewater. Air boats are expensive to buy and maintain. In addition, they are also rather large craft, which limits their deployment flexibility. However, as an all-purpose flood (as opposed to swiftwater) rescue boat they are hard to beat.

Choosing a rescue craft is never easy. Frequently, the deciding factor is the agency budget; however, expense need not compromise effectiveness. Most of the rescues described in this book are well within the capabilities of a four-person paddle raft and an experienced crew.

Communications

Communications are essential for any team operations. The most basic are voice, whistle, and hand signals, which are discussed in the chapter on incident command. Some of the more technical communications systems are discussed below.

Radios. Radios are vital for incident command. One of the major, recurring problems of river rescue is that of on-site communication. The river itself is a wet, abusive, high-noise environment. On the positive side, river rescue sites (as opposed to search sites) are usually relatively compact. Fortunately for rescuers, the cost of hand-held FM radios has dropped dramatically in the last few years, ensuring that most agencies can now afford them. The problem has been the lack of a reliable, waterproof, voice-operated radio that can be worn by a rescuer while in the water. One solution has been flexible plastic waterproof housings (e.g., Ewa) for on-river use. Another promising innovation has been a waterproof tactical radio pouch with headset (Television Associates' "Occasional Swimmer's Kit"). While expensive, it does allow sustained underwater operation at depths of up to 25' (7.6 m).

Cellular Phones. Cell phones have become the communications system of choice in many urban areas. New models are tiny, and a few are waterproof. For those that are not there are waterproof plastic housings. One advantage of a cellular phone is that rescuers may dial each other directly without making a net call, as well as being able to dial off-site. Another is that it is harder for other parties to eavesdrop, and there is less tendency for several people to overload the net. With the advent of digital satellite communications systems, it may be possible to use cellular phones anywhere in the world. Cell phones have their disadvantages as well. They are frequently overloaded during emergencies and may stop working altogether, making them more suitable for supplemental rather a primary mode of rescuer communication.

Fig 3.24 One solution for on-water communications: a waterproof boom mike, and a radio pouch.

Bullhorns. The bullhorn is a mainstay for high-noise environments. Radios or cellular phones work better for contacting rescuers, but a bullhorn is often essential for warning off spectators and upstream watercraft.

Smoke and Mirrors. In addition to communicating with each other, rescuers must sometimes mark their position for others, particularly aircraft. Pyrotechnics, mirrors, strobes, dye markers and marker panels afford excellent marking capability and can also be used for crude ground to air communications.

Pyrotechnics include colored smoke and flares. Smoke is most effective during the day. Some aerial flares may be visible for up to 20 miles under ideal conditions (5–10 miles/8–16 km, is more common); however, they may be dangerous to use, particularly in dry conditions. Ground or hand-held flares will also work for marking.

Dye markers can be sprinkled on snow or thrown in the water to create a high-visibility spot. Panel markers, usually plastic 4' x 8' international orange panels, also work well.

Mirrors are very effective for signaling aircraft. They are cheap, lightweight, and require no power. Signal mirrors have an aiming grid in the center and can be used at distances of up to 20 miles (32 km). Glass mirrors work best but are fragile; plastic and metal mirrors last longer under adverse conditions. If you have a bright light source, mirrors can also be used at night.

The price of *strobe lights* has come down enough for almost any team to be able to afford them. The lightweight, waterproof models now on the market make marking easy at ranges of up to three miles (5 km). Strobes can be distracting for search use, but are ideal for signaling aircraft.

When considering signaling options, don't overlook ordinary flashlights and car headlights. Chemical light sticks work well for marking areas like landing zones at night as well as team members.

Line-Throwing Devices

Another recurring problem in river rescue is getting a line across the river. It is simple to say, hard to do. Line ferries will be discussed in more detail in the boat handling chapter. For now, it is enough to look at some of the specialized equipment devised for the purpose. Almost all these devices work in similar fashion, using a technique that has been around at least since the days of sailing ships. A light line (the *messenger line*) is somehow (this is the hard part) passed over to the other bank, then used to pull over a heavier line, or series of lines.

Line Guns. Line guns shoot a line-trailing projectile across the river. They range from units shooting a lightweight plastic projectile 250' (75 m) with a .22 blank to a .45-70 rifle firing a brass rod more than 800' (250 m), to "line cannons" with a reported range of well over a mile. Mossberg makes a line launcher kit for its Model 500 shotgun with a reported range of up to 700'. Rescuers have adapted tear gas launchers and hunting dog decoy trainers, and Res-Q-Max makes an air-powered line rocket with a range in excess of 400' (122 m). For very wide rivers these guns can be invaluable, but they are expensive and some of them require a federal firearms license.

Slingshots, Bows and Arrows. Hunting slingshots can fire lead fishing weights over 100' (30 m) with a light fishing line. Bows shooting heavy hunting

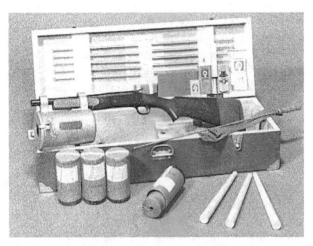

Fig 3.25 This line gun is made from a sawed-off 45-70 rifle. It fires a brass rod about 800' (250m).

shafts will also work, but crossbows are better and are capable of shooting a line-carrying bolt up to 150' (46 m).

Line Throwers. Not all line throwers are guns.

The B.E.L.L. Balcan Life Line resembles an old-fashioned potato masher. It has 130' (40 m) of light braided twine inside, and was designed for passing lines between ships. Effective range with the added "flick stick" is just over 100' (30 m).

Fig 3.26 This line gun uses a .22 blank to throw a soft projectile about 250' (75m).

Choosing The Right Messenger Line

A messenger line is a light line shot or ferried across the river and used to pull over a heavier line. In swiftwater, some special considerations apply. Many systems use a light twine or monofilament line as the messenger line. Since the force on the line can easily reach several hundred pounds/kgs in heavy current, these may not be adequate. Small-diameter line can also be very hard on cold hands, even with gloves. Parachute cord (tensile strength of about 550 lbf/2.45 kN) works much better than twine or monofilament and is very compact, although it is still hard to handle. Another solution is to use a ¼" (6.5 mm) Spectra™ line to pull the heavier lines across. You can also use a succession of lines: first, a light monofilament, then a ¼"(6.5 mm) line, and finally a ½" (13 mm) kernmantle rope.

Expedient Line Throwers

For those truly strapped for funds, here are some workable expedients.

● A water balloon slingshot throwing a softball. The softball has an eyehook screwed into it and a light line (such as surveyor's twine) attached. Variations include smaller slingshots throwing baseballs or even golf balls. If no slingshot is available, you can always just throw the ball across.

● Take an empty plastic milk jug and punch a hole in the bottom. Using a light line (100–120'/45–55 m of parachute cord or cheap poly), tie a knot through the hole and stuff the line inside. Put some water in the jug for better range.

Line Capture Devices

Line capture devices allow rescuers to catch lines floating in the water, even in swift current. This is very helpful, for instance, when making a rescue from one side of the river only, either because access is limited or the river is wide. A rescuer may throw a line upstream past a victim, allow it to float down past him and then snag the line with a capture device to pull it to shore. It may also be a great help when rigging a pinned boat in mid-stream.

Two current examples are the Crossline Reach System and the Snag Plate. The Reach System is a triangular metal device with spring clips that will grab a floating or tensioned line when pulled over it. It works well but is rather expensive and can snag on obstacles. The Snap Plate is a simple, inexpensive toothed plastic wheel that fits inside a throw bag. When pulled over a floating line the teeth snag it and allow it to be pulled ashore.

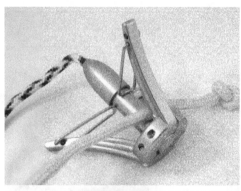

Fig. 3.27 *Line capture devices: Crossline Reach System (top) and Snag Plate (bottom)*

Fire Equipment

Fire companies often have many useful things for the river rescuer; however, each piece of equipment should be carefully evaluated before use. While hose inflation devices and fire ladders have proven applications, firefighters should not attempt swiftwater rescues in turn-outs, nor try in-water rescues with inappropriate equipment like Pompier belts.

Hose Rescue Devices. One of the most useful pieces of equipment for firefighters is the common fire hose. When fitted with an airtight cap on one end and an air inflation fitting on the other, the fire hose becomes a floating rescue device that can be extended to victims in the water. The 2" (6.5 cm) diameter standard fire hose, inflated to 25–30 psi (172-207 kPa) with air or nitrogen, works well for this application. Larger hose sizes provide more flotation but are more difficult to maneuver. Hose inflation devices can either be fabricated locally or purchased commercially. Such a device is simple, cheap and effective, and should be on every fire truck that has any possibility of being involved in a river rescue. Their use is discussed in more detail in Chapter 11.

Pieces of retired fire hose make excellent edge pads for technical rope operations.

Ladders. Another handy, yet common, rescue device for firefighters is the fire ladder. Ladders make excellent "reach" devices for shore-based rescues.

They are also very handy for getting up and down steep, slippery banks and can be extended down from bridges. An inflated fire hose can also be interlaced in the ladder's rungs to give it more flotation on soft surfaces.

Aerial Ladders. A truck-mounted aerial ladder extended to a victim can make a very effective "reach" device. However, not all aerial ladders can be used this way. When the ladder is fully extended at or below horizontal, it creates a huge lever that puts pressure on the traversing and elevating mechanism that the designers may not have anticipated. When you add the weight of one and often two persons on the end of the ladder, the strain becomes great indeed. Any fire company that anticipates using aerial ladders in this way should carefully check the manufacturer's specifications first.

Pike Poles. The common fire department pike pole also makes an excellent "reach" device for river rescues. Adding a loop of webbing on the end helps a victim with cold hands hang on to it.

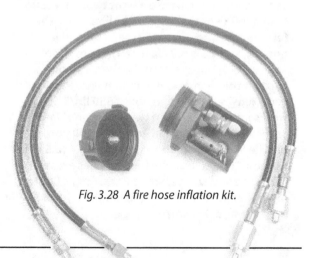

Fig. 3.28 A fire hose inflation kit.

Rigging for River Rescue I
The Basics
Chapter 4

This chapter covers the basics of technical rescue: the equipment, including ropes and technical gear, and knots, slings, and improvised harnesses. While this is not intended as a treatise on technical ropes, it soon became apparent when writing that it was just not enough to cover only those areas where swiftwater practice differed from other technical rope applications. This is especially true since technical ropes have themselves become a hot and somewhat contentious subject in the SAR community in recent years, with rival schools promoting their own solutions and attacking those of others. I have tried to avoid endorsing any one school of ropes, but rather to set out the alternatives and let the user decide what's best. It's been my experience that in swiftwater applications, practical solutions are often of the "quick and dirty" type. River rope rescues differ substantially in many ways from conventional high-angle practice, and most closely resemble those used in caving. The ropes (and indeed, all gear) are usually wet and muddy, and the forces encountered can be surprisingly high. River rope systems, even high lines, are seldom very high above the water; riggers must constantly be alert for the danger of entanglement if the rescuer or victim somehow falls into the current.

Ropes have many uses in swiftwater rescue. These include:
- Throw ropes
- Moving casualties and rescuers up and down river banks
- Moving litters and personnel across rivers
- Unpinning rafts and boats
- Rigging for helicopter operations
- Bridge rescues
- Positioning rescue boats in the river

In later chapters, we will consider the differences between low- and high-angle rescue, methods of rigging and recovering boats and rafts, and basic high lines. Still further along, we'll look at rigging for helicopter operations, bridge rescues, and tethered boat rescue systems. Since rope equipment is a rather specialized field, we have put the discussion of it in this chapter rather than in the equipment chapter.

Ropes and Webbing

Ropes. Humans have made and used natural fiber ropes for thousands of years. In the years since WWII, however, synthetic ropes have gradually displaced natural fibers and have come to be used almost exclusively because of their superior strength and resistance to rot. Today's ropes are literally marvels of technology, and outright failures are virtually unknown. While the coils in the store may look similar, they are in reality quite different.

The ideal rope for technical rescues in swiftwater:
- Should have low stretch, absorb shock loads, (1–3% elongation under typical working loads), be very strong, and resist abrasion as well as environmental factors like ultraviolet deterioration.
- Should not absorb water, lose strength, or become excessively slippery when wet. Ideally, it would float and be colored so as to be visible on the water.

Unfortunately, there is no rope currently manufactured that fits this description. Nylon fits most of the requirements in the first category, except that it is subject to UV deterioration. Spectra™, a recently introduced synthetic material, is stronger but much more expensive, has a lower melting point and poor shock-absorption characteristics. Nylon rope does

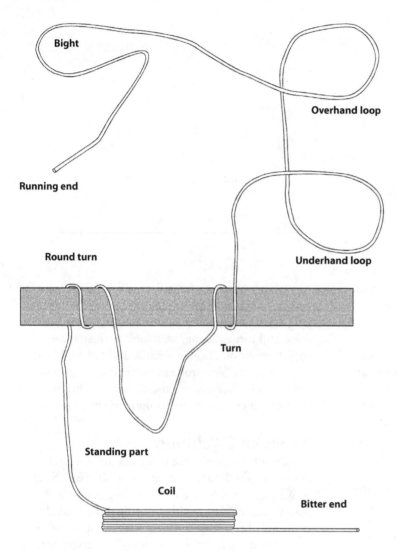

Bight

Overhand loop

Running end

Round turn

Underhand loop

Turn

Standing part

Coil

Bitter end

Fig 4.0 Rope Nomenclature

not score well at all in the second category. It absorbs water, which can reduce its strength up to 15%, and does not float.

Polypropylene ropes, on the other hand, score well in the second category but are simply not strong enough for serious technical work. A typical ½" (13 mm) nylon rescue rope has a breaking strength of about 10,000 lbf (45 kN), whereas a poly rope of the same diameter is rated at about 4,500 lbf (20 kN). Poly ropes also have much more stretch than nylon, a lower melting point, and less resistance to abrasion. For these reasons we will almost always choose a *low-stretch*, or *static*, nylon rescue rope for technical river rescues. Poly ropes will be relegated to throw lines and emergencies. While we are on the subject of stretch, it is worthwhile to note that we do want *some* stretch in any rope to help absorb the inevitable shock loads.

Thus we do not want a rope with no stretch at all (e.g., steel cable) nor do we want a *high-stretch*, or *dynamic*, rope such as those used for lead climbing.

Rope Construction. *Laid* ropes have bundles of fibers woven into strands, which are then twisted together around one another. This is the traditional way of making a rope. Ropes made this way often twist under load and are vulnerable to abrasion. Braiding (*solid braid*) the rope solves the twist problem, but does not improve abrasion resistance. Adding a separate braided outer sheath over a braided core (*braid on braid*) helps protect the inner braid from abrasion. Both sheath and core bear part of the load. Modern rescue and climbing ropes take this concept a step further where the *kern*, or core, is surrounded by a braided *mantle*, or sheath, to make a kernmantle rope. It differs from braid on braid in that the core is not braided and bears most of the load. The mantle's function is primarily abrasion protection. For dynamic (high-stretch) ropes, such as those meant for lead climbing, the core is twisted into a spiral pattern. This feature allows it to elongate under sudden loads like leader falls. A static or low-stretch rope, however, has the core strands laid parallel to each other, resisting elongation under load. Any rescue rope used for life support should have *block creel* construction, which means that the fibers run from one end of the rope to the other with no splices. When we talk in this chapter about ropes used, we mean a low-stretch nylon or Spectra™ kernmantle rope.

Rope Strength. When talking about rope strengths, we need to distinguish between the *tensile strength* of the rope and its *working load*. The tensile strength of the rope is the force required to break it. This can be readily determined by testing. However, in the real world, particularly on the river, factors begin to intervene that reduce the strength of any rope, sometimes drastically. These include knots, age, environmental deterioration, abrasion, harmful substances, water, and prior use.

Obviously, we don't want the rope to break, particularly if we are hanging from it, so we factor in these variables (which are difficult to actually measure) by using a more or less arbitrary figure to reduce the actual load we put on the rope. This figure is the *working load*, and the ratio between the working load and the tensile strength of the rope is the *safety factor* or *safety margin*. For example, if we have a rope rated at 6,000 lbf (27 kN) and wish to maintain a 10:1 safety margin, the maximum working load would be 600 lbf (2.7 kN).

The safety margin varies with the type of work done. If we are using a haul system only to recover a boat, a lower margin is acceptable. For general hauling and lifting, where the rope is not a lifeline, a factor of 4-5:1 is generally considered adequate.

For lifelines (i.e., where a person's life depends on the integrity of the rope), we use a higher figure. While cavers and mountaineers generally accept a safety factor of 10:1 as adequate, the National Fire Protection Association (NFPA) recommends a 15:1 safety factor for life safety ropes. However, safety margins are often very difficult to apply in river work where equipment is wet and loads unpredictable.

Rope Care. Because of their critical function and complex construction, there are special considerations for the care of rescue ropes. Since their lives may depend on them, rescuers are well advised to take special care of any ropes used as lifelines.

● Protect the rope from physical abuse when using it. *Do not step on the rope!* Use edge protection to reduce abrasion and don't run the rope over sharp edges or around tight bends. Don't drag the rope on the ground if you can avoid it. Don't let anything fall on the rope, and don't overload it.

● Avoid overheating a rope when using it. Nylon begins to melt and deteriorate at relatively low temperatures (440–460° F/226–238° C). Running nylon over nylon, whether rope or webbing, quickly generates enough heat to melt or damage the fibers. Rappelling devices may also overheat ropes (e.g., "flash" rappels).

Types of Rope

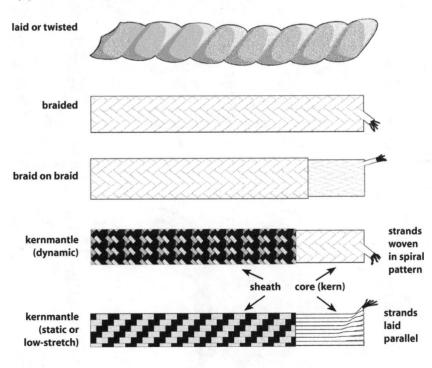

laid or twisted

braided

braid on braid

kernmantle (dynamic)

kernmantle (static or low-stretch)

strands woven in spiral pattern

sheath   core (kern)

strands laid parallel

Fig 4.1

What's in a Load?

The performance of any piece of rope-related equipment depends to a large extent on how it is loaded. Various groups have, for planning and testing purposes, attempted to define a standardized load that might be expected in a typical operation. Much of the rope equipment and techniques used in mountaineering and sport climbing are similar to those used in rescue operations. However, a *climbing load*—the weight of a single sport climber and his gear—is much less than that expected in a technical rescue situation, where the *rescue load* might be as much as one victim, two rescuers, and associated rescue gear such as a Stokes litter. Thus, equipment like slot-type belay devices (e.g., Stitch plate), that work quite well to catch a lead climber, will not hold a rescue load. When selecting equipment and figuring safety margins, these differences in design and performance must be taken into account.

There are, however, no universally accepted numbers. Some rescue authorities use 80 kg (176 lb) for a "climbing" load and 200 kg (448 lb) as a mountain rescue load (one patient, one rescuer, and gear). The National Fire Protection Association, whose standards are widely used in the fire-rescue community, defines a two-person "G" rescue load as 600 lb (272 kg). Any rescue organization is well-advised to determine what their standard rescue loads will be and select equipment and techniques accordingly.

- Clean the rope after use, especially when it's wet and muddy. Commercial rope washers are available for this purpose. Do not use solvents or harsh detergents on ropes; follow the manufacturer's directions for cleaning (most recommend cold water and a mild detergent) instead. Allow the rope to dry thoroughly before storing.

- Inspect the rope before and after each use. If it displays physical damage (cuts, abrasions, glazed or melted areas, "boogers," etc.), do not hesitate to cut off the damaged section or retire the rope altogether. Keep an accurate log of when and what kind of use it's had. Mark your ropes so that you don't get them mixed up. Use your newest and best ropes for lifelines, and periodically downgrade them to training or utility purposes.

Fig 4.2 Rope and equipment bags

Rope Storage. Remove any knots, then coil, chain, or bag the rope carefully and store it in a dry, well-ventilated place away from direct sunlight and chemicals like battery acid and exhaust fumes. Webbing should be treated with the same respect due rope. It is usually chained rather than coiled.

A rope bag provides the best protection for rope transportation and storage, and can be used for edge protection in a pinch. It also makes for speedy rope deployment, as long as you remember that the way that you put the rope into the bag is the way it will come out. Rope bags range from the very simple to elaborate bags with shoulder straps and outside pockets for associated gear.

If no bag is available, ropes can be coiled. The *climber's coil* and *caver's coil* are the two most common ways of coiling a rope. If you use a rope on a boat in swiftwater, take special care that you store it so

Fig. 4.3 Boatman's Coil

Fig. 4.4 Caver's Coil

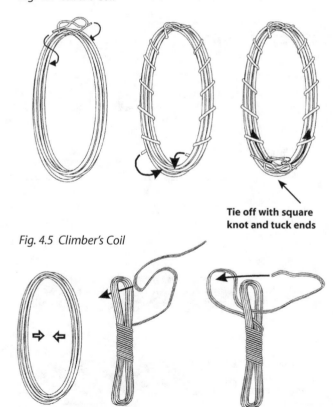

Tie off with square knot and tuck ends

Fig. 4.5 Climber's Coil

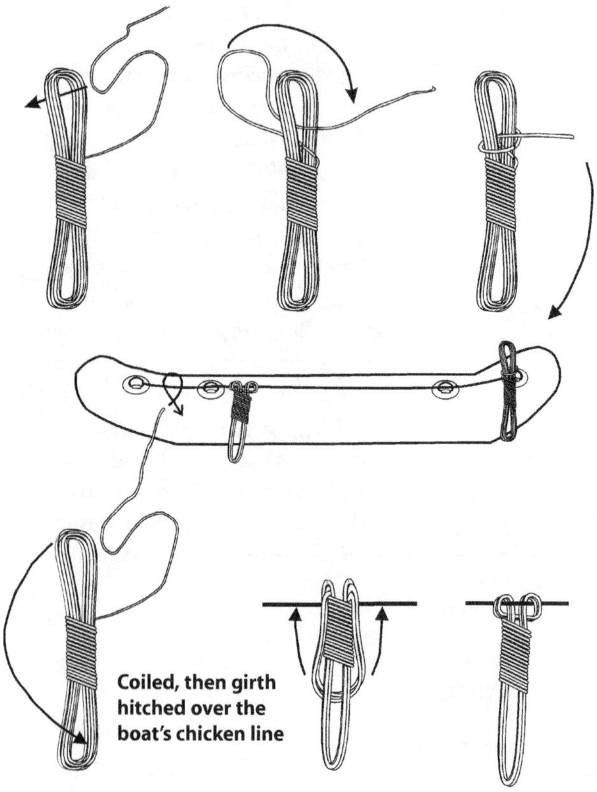

Coiled, then girth hitched over the boat's chicken line

Before using a rope, make sure that it will not tangle when you try to use it. One way to do this is to flake or stack the rope into a pile just before using it.

Fig 4.6 *Flaking or stacking rope into a pile prior to use.*

that does not come loose if the boat flips. One way of doing this is to use a modified climber's coil, with the rope coiled up to its attachment point at the boat and then secured, leaving as little slack as possible.

Turning the corner: How sharp the bend?

The sharper the bend we force a rope to go around, the more strength it loses, because most or all of the load is then transferred to the fibers on the outside of the bend. In extreme cases this can permanently damage the rope. Therefore, when changing directions with a loaded rope, we want to use as large a radius as possible. Opinions vary somewhat on the subject, but most authorities suggest that a radius of less than 4:1 (that is, a bend less than four times the diameter of the rope) begins to cause a significant strength loss. Others hold, however, that modern nylon ropes can go down to 2.5:1 without significant strength loss, while still others maintain that an 8:1 ratio is desirable. In practical terms this will be most important when selecting a pulley. Thus, for a 1/2" (13 mm) rope, we would need a pulley with a sheave diameter of 2" (5 cm) to maintain a 4:1 radius. When you measure this, be sure that you use the *inside* diameter, or *tread*, of the pulley sheave, since this is where the rope actually runs. Many two-inch pulleys, for example, have an actual tread diameter of 1.5" (3.8 cm).

Edge Protection. One of the primary causes of rope failure is the rope being cut or abraded when it is run over a sharp edge. The more heavily loaded the rope, the more likely it is to fail. Hence, edge protection assumes major importance in both technical rope work and climbing.

The most basic method of edge protection is to pad all sharp edges. This can be done with a wide

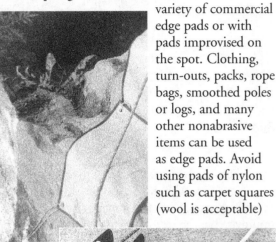

variety of commercial edge pads or with pads improvised on the spot. Clothing, turn-outs, packs, rope bags, smoothed poles or logs, and many other nonabrasive items can be used as edge pads. Avoid using pads of nylon such as carpet squares (wool is acceptable)

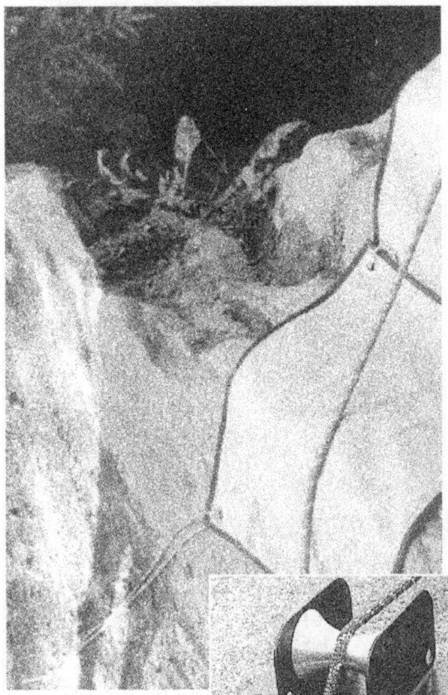

Fig 4.7 *Commercial edge pad.*

Fig 4.8 *Edge rollers (l) and roof roller (r).*

or nylon clothing, since these may melt. Edge pads do need to be carefully watched, however, to ensure that they do not shift or wear through. Usually they will need to be tied or hung in place.

Unlike pulleys, edge rollers, roof rollers and other similar devices, edge pads do not reduce friction on moving ropes, and ropes may wear through the pads.

Another method of edge protection is to avoid touching the edge at all by suspending a pulley over it. The pulley can be slung from a convenient object like a bridge railing or the rescuers can buy or fabricate an A-frame or a tripod from which to sling the pulley. A large pulley can also serve as edge protection where a rope is forced against an obstacle.

TIP: Fireman, spare that hose! Retired fire hose makes excellent edge pads. Don't throw it away.

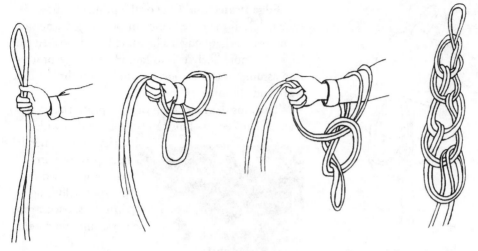

Fig 4.9 Daisy chain.

things as well. They can also substitute for pulleys in an emergency and can hold your keys. Carabiners come in an astonishing variety of sizes, shapes and materials, but generally they can be divided into two major categories: *locking* and *nonlocking*. A locking carabiner has a screw lock that can be run down over the gate to positively lock it. A variation of this design is the autolocking carabiner, in which a spring-loaded gate lock closes automatically when released.

Nonlocking carabiners are generally used in sport climbing and offer less security. *If you use nonlocking carabiners in a critical application, use them in pairs with the gates opposite and opposed.* A non-locking carabiner can also be clipped in alongside a locking carabiner for added security.

Carabiners are usually made from either aluminum or steel. Steel is stronger and less liable to wear and stress fractures, but aluminum is considerably lighter. If you have a truck to carry your gear, you

Webbing. It is hard to have too much webbing at a rescue site. Webbing serves a multitude of purposes in rescues, including anchor and gear slings, improvised harnesses, and many more. Webbing comes in many widths from ½"(13 mm) to 3" (76 mm), although 1" (25 mm) webbing is the size most commonly used. Most webbing is nylon, although more exotic (and expensive) materials like Kevlar and Spectra are coming into use for things like sewn runners. *Flat webbing* is a single layer of material (e.g., seat belts), while the more commonly-used *tubular webbing* is woven into a tubular shape and then pressed flat. *Spiral weave* tubular webbing, which is woven as a single unit, is generally preferred to *edge-stitched* webbing, where the web is folded over and stitched together into a tube. Since webbing is a key link in a very important chain, use the same care in choosing webbing that you would in selecting a rescue rope. For rescue, the preferred material is a high-quality 1"(25mm) tubular webbing that has a tensile strength of about 4,000 lbf (1,800 kgf). It is soft, knots easily, and has good resistance to abrasion—all qualities that recommend it for anchor slings.

Webbing also comes as presewn *runners* and slings, or in *daisy chains* with a series of presewn loops. Some manufacturers also make heavy duty anchor slings of flat webbing with sewn-in loops or D-rings. These eliminate the need for knots and can considerably speed up rigging.

Hardware

Carabiners. Carabiners ("snap links," "biners," "krabs") are another indispensable link in the rescue chain. They are roughly oval-shaped metal links with a spring-loaded *gate* that can be opened to clip in various objects, principally ropes but many other

Carabiners

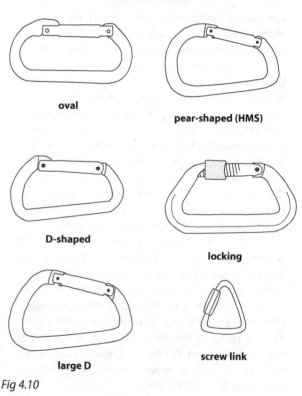

oval

pear-shaped (HMS)

D-shaped

locking

large D

screw link

Fig 4.10

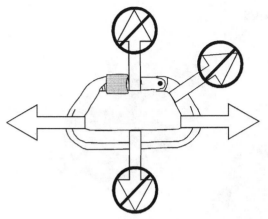

Do not cross-load carabiners.

Do not load carabiners with the gate open.

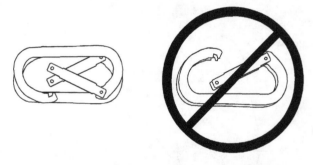

When using non-locking carabiners in critical applications, make sure the gates are opposite *and* opposed.

Fig 4.11

may opt for the extra strength of steel; if you have to hump your gear long distances over rough terrain, you will undoubtedly prefer aluminum.

Carabiners should be loaded in only one way: along the long axis, with the gate closed and locked. "Cross-loading" dramatically reduces their strength. For example, the Petzl "Attache" locking carabiner is rated at 22 kN (5,000 lbf) when closed and locked, but only 8 kN (1,800 lbf) when cross-loaded, and 6 kN (1,350 lbf) with the gate open.

Carabiners come in many shapes. The basic oval was quickly supplemented by the "D" shape, which

had the effect of moving the load closer to the spine of the carabiner and increasing its strength. The next step was a pear-shaped carabiner with a larger oval at one end, which worked well for passing knots and rigging Münter hitches (these are sometimes called HMS carabiners). Rescue and fire services demanded larger carabiners to fit over litter rails and ladder rungs.

Today, any rescue organization is likely to have a mixed bag of carabiners, which is fine as long as everyone understands how they work and what they are to be used for. There are a few "gotchas" that rescuers should keep in mind: different manufacturers make carabiners that lock in opposite directions, and the newer "superlight" lead climbing carabiners work fine for climbing loads but may not be adequate for rescue loads. Like ropes, carabiners should be checked after each use and retired if they show signs of damage. Some teams go so far as to X-ray their hardware periodically.

Another piece of gear that looks like a carabiner (but isn't) is the *screw link*. Unlike a carabiner, a screw link does not have a spring-loaded gate. Instead, it has an opening that screws open and closed. As a result, it is slower to operate but tolerates cross-loading much better. They have about the same strength as carabiners and are often used where two- or three-way loading occurs, such as on a sit harness.

Pulleys. Pulleys change the direction of a rope. Properly rigged, they reduce friction and add mechanical advantage. The rope runs over the pulley's *sheave*, which in turn runs on either a *bearing* or *bushing* held in place with an *axle*. Metal *side plates* hold the axle, protect the rope, and provide an attachment point for carabiners. Most pulleys have side plates that move independently, allowing the pulley to be fitted over the rope at any point. Some pulleys also have an additional rigging hole below the sheave called a *becket*.

Since pulleys take a lot of stress—up to twice the load on the rest of the system—it's worthwhile investigating what you buy. Stick with heavy-duty hardware designed for rescue use. Cheap pulleys with plastic bushings and thin side plates may be adequate for light gear-hauling chores, but should not be trusted with rescue loads. You will be presented with the choice of pulleys with bushings or bearings. For river use, pulleys with sealed bearings are generally a better choice, since they are less likely to become contaminated with dirt and grit. Another welcome trend in recent years has been pulleys with larger and stronger side plates (which give more protection to the rope) and bigger openings, permitting the use of multiple carabiners.

Rescue Pulleys

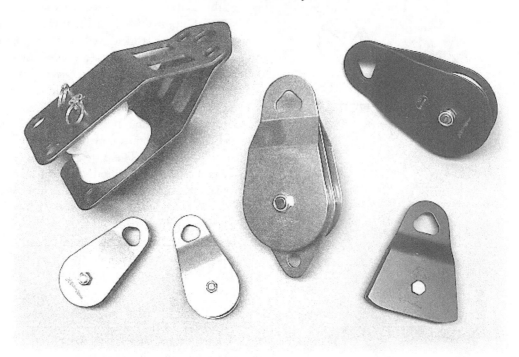

Fig 4.12 A large rescue pulley (top left) has a throat wide enough to pass knots or multiple tracklines. The center pulley has a double sheave and an additional rigging point called a becket below it. The bottom right pulley is a Prusik-minding pulley, with square edges to keep the Prusik knot from being pulled into the throat.

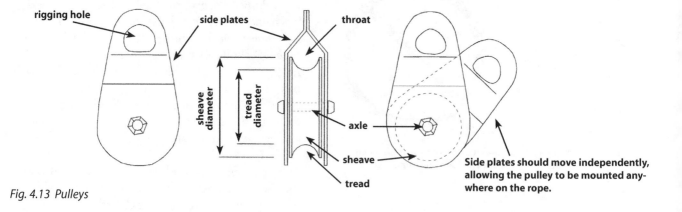

Fig. 4.13 Pulleys

Some specialized designs have also appeared. The *Prusik-minding pulley* has large side plates, squared-off at the bottom. These "mind" or catch the Prusik knot and keep it from getting jammed between the pulley's side plates and sheave. *Knot-passing pulleys* are large pulleys with wide throats that allow knots to pass through without hanging up. The *Kootenay carriage* is a unique design that functions as a knot-passing pulley, a high line carriage capable of accepting multiple tracklines, and, with the attached locking pins, a high-strength tie-off.

There are also pulleys with double and triple sheaves for rigging more complex haul systems.

Another very specialized type of pulley is the *edge roller*. Here the pulley is mounted on an edge like the side of a building. Its purpose is to change the direction of the rope with a minimum of friction and abrasion.

Carabiners can be substituted for pulleys in an emergency. However, this greatly increases friction and puts a very sharp bend on the rope, reducing its strength and possibly even damaging it. One

Fig. 4.14

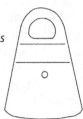

A Prusik-minding pulley has squared-off side plates

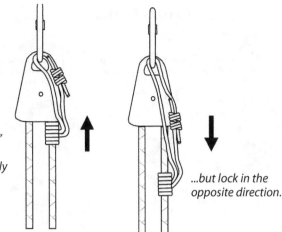

These catch or "mind" the Prusik, allowing the rope to move freely in one direction...

...but lock in the opposite direction.

company makes a nylon roller that fits on a carabiner to effect a makeshift pulley.

Rope Clamping Devices. A recurring need in rescue rope work is the need to securely grip the working rope without damaging it. There are a host of reasons: ascending, hauling, and belaying are the primary ones. The many available devices can be divided into two categories: hard clamps and soft clamps.

Hard clamps are generally some kind of mechanical ascender that works by pushing a metal cam against the rope when it is pulled. These devices were originally designed to grip a rope when a climber put his body weight on it. Using a pair of these, a climber could ascend the rope, even if it was wet or icy. The principle was later adapted to grip the rope for hauling systems or belays. While the metal clamp provides a very positive method of gripping the rope, it may also damage and even cut the rope under the right circumstances, particularly if the system is shock loaded. As things stand now, handled ascenders (i.e., those with handles, sometimes generically called Jumars) are not recommended for any purpose other than what they were designed for—ascending fixed ropes with a single person's body weight. The Gibbs ascender, however, has been widely used not only for ascending but also as a general-purpose rope clamp. Some groups decry this practice, while others continue to use the Gibbs routinely. All agree, however, that if the system is overloaded or shock loaded, the Gibbs will damage the rope. Later variants have been redesigned to be kinder and gentler to the rope. The Rock Exotica Rescuscender™ and Microscender™ are designed specifically to reduce the chance of damage to the rope by using a gently curved cam and rounded teeth. While these design features certainly minimize the problem, they do not entirely eliminate it.

Soft clamps are various types of cinching knots like the Prusik and Bachmann that grip the rope when tightened. These are normally tied with 6–8 mm (¼–⁵⁄₁₆") nylon accessory cord (smaller versions of kernmantle climbing ropes). Cinching knots are less likely to damage the main rope than metal ascenders since they will slip under heavy loads. However, this introduces its own problem—a nylon Prusik slipping on a nylon rope generates friction, and it will eventually melt through and fail.

Mechanical Ascenders

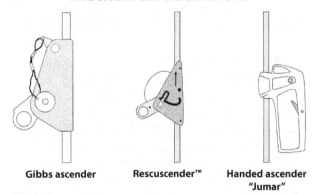

Gibbs ascender   **Rescuscender™**   **Handed ascender "Jumar"**

Fig 4.15

Friction Devices. Unlike clamping devices, friction devices do not actually grip the rope. Instead, the rope is run over or through them, and their task is to impart a controlled and sometimes variable amount of friction. Normally, this friction is used for a controlled lower (belaying or descending) of either a rescuer/climber or of a load. There are many friction devices, and as with other systems, some are adequate for rescue loads and some are not. The performance of various devices, especially for belaying rescue loads under shock loading, continues to be one of the most hotly debated areas of technical rescue.

Figure Eights

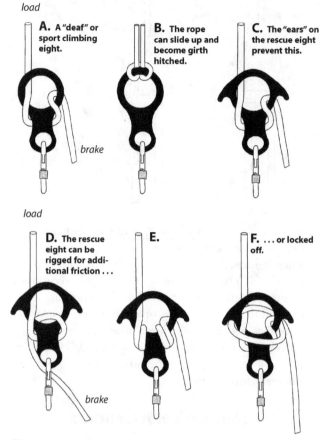

A. A "deaf" or sport climbing eight.

B. The rope can slide up and become girth hitched.

C. The "ears" on the rescue eight prevent this.

D. The rescue eight can be rigged for additional friction . . .

E.

F. . . . or locked off.

Fig 4.16

The most common friction device is the *figure eight*. The smaller version, sometimes called a "deaf" eight (because it has no ears), is commonly used in climbing. However, it is possible for the rope to girth hitch itself on the device, stranding the climber. The larger version, generally called a rescue eight, has "ears" that stick out from the body of the device to prevent girth hitching and allow extra rope wraps for more friction. The rescue eight is physically larger for better heat dissipation and to accept larger diameter ropes. The friction on the rescue eight can be varied by increasing or decreasing the number of wraps around the body, which allows adjustment for the weight of a victim (although this cannot be done in the middle of a lower). While most figure eights are made from aluminum, which is lighter and gives better heat dissipation, they are also available in steel, which has less friction and poorer heat dissipation, but which wears better and can be used as a rigging spider. The figure eight is simple, light, and inexpensive, but has two major disadvantages: it twists the rope and the amount of friction cannot be adjusted during use. Because of this, it is not recommended for long (over 150'/46 m) rappels or rescue loads.

Brake bar racks (rescue racks, rappel racks) solve many of the problems of the figure eight. It is a "U"-shaped metal rack with several (the number varies) crossbars made from steel or aluminum. The rope is laced through the bars, and the amount of friction can be adjusted while the device is in use by adding or subtracting bars or by sliding them closer together or farther apart. It does not twist the rope and can be used with doubled ropes. Racks have been used for rappels in excess of 3,000' (914 m). The disadvantage of the rack is that it is rather bulky and expensive, although several recent designs are considerably smaller.

Brake Rack

less friction

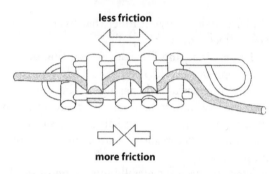

more friction

Fig. 4.17 The friction in a brake rack can be increased by sliding the bars closer together or by adding more bars against the rope. Friction can be reduced by sliding the bars apart or removing them.

wrap 2-3 times

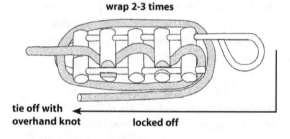

tie off with overhand knot   locked off

The brake bar can be locked off by sliding the bars together, wrapping the rope 2-3 times around the rack, and tying it off with an overhand knot above the rack.

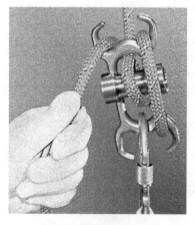

Fig. 4.18 The Scarab Rescue tool.

One such is the Scarab (R) which weighs only 385g (13.8 oz) and combines a single bar with "horns" to increase and vary friction.

If no friction device is available, one can improvised from carabiners. The *carabiner brake* is similar in function to the brake bar rack. It is made by clipping one or more carabiners across a rope loop through another carabiner. Locking carabiners are preferable, but an opposed pair of non-lockers can also be used. If more friction is needed, more brakes can be added to the first one. A large pear-shaped carabiner rigged with a Münter hitch will also work as an improvised friction device.

Adjustable Friction Devices

In recent years manufacturers have developed a whole series of adjustable friction devices. Some, like the Petzl Stop, rely on a pair of wheels, or bobbins, which can be pulled together with a lever to increase rope friction and slow a rescuer's descent down a rope. Later models, such as the Petzl Grigri, use a stainless steel cam to press the rope against a plate. With the friction controlled by a lever, these devices can be used for rappelling as well as belaying and lowering climbing loads. Newer designs like the Petzl I'D can handle heavier rescue loads, have a "panic" handle, and can be used for limited ascending as well.

Another innovative device is the 540 Rescue Belay, which has an eccentric cam inside a housing similar to that of a pulley. When the cam catches on belay it forces the rope against a stationary wedge on the housing, slowing it. Unlike the Grigri and I'D, it is a single purpose device.

Fig. 4.19 The Petzl I'D combines the function of descender and belay device, and features an "anti-panic" stop system.

Harnesses

The link at the rescuer's end of the chain is the harness. These are sit harnesses, which fit around the thighs and waist; chest harnesses; and full body harnesses, which are a combination of the two. The sit harness is less restrictive and expensive than the full body harness. It is adequate for most purposes, and

a chest harness can be added for additional safety. However, a full body harness does give more overall security. John Weinel, who makes and markets harnesses, says "a rescue harness should be comfortable to hang in for long periods of time. It must be robust in design and capable of withstanding high stress and adverse conditions. The harness should have gear loops or provisions for them so that you can carry the tools of the trade with you. It should be quick and easy to don on yourself or on the subject you are there to rescue. The wearer should feel secure and not have to expend a lot of energy to maintain an upright stance." Firefighter Gary Seidel cautions, however, that "if rescuers are going into the water with harnesses on, they should not have gear loops. The harnesses should be loop-free to prevent getting hung up."

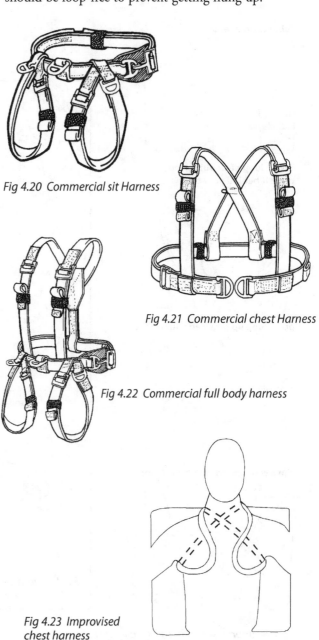

Fig 4.20 Commercial sit Harness

Fig 4.21 Commercial chest Harness

Fig 4.22 Commercial full body harness

Fig 4.23 Improvised chest harness

Rope rescue harnesses are designed so that the rescuer does not part company with the rope, no matter what. However, *river rescue harnesses* routinely incorporate some form of quick release so that the rescuer can get free of the rope in an emergency. Rescuers in technical rope systems over water face a dilemma here. Keep in mind the following:

- Do not use a rescue/climbing harness for an in-water rescue. By the same token, a river rescue harness/buckle system is not suitable for rappelling or technical rope work.
- In an over-water technical rope system in which the rescuer wears a climbing-type rescue harness, either rig a run-free system (i.e., with no knots to keep the rescuer from washing free) or provide a means of cutting the rope.

Tying the diaper seat:

Take a loop of webbing and place it behind your thighs . . .

pull a loop up between your legs . . .

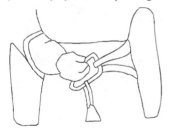

clip a carabiner through the loops.

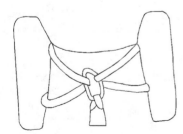

Fig 4.24 *Improvised sit harness – diaper seat*

Although "store-bought" harnesses definitely have the edge in terms of security, strength and comfort, there may be times when they are not available. Every rescuer should, therefore, be able to rig a simple improvised harness out of webbing for either himself or for a victim. Improvised harnesses, while they can be quite effective, are less comfortable even when made from the widest available webbing, and can cut off circulation after even a short period of time. It is possible to tie and use them incorrectly.

The *diaper seat* is one of the simplest improvised seats to tie. It is simply a large loop (roughly the length of a person's outstretched arms) tied or sewn together. Put the loop against the small of your back, then pull the loop around the front of the body. Bring the lower part of the loop up between your legs, and lock the loops together in front of the body with a large locking carabiner or a screw link. The diaper seat is reasonably secure, easy to remember, and requires a minimum of equipment to rig. However, it does give the wearer a rather low center of gravity. To increase the security of this or any sit harness, you can add a simple chest harness.

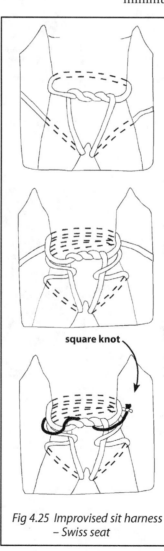

square knot

Fig 4.25 *Improvised sit harness – Swiss seat*

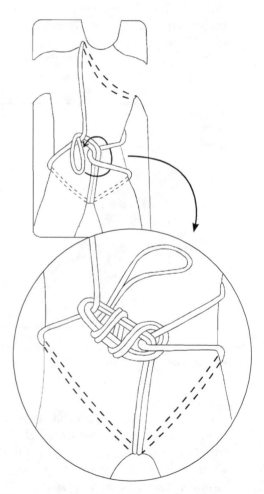

Fig 4.26 *Improvised body harness – Hansen harness*

The *Swiss seat* (G.I. rig, quick-don harness, swami seat harness) is also simple and quick to tie but is less secure. The web is passed around the waist and looped into a double overhand knot, then run between the legs and behind the thighs. The web then goes under the leg loops on the front and back around the body to be tied off on the side with a square or water knot.

The *Hansen harness* is a combination chest and sit harness. Essentially a body girth hitch, it was originally used in Vietnam to extract soldiers from the jungle via helicopter. It is not particularly comfortable but it is more secure than either the Swiss seat or the diaper seat. To tie, take a large loop of webbing and place it over your shoulder. It should be long enough that the ends touch the floor. Reach down and pull a bight up between your legs and wrap it through the web coming down from your shoulders. Wrap any remaining slack around this web and clip a carabiner into the end. The Hansen harness is quick to tie and functions equally well for victims or rescuers.

A variation of the diaper seat makes an improvised victim harness that is relatively simple for a rescuer to tie on an uninjured or lightly injured victim.

> **TIP:** Keep a Prusik loop, or two, in your pocket when up on a rope. A safety knife or seat belt cutter isn't bad idea either.

To tie it, the rescuer, using a loop similar to that used for the diaper seat, reaches between the victim's legs, holding the top of the loop with his chin. He then pulls two bights under the leg loops formed in the first step and clips them together with a carabiner. If possible, use a chest harness for added security, since the chances are good that a victim will not have much experience in high-angle work.

Other Rigging Paraphernalia: A good pair of leather *gloves,* with reinforced palms, comes highly recommended for rope work. A knife can also be very handy. There are special *rigging knives:* these are folding knives that have a built-in marlinespike for loosening stubborn knots as well as a blade. Some rescuers like to carry a knife during technical rope maneuvers, especially rappelling, in case something (hair, clothing, etc.) gets caught or jammed. However, since an open blade isn't a good idea around loaded ropes, they use either a *safety knife,* a *seat belt cutter,* which have protected blades, or *EMT shears.*

A steel *rigging ring* (pear ring, "O" ring, connecting ring, rigging spider) is used to gather lines together for litter spiders, lowering systems, etc. These rings are much stronger than the aluminum figure eights sometimes used for this purpose and resist side loading better than carabiners. *Rigging plates* are steel

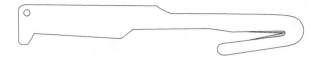

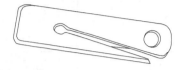

Fig. 4.27 Safety knives have protected blades

or aluminum plates with multiple openings to accept carabiners and other rigging devices. Made in almost every conceivable shape and size, they allow a neater, cleaner installation and reduce the likelihood of tangling and cross loading.

Fig. 4.28 A portable rescue winch used by Grand Canyon National Park Rangers.

Winches, both power and hand cranked, provide a lot of pull in a small package some with mechanical advantages of up to 48:1. *Come-alongs* are a somewhat simpler version with a much more limited range of pull. The Griphoist/Tirfor hand winch, which uses cable instead of rope, is capable of lifting very heavy loads. Fire and rescue trucks often have power winches mounted on them. While power winches can be very useful, it is also possible for them to overload a system very quickly.

Sport climbers have invented and used a bewildering variety of artificial rock anchors to protect themselves on difficult routes. Although they can be used to advantage in technical rope systems, it is far beyond the scope of this book to do more than a quick review. *Pitons* and *bolts* are fixed hardware meant to remain in place. A piton is a thin metal spike driven into a crack in the rock, while a bolt is an expansion device that is first drilled into the rock. A *hanger* is a metal extension at the end of the bolt with a provision for attaching a rope or carabiners. Passive *chocks* (tapers, nuts,

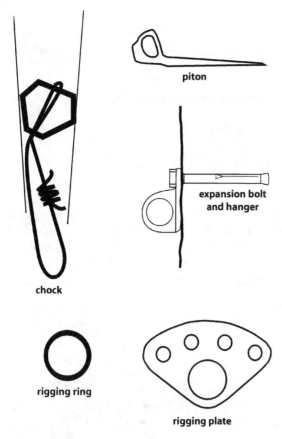

piton

expansion bolt
and hanger

chock

rigging ring

rigging plate

Fig 4.29 Rigging accessories

stoppers, cams, hexes, etc.) insert into a rock crevice and are held there by their shape. Active chocks, on the other hand, are mechanical devices that physically expand to grip the crack. The most common types (e.g., Friends™) are commonly called spring-loaded camming devices, or SLCDs. Anyone interested in a more in-depth review of this subject should consult the books listed in the bibliography.

Knots

Knot is a generic term that covers the entire area of tying ropes. We will use it as such here, although for the technically-minded the following definitions are offered. A *bend* joins two ropes together (e.g., double fisherman's). A *hitch* incorporates another object (such as a carabiner, pole, etc.) as a functional part of it, so that if the object is removed, the hitch falls apart (e.g., clove hitch). A *knot* ties a rope to something or makes an attachment point like a bight (e.g., figure eight on a bight).

Whole books have been written on knots. Although different knots can be used for the same purpose, some work better than others. For rescue use, a knot should be strong and secure. Recall that we said that when we bend a rope sharply, we reduce its strength. This is also true with knots. Any knot

will reduce the strength of the rope—a good rule of thumb is about 30%. CMC Rescue tested the strength of some common knots, and the results are shown below. By *secure,* we mean that the knot will not capsize or slip if it is loaded from another direction or snags against something. The "knotability" of the rope also affects this equation, since an extremely stiff rope does not hold a knot very well when unloaded.

Before use, each knot should be dressed and set. By *dressing,* we mean that the strands of rope should be smoothly laid against each other in the shape of the knot and not laid across each other randomly. Failure to dress a knot will reduce its strength as well as make it much harder to untie. The knot is *set* when all the slack is taken out of it and the parts of the knot touch each other for proper functioning.

The figure eight family of knots can serve almost any rescue task. They are strong and secure, but, more important, since they are all tied similarly, they are easy to teach and remember.

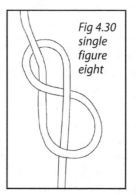

Fig 4.30 single figure eight

This can be an important consideration for anyone under the stress of rescue work. The figure eight is also very easy to visually identify—one can tell at a glance whether it is correctly tied or not. Do not confuse the figure eight with an *overhand* knot, (Fig. 4.32) which is very difficult to untie after being loaded.

The *single figure eight* (Fig. 4.30) is the basis for all other knots in the family. It is used primarily as a stopper knot on the end of a rappel rope.

The *figure eight loop* (Fig. 4.31) or figure eight on a bight, provides a rope with a loop at the end of it— ideal for clipping the rope to something else.

The *double-loop figure eight* (Fig. 4.36) gives two loops instead of one. This knot is useful for backing up anchors and forms the basis for a self-equalizing anchor system.

The *in-line figure eight,* (Fig. 4.35) or directional figure eight, gives a midline knot with the loop pointing down along the rope.

The *figure eight bend* (Fig. 4.33) joins two ropes together.

The *figure eight follow through* forms a loop around an object like a post or tree (Fig. 4.34). This and the figure eight bend are sometimes called *tracer knots,* since both knots start with a single figure eight and have the rope traced back through the same way as the original single knot.

Fig 4.31 Figure eight loop

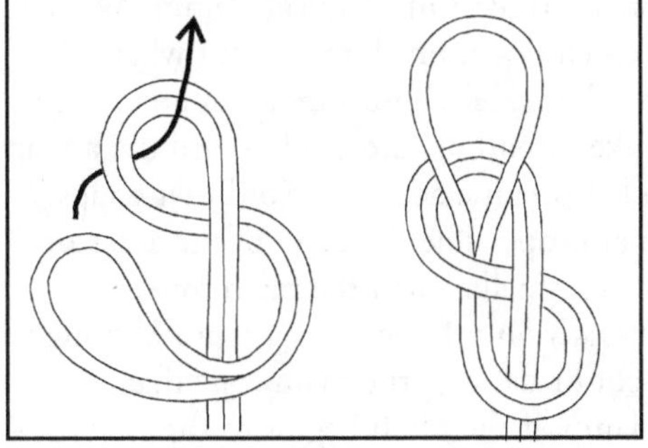

Fig 4.33 Figure eight bend

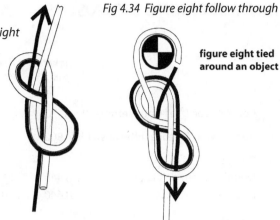

Fig 4.34 Figure eight follow through

figure eight tied around an object

Do not confuse the figure eight (Fig. 4.31) with an overhand knot, (Fig. 4.32) which is very difficult to untie after being loaded.

Fig 4.32 Overhand knot

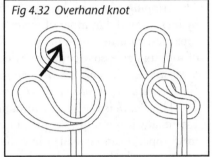

Fig 4.35 Directional or In-line figure eight

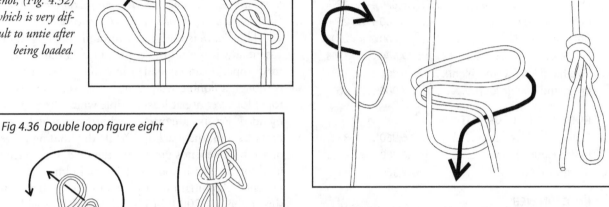

Fig 4.36 Double loop figure eight

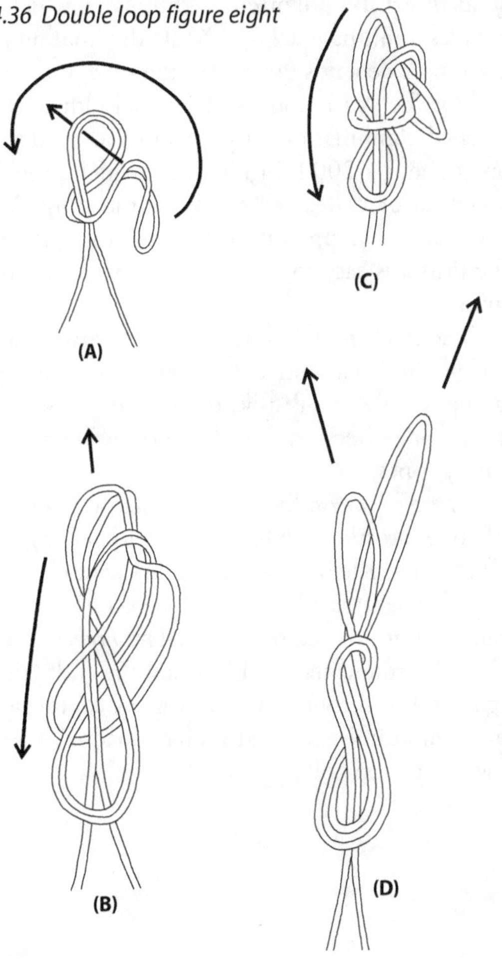

(A)

(B)

(C)

(D)

The *load-sharing figure eight,* (Fig. 4.37) which incorporates a locking carabiner into the body of the knot, provides a quick way to tie a load-sharing anchor system.

Fig 4.37 Load-sharing figure eight

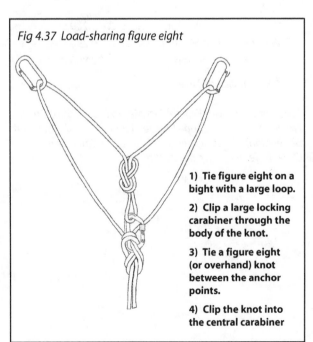

1) **Tie figure eight on a bight with a large loop.**

2) **Clip a large locking carabiner through the body of the knot.**

3) **Tie a figure eight (or overhand) knot between the anchor points.**

4) **Clip the knot into the central carabiner**

KNOT BREAKING STRENGTHS

Fig 4.38 Knot breaking strengths

	STRENGTH IN lbf	% STRENGTH LOST
Control rope (¹/₂" static kernmantle)	10,705	
Control web (1" spiral tubular web).	4,800	
BENDS		
Double fisherman's knot	8,440	21%
Figure 8 bend.......................	8,640	19%
LOOPS		
Figure 8 loop (with a bight)	8,560	20%
Figure 8 loop (follow-through)	8,640	19%
Double Figure 8 Loop	8,820	18%
In-line Figure 8 Loop	8,000	25%
Butterfly knot......................	8,000	25%
Bowline............................	7,180	33%
Overhand loop (with a bight)........	9,060	15%
Overhand double loop..............	7,900	26%
ROPE WITH A LOOP IN IT *		
Figure 8 Loop	6,960	35%
In-line Figure 8 loop................	6,280	41%
Butterfly loop	7,360	31%
KNOTS IN WEB		
Water knot.........................	3,060	36%
Overhand loop.....................	3,120	35%
Figure 8 loop (with a bight)	3,360	30%
Figure 8 loop (follow through).......	3,560	26%
Web slings:		
Water Knot-Single loop	5,700	
Water Knot-Double loop	12,290	
Water Knot-Triple loop	22,860	

* *Rope pulled end to end.*

Test conducted with Federal Test Method 191A, 6016 (from Frank & Smith, CMC Rope Rescue Manual, 2nd ed., used with permission). These figures are for static kernmantle rope. Other ropes may give different results.

Cinching knots (soft clamps) serve much the same function as do hard mechanical clamps. When loaded, they cinch down on the rope, but when unloaded they slide freely. Thus, they can be used for applications like ascending and hauling. All use a loop of soft nylon kernmantle accessory cord that wraps around the main rope. The diameter of the accessory cord is somewhat smaller than the main rope, commonly 8 mm (⁵⁄₁₆") on a 13 mm (½") rope. Although less likely to cut or damage the main rope than a mechanical clamp, if the hitch begins to slip, the friction generated may burn through the accessory cord. The gripping power of these knots can be improved by increasing the number of wraps around the main rope, and by using them in pairs. Since the properties of rope and accessory cord vary widely, *it is imperative that users test the compatibility and friction qualities of both before using them in critical applications.*

The *Prusik* is the most common cinching knot. Essentially, it is a multiple girth hitch around the main rope. It can be loaded in either direction. Although a double-wrap Prusik will work on dry rope, for river use at least a triple wrap is recommended. Prusiks are often used in pairs, which provides a mutual backup. While the holding power of a Prusik depends greatly on the type of cordage used for the Prusik loop and the main line, a well-matched combination (triple-wrap Prusik) begins to slip at about 1,200 lbf (5.34 kN), which provides somewhat of a check on the system loading. A carabiner can be wrapped into the hitch for a handle. The Prusik is easy to tie and can be tied with one hand.

The *Bachmann* is similar to the Prusik, but incorporates a carabiner. This provides a convenient handle. Unlike the Prusik, the Bachmann works in only one direction, but does grip well on wet or muddy ropes.

The *Kleimheist*, like the Bachmann, works in only one direction. Unlike the other cinching knots, however, the Kleimheist can be tied with either webbing or rope. It is a handy knot if you have webbing available but no accessory cord. The *French Prusik* is looped around the haul line and the ends clipped together. It can function as a progress-capturing device and can be released under a load with a sharp blow to the top coils.

Fig 4.39 Cinching knots

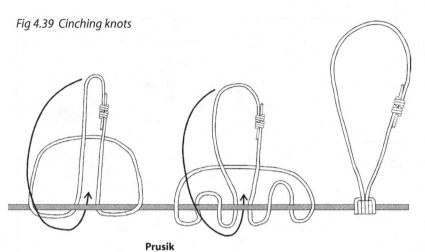

Prusik

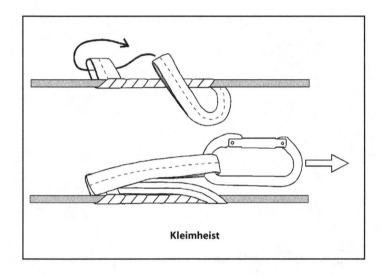

Kleimheist

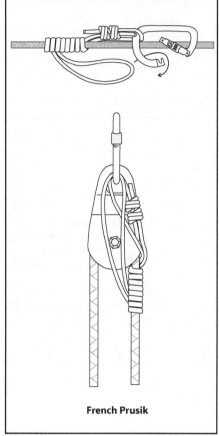

French Prusik

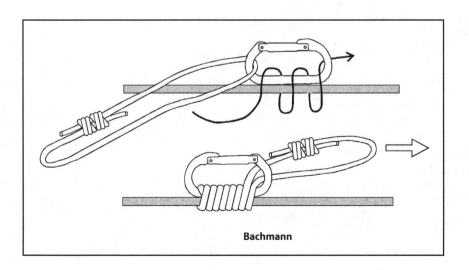

Bachmann

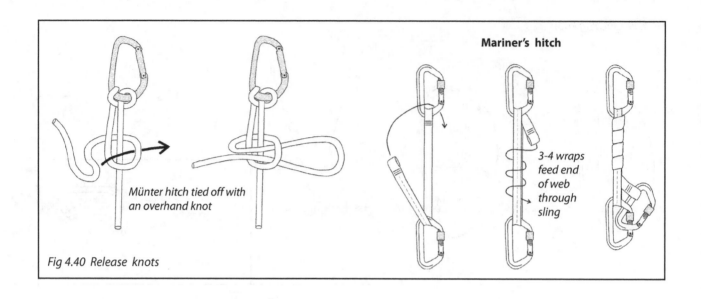

Mariner's hitch

Münter hitch tied off with an overhand knot

3-4 wraps feed end of web through sling

Fig 4.40 Release knots

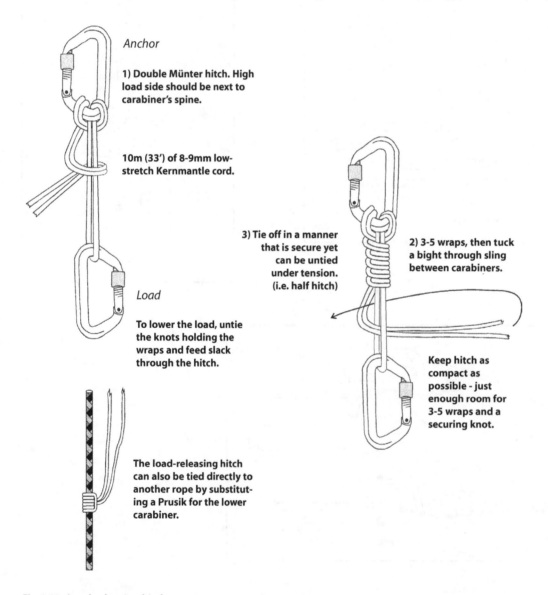

Anchor

1) Double Münter hitch. High load side should be next to carabiner's spine.

10m (33') of 8-9mm low-stretch Kernmantle cord.

Load

To lower the load, untie the knots holding the wraps and feed slack through the hitch.

The load-releasing hitch can also be tied directly to another rope by substituting a Prusik for the lower carabiner.

3) Tie off in a manner that is secure yet can be untied under tension. (i.e. half hitch)

2) 3-5 wraps, then tuck a bight through sling between carabiners.

Keep hitch as compact as possible - just enough room for 3-5 wraps and a securing knot.

Fig 4.41 Load-releasing hitch

Release Knots. Another recurring need in rescue operations is the need to release a load under tension. Ideally this should happen gradually so as not to shock load the system. A *Münter hitch* makes a quick-and-dirty release knot if it is tied off with an overhand knot to the main rope, but is not considered suitable for holding rescue loads. The *mariner's hitch* uses a web loop or sewn runner wrapped around itself to provide enough friction for a controlled release although it, too, is not considered suitable for rescue loads. The *load-releasing hitch* is similar but uses 10 m (33') of 8 mm (⁵⁄₁₆") accessory cord. It starts with a double Münter hitch, then is followed by multiple wraps around itself. The Radium Release hitch resembles the Load-releasing hitch except that it uses a three to one instead of a two to one arrangement. This increases its load capacity and reduces lowering effort but restricts the amount of rope available for it.

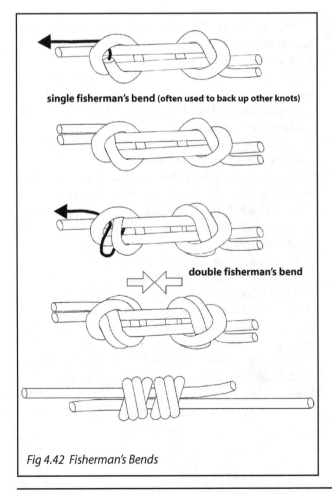

Fig 4.42 Fisherman's Bends

single fisherman's bend (often used to back up other knots)

double fisherman's bend

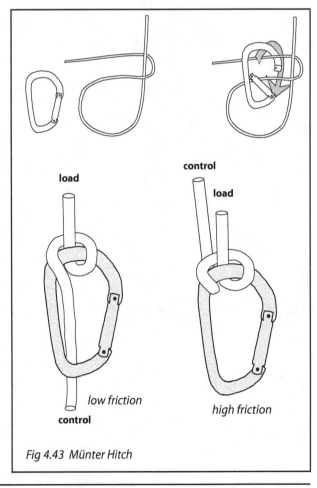

Fig 4.43 Münter Hitch

load

control

load

low friction

high friction

control

Fig 4.44 Radium release hitch

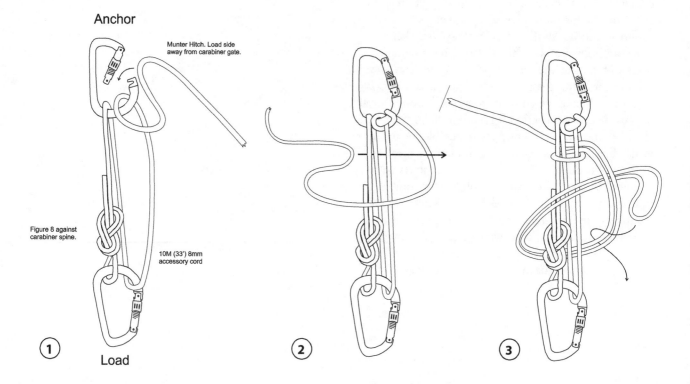

Anchor

Munter Hitch. Load side away from carabiner gate.

Figure 8 against carabiner spine.

10M (33') 8mm accessory cord

Load

① ② ③

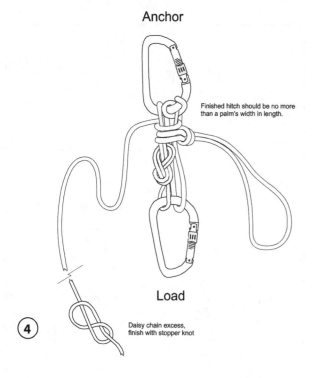

Anchor

Finished hitch should be no more than a palm's width in length.

Load

Daisy chain excess, finish with stopper knot

④

Other Knots. The *double fisherman's bend* (grapevine knot, barrel knot) is a very strong and secure knot used for tying two ropes of the same diameter together. It is somewhat more compact than the figure eight bend, and is therefore, preferable when a knot pass is necessary. After being loaded, however, it is virtually impossible to untie.

The single fisherman's bend also functions as a *safety* or *backup knot*. Tied on the free end of a knot, it provides extra security for any knot, but especially those like the bowline that are prone to work loose when unloaded. All knots used in critical applications, especially lifelines, should be backed up with a safety knot. *However, for purposes of clarity, the safety knot backup will not always be shown in this book.*

The *Münter hitch* (Italian hitch) is a handy and versatile hitch that can be used as a friction/belay device. The Münter is bi-directional: that is, a pull on either end will flip the knot over to work in the other direction. It is simple and easy to tie, works with doubled ropes, and can be tied off and released under load. It does require a large pear-shaped locking carabiner for best functioning, and twists the rope during operation. It can be doubled for increased friction.

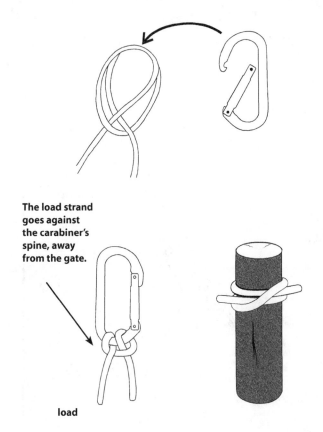

The load strand goes against the carabiner's spine, away from the gate.

load

Fig 4.45 Clove hitch

The *clove hitch* is a simple and effective knot that should be part of any rescuer's repertoire. It can be used to secure a rope around an object like a post, or to a carabiner (it slips at about 1,000 lbf/4.45 kN). A rescuer can put a clove hitch over his own arm and slide the coils down over the arm of a victim for stabilization or recovery (see figure 15.4).

The *butterfly* is a strong, secure, and useful midline knot that can be loaded from three directions.

The *bowline,* long a staple knot for rescue, has fallen out of favor in recent years. It is not quite as strong as the figure eight and—worse—tends to loosen when unloaded, especially with a stiff rope. For this reason, *any* bowline should be backed up with a safety knot. Still, when a fixed loop is needed, it is simple to tie and effective, and is still widely used.

The *water knot* (ring or water bend) is used to join webbing. It is an overhand knot tied with both pieces of webbing together. Since webbing sometimes slips under load, it is recommended that at least a 3–4" (8–10 cm) tail be left free on each end of the knot. For critical applications, tie off the free ends with an overhand safety knot. Do not use the water knot for rope.

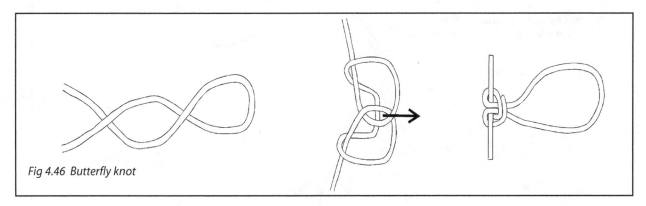

Fig 4.46 Butterfly knot

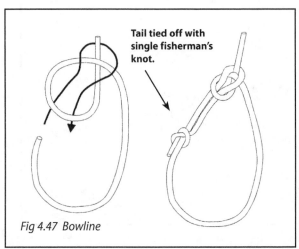

Tail tied off with single fisherman's knot.

Fig 4.47 Bowline

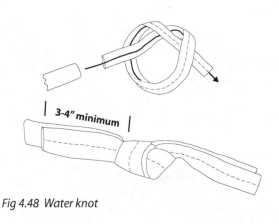

3-4" minimum

Fig 4.48 Water knot

Conclusion

A thorough understanding of the properties of the various different types of ropes used in swiftwater rescue is essential, as are the correct ways of employing, protecting, and storing rope. Equally essential are the various knots used in technical rescue, as well as the characteristics of an ever-growing array of technical rescue gear.

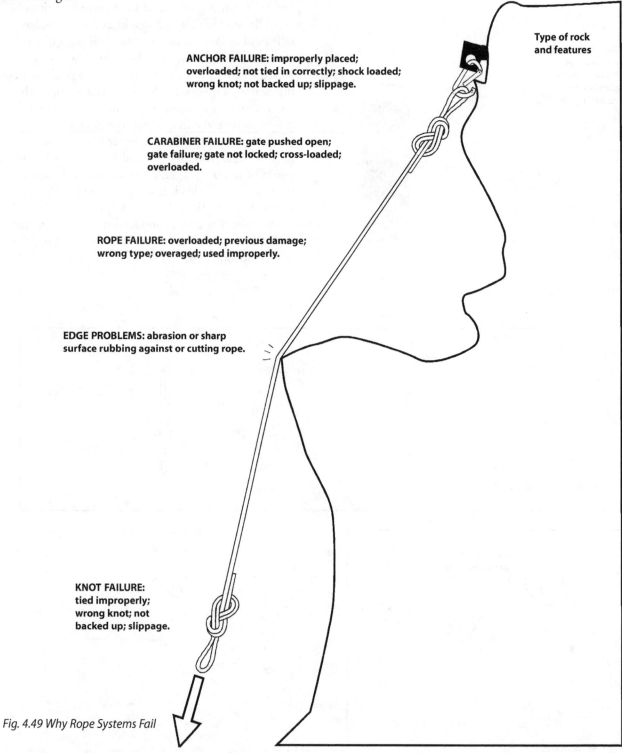

Type of rock and features

ANCHOR FAILURE: improperly placed; overloaded; not tied in correctly; shock loaded; wrong knot; not backed up; slippage.

CARABINER FAILURE: gate pushed open; gate failure; gate not locked; cross-loaded; overloaded.

ROPE FAILURE: overloaded; previous damage; wrong type; overaged; used improperly.

EDGE PROBLEMS: abrasion or sharp surface rubbing against or cutting rope.

KNOT FAILURE: tied improperly; wrong knot; not backed up; slippage.

Fig. 4.49 Why Rope Systems Fail

Rigging for River Rescue II
Anchors, Haul Systems, and Belays
Chapter 5

This chapter covers anchor systems, including multiple, load-sharing and load-distributing systems; haul systems and mechanical advantage; and belay systems. A solid anchor is essential for any rope system, and the basic techniques of raising and lowering loads, as well as releasing them when necessary, are essential skills for any rope rescuer to master. The basics of rappelling are also covered at the end of the chapter.

Anchors

The last link in the rescue chain is the anchor to which the rope system is attached. Many rope system failures can be attributed to anchor failure, in part because anchor strengths and placements are not as much of an exact science as figuring the working strengths of rope and rescue gear. Our primary considerations for choosing an anchor point are that it should:

- be in the right place
- not fail under the greatest anticipated load

A super-strong anchor is often referred to as bombproof or simply as bomber. Obviously these two requirements will not always coincide, and anchor points may have to be created where they are needed. Another current point of debate among rope systems proponents is the desirability of using self-equalizing multi-point anchors, since these can shock load the system if a single anchor point fails.

Anchors can be either natural or artificial. Natural anchors include:

- Rocks and trees (sometimes called BFRs and BFTs)
- Buildings, poles, guard rails, or anything in the right place that is able to bear the load without failure
  Artificial anchors include:
- Rock anchoring devices like SLCDs, bolts, chocks, hexes, nuts, and pitons
- Vehicles
- Stakes, pickets and deadmen

The *no-knot* (tensionless anchor high-strength tie-off) is very strong and secure. It is a series of 3–5 wraps around an object such as a tree. The line is then clipped or tied back on to the main line. Since friction does the work, this is the only "knot" to retain the full strength of the rope. The no-knot works best on a round object. As the object becomes more square, the bend becomes sharper and rope strength is lost. In general, tie into an anchor like a tree or a post as low as possible to avoid creating a lever effect, and be sure to pad any sharp edges.

If you are unable to use a no-knot to tie the rope directly to an anchor point, you will have to either clip directly into the anchoring device (e.g., a piton) or use an *anchor sling*. Anchor slings are usually made from webbing, although rope can be

Fig. 5.0 No-knot or tensionless anchor.

used. The working rope can then be clipped directly into the anchor sling, or a release knot can be added in between the two if it's necessary to release the load gradually. The strength of an anchor sling depends on the type of webbing (or rope) used and the sling configuration. The 1" (25 mm) tubular webbing commonly used for anchor slings has a breaking strength of about 4,000 lbf (17.8 kN), but this can be greatly increased by using multiple wraps in either a three-bight or a "wrap three, pull two" configuration.

Vehicles can also make good anchors, especially for low-angle evacuations up river banks. They are mobile and often can be moved where no natural anchors exist. If you use one, consider the following:

- The bigger the vehicle, the better the anchor. A large fire truck works better than a motorcycle.
- Rig the anchor as low on the vehicle as possible. A high rigging point may cause the vehicle to move or even overturn. Running a rope over a vehicle may damage both.
- Make sure the vehicle doesn't move. Put it in gear (or PARK), engage the emergency brake, and chock the wheels. Set the angle of pull perpendicular to the vehicle's rolling axis, and give the keys to the safety officer.

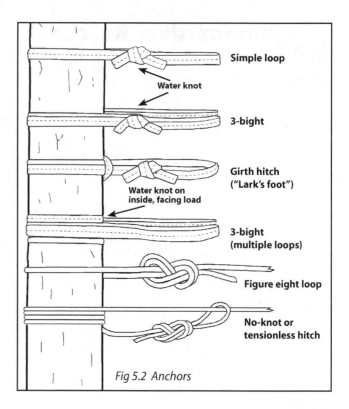

Fig 5.2 Anchors

- Watch what you tie on to. Select a part of the vehicle that won't pull off, such as the frame or suspension rather than bumpers or external hooks. Try to avoid getting anchor slings or ropes greasy or from coming into contact with chemicals like battery acid. Avoid sharp or abrasive edges and, if you can't, pad them.

Stakes and *pickets* can make effective anchors in soil, sand, and snow. They can be either improvised locally or bought as prefabricated metal stakes. Stakes are typically 4–5' (1.2–1.5 m) long, driven into the ground for half to two-thirds of their length at a 10–15° angle away from the load. They are normally employed in series with a rope or web loop run from the bottom of one stake to the top of the one behind it. To tension the loop, a smaller stake is twisted in the center of the loop and then driven into the ground. The load is anchored to the bottom of the first picket. A useful variation is to put in two rows of stakes holding a log or pipe for an anchor point. Picket anchor systems are cheap, easy to transport, and relatively simple to install. They will work where just about nothing else will.

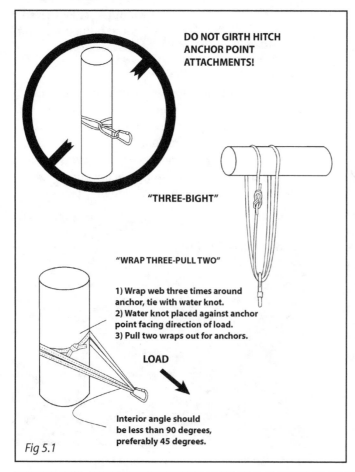

DO NOT GIRTH HITCH ANCHOR POINT ATTACHMENTS!

"THREE-BIGHT"

"WRAP THREE-PULL TWO"

1) Wrap web three times around anchor, tie with water knot.
2) Water knot placed against anchor point facing direction of load.
3) Pull two wraps out for anchors.

LOAD

Interior angle should be less than 90 degrees, preferably 45 degrees.

Fig 5.1

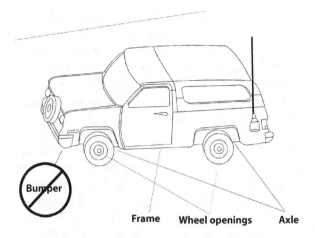

Fig 5.3 Vehicle Anchor Points – All sharp surfaces must be padded. Ropes and slings must be protected from oil, grease & especially battery acid!

However, they are slow to set up and hard to move. While the actual holding power of a stake system depends on a number of factors (e.g., soil type, density, and stability), the following figures are adequate for rough planning purposes:

Single	**700 lbf (3.1 kN)**
Double (1-1)	**1,400 lbf (6.2 kN)**
Triple (1-1-1)	**1,800 lbf (8 kN)**

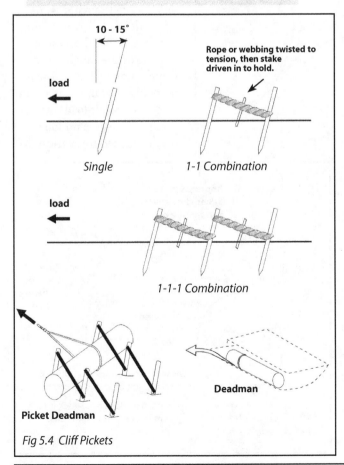

Fig 5.4 Cliff Pickets

A *deadman* is any object buried and used as an anchor point. Almost any object can be used, but the most common is a pipe, log, or tire rim. The deadman is buried in a vertical trench, and a channel dug out for the anchor rope so that the deadman is pulled against undisturbed soil.

Backup/Multiple Anchors

It's a good idea to *back up* any anchor, and mandatory for life lines. An anchor is considered backed up when:

- there is more than one anchor in the system
- each anchor will hold the entire load

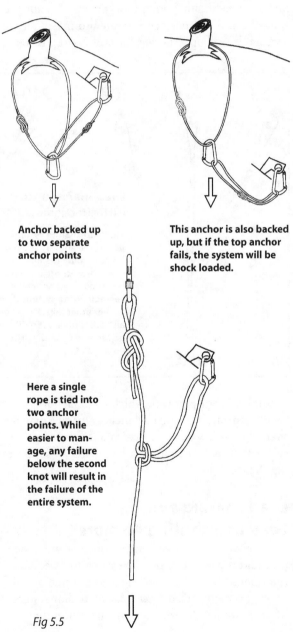

Anchor backed up to two separate anchor points

This anchor is also backed up, but if the top anchor fails, the system will be shock loaded.

Here a single rope is tied into two anchor points. While easier to manage, any failure below the second knot will result in the failure of the entire system.

Fig 5.5

Any live load (i.e., any rope from which a person hangs) should be supported by at least two independent anchors, and it's a good idea for any system. Backing up an anchor does not necessarily mean that the backup anchor shares the load under normal conditions—it may only be loaded if the first anchor fails. Anchor failure may cause a shock load, which is any sudden load on the system caused by an unexpected drop or shift of the load. Since this may induce forces far beyond those anticipated, riggers try studiously to avoid it while at the same time taking precautions to contain it if it does occur. One of the basics is to keep unnecessary slack out of the system. In the case of anchor backups, if the primary anchor fails the system should not shift or drop enough to shock load the backup anchor. If an anchor point is known to be totally bombproof, rigging two anchor slings from the same anchor point also makes an acceptable backup. Another form of backup is to use

subject of much testing and heated debate, and testing indicates that "equalizing" anchors don't do the job very well. In the real world rope stretch on the longest legs means that the shortest anchor legs carry proportionally more of the load, and even a small change in the direction of pull can dramatically shift the load from one anchor to another. This is not a problem *if* all anchors can hold the entire load, and in fact offers a quick way to back up anchors. On the positive side, anchor failure on a load-sharing system usually induces little shock loading if an anchor fails providing the rigger keeps slack out of the system. Load-sharing anchor systems like the *cordelette* work well for anchor backups and do not induce a shock load if an anchor fails.

Sometimes, however, only marginal anchors, i.e. ones not capable of reliably holding the entire load, are available—a problem often encountered by sport climbers and sometimes by rescuers as well. Until recently the preferred solution was the *load distributing* or *self-equalizing anchor*, a connecting loop that could self-adjust for changes in the angle of pull as well as for a failed anchor. In theory, even in the event of an anchor failure, no single anchor would be overloaded. Again, testing has shown that this is simply not the case—equalization is poor and anchor failure is quite likely to induce a shock load. As a result load distributing anchors are no longer recommended for vertical or

70 #   **70 #**   **70 #**   **210 #**   **0 #**   **0 #**

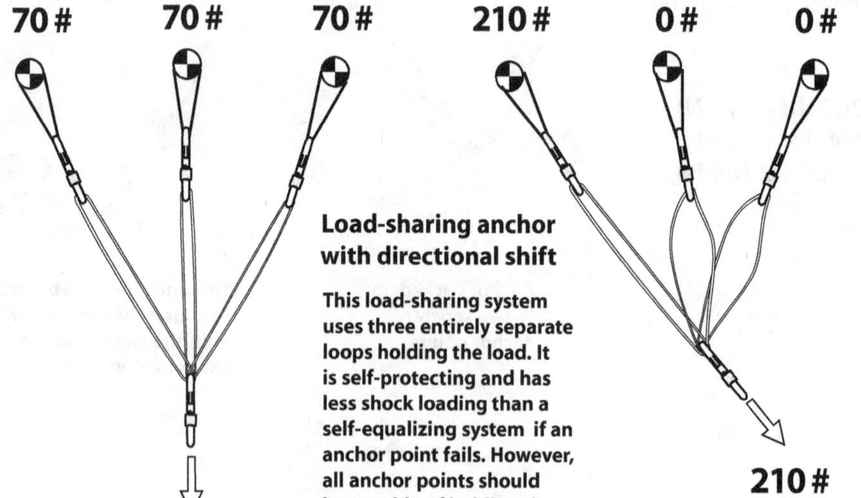

Load-sharing anchor with directional shift

This load-sharing system uses three entirely separate loops holding the load. It is self-protecting and has less shock loading than a self-equalizing system if an anchor point fails. However, all anchor points should be capable of holding the entire load.

210 #

Fig 5.6   **210 #**

multiple anchor points in which each anchor is a totally separate system. The anchors may or may not share the load. With multi-point anchors the general rule is that each anchor must be capable of supporting the *entire* load.

Load Sharing and Load Distributing Anchors

If two or more anchors are loaded at the same time and the load is allowed to swing to its equilibrium point, they will share the load to some extent, an arrangement called a *load sharing* anchor system. Multi-point load sharing anchors have been the

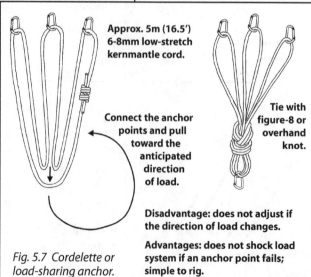

Approx. 5m (16.5')
6-8mm low-stretch kernmantle cord.

Connect the anchor points and pull toward the anticipated direction of load.

Tie with figure-8 or overhand knot.

Disadvantage: does not adjust if the direction of load changes.

Advantages: does not shock load system if an anchor point fails; simple to rig.

Fig. 5.7 Cordelette or load-sharing anchor.

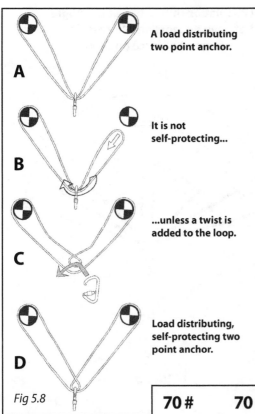

A load distributing two point anchor.

A

It is not self-protecting...

B

...unless a twist is added to the loop.

C

Load distributing, self-protecting two point anchor.

D

Fig 5.8

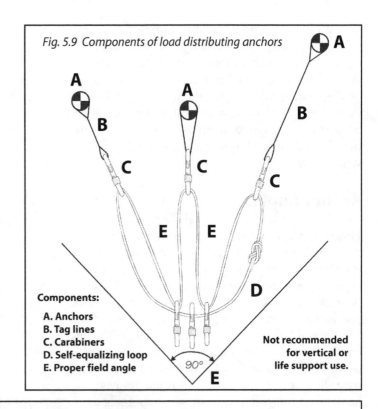

Fig. 5.9 Components of load distributing anchors

Components:

A. Anchors
B. Tag lines
C. Carabiners
D. Self-equalizing loop
E. Proper field angle

90°

Not recommended for vertical or life support use.

life line work. However, they still have a place on the river where loads move much more slowly, no one is hanging from a rope, and shock loading is much less of a problem. In many places in the Western US, for example, a rescuer often encounters a situation where a pinned raft has

70 # 70 # 70 # 70 # 70 # 70 #

Load shift - load distributing anchor

The load distributing anchor adjusts for a changing angle of pull, even though it equalizes poorly and may shock load the system if an anchor fails. A third carabiner inside the loop is required for a self-protecting setup. Not recommended for vertical or life support use.

Fig. 5.11

210 #

210 #

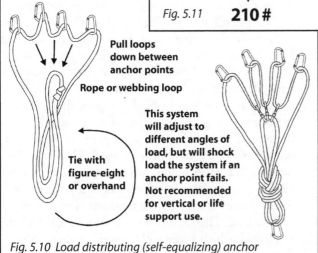

Pull loops down between anchor points

Rope or webbing loop

This system will adjust to different angles of load, but will shock load the system if an anchor point fails. Not recommended for vertical or life support use.

Tie with figure-eight or overhand

Fig. 5.10 Load distributing (self-equalizing) anchor

to be hauled from some very marginal brush and rock anchors. The load distributing anchor is still the only system that can adjust for a direction of pull that changes over a wide angle, which often happens when rescuers pull a raft off an obstacle and have it swing round into an eddy. The pinned watercraft is also a weak point and often it's necessary to spread the load, however imperfectly, over several marginal anchors (e.g. the D-rings on a raft as described in Chapter Six). When the consequences are no more than popping a few D-rings or at most losing a raft, the load distributing/self-equalizing anchor is still an acceptable solution.

When setting up a load distributing anchor it's preferable to keep the equalizing loop as small as possible. If the anchors are not equidistant, extend the anchors to the equalizing loop using static rope or slings rather than making the sling larger to reach them. Load distributing systems should be rigged as a *self-protecting* system so that if an anchor points fails the sling will not pull through and cause the whole system to fail.

Anchor Angles

Hanging a load from a single line is easy. However, when we use multiple anchors, we must carefully consider the angles between them. The simple expression of the principle is: the greater the angle between the anchors, the greater the force on them. You can prove this to yourself by filling a milk jug with water, running a piece of rope through the handle, and holding an end in each hand. With your hands almost together and the milk jug hanging down between them, it's not too hard to hold. Now try pulling your hands apart. It gets progressively harder, and you will find it virtually impossible to pull the line completely straight. It is the same with anchors. If we connect two anchors directly and load the rope between them, the force on each anchor is greatly multiplied. For this reason we prefer to *keep the angle of loading between any two anchors at 90° or less*. Remember this principle—we will see it again.

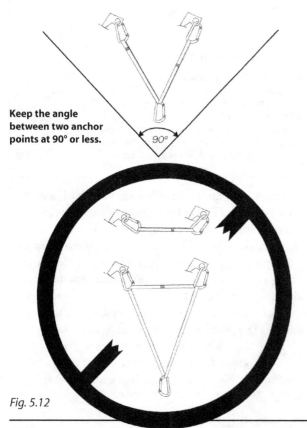

Keep the angle between two anchor points at 90° or less.

90°

Fig. 5.12

The relationship between the angle between anchors and the resulting amount of force on them

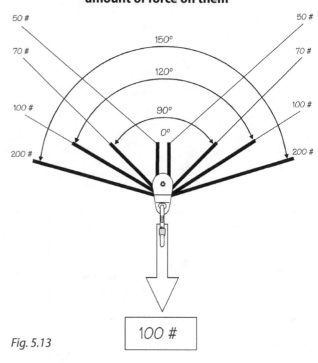

50 #   50 #
70 #   70 #
150°
120°
100 #   100 #
90°
0°
200 #   200 #

100 #

Fig. 5.13

A Quick Course in Mechanical Advantage

Figuring anchor loads and mechanical advantage causes a lot of confusion. Fortunately, it's not mysterious, and basic principles need to be understood by anyone who wants to learn rigging. We've already learned one of the most important principles: if we suspend a load from several ropes, they will share the load equally. If we suspend a 50 kg load from two ropes, each anchor will bear 25 kg. With four ropes each anchor holds 12.5 kg. Now let's put a pulley *on the load* and pull on one of those two anchor ropes. Both ropes still bear the load, so if we pull on one of the ropes with a force greater than 25 kg, the load will begin to move. We are now lifting a weight of 50 kg with only 25 kg of pull. In effect we are doubling our strength with a *mechanical advantage* of 2:1. There is, however, a price to be paid. To move the load one meter we must pull two meters of rope. Now look at the second example: using four ropes reduced the load on the anchor to one-fourth the load. Therefore, if we have four ropes pulling we have to pull with only 12.5 kg, but must pull four meters of rope to move the load one meter. This is a mechanical advantage of 4:1. Naturally, there will be some friction losses in the real world, which is generally figured at about 10% from each pulley.

Not all pulleys add mechanical advantage. In order to do so, the pulley must move with the load. If the pulley does not move, it just changes the direction of the pull without adding any advantage to the haul system. However, this does not mean that there are no forces on a change of direction pulley. Instead of putting a mechanical advantage on the load, a change of direction pulley puts it *on the anchor*. For this reason the haul system may put a load on an anchor that is greater than the weight being lifted. In some cases, the load on the anchor may be double the weight being lifted. Why? Because we have simply reversed our 2:1 haul system into a directional change system.

We can stack pulley systems—that is, have one system pull another one—for a *compound pulley system*. Here we multiply, rather than add, the mechanical advantages (MAs) of the two systems. For example, a 2:1 system pulling a 3:1 system gives a total MA of 6:1, not 5:1. We can also use pulleys with multiple sheaves for increased MAs. However, we quickly reach a point of diminishing returns, since for every increase in MA, friction increases and we must pull more rope though the system to get the load to move.

Three quick rules for figuring mechanical advantage are:

- Count the number of lines coming *from the load*. Three lines means that the MA is 3:1.
- If the rope is attached directly to the load, the MA is odd (e.g., 3:1). If attached to the anchor, it is even.
- If the last pulley in the system is attached to the anchor, it functions only as a change of direction and does not add MA.

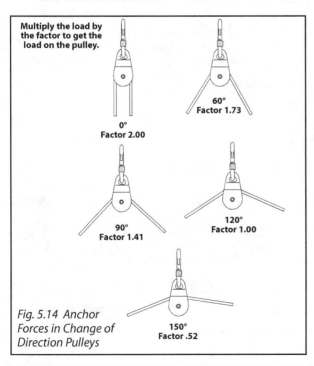

Fig. 5.14 Anchor Forces in Change of Direction Pulleys

Multiply the load by the factor to get the load on the pulley.

0°
Factor 2.00

60°
Factor 1.73

90°
Factor 1.41

120°
Factor 1.00

150°
Factor .52

Haul Systems

A haul system allows us to pull or raise something. The simplest haul system is the direct pull. This can be several people pulling on a rope. Great monuments have been built with no more than this. Austrian engineer Peter Reithmaier found that the most important factor in how hard a person can pull on a rope is not strength, but weight. A person can, using a shoulder belay, generate a pull equal to his body weight, with a maximum "jerk" pull of 120% of body weight. When holding the rope in his hands, however, the maximum pulling force is only 60% of the person's body weight. Using a smaller diameter rope (6 mm vs. 8 mm) also reduced the amount of pull by about 10%. Rigging a triple harness that allows all persons pulling to use a shoulder belay effectively increases the force available in this simplest of haul systems. There are other direct haul systems as well. We may use a hand winch, a come-along, the power winch on a vehicle, or pull with the vehicle itself.

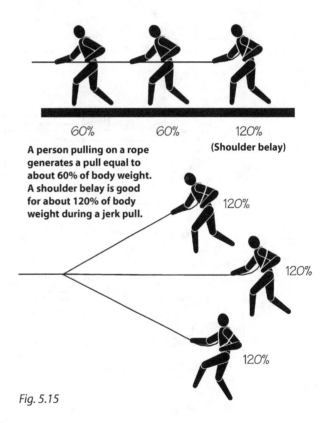

60%   60%   120%
(Shoulder belay)

A person pulling on a rope generates a pull equal to about 60% of body weight. A shoulder belay is good for about 120% of body weight during a jerk pull.

120%

120%

120%

Fig. 5.15

By using the principle of mechanical advantage we can construct very powerful haul systems. In the real world of swiftwater rescue and technical rope rescue, we will seldom need a system more powerful than three or four to one. A system with this MA, pulled by four to six beefy adults, can generate a force of over 1,000 lbf (4.4 kN), or about the same as a mechanical hand winch.

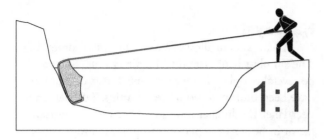

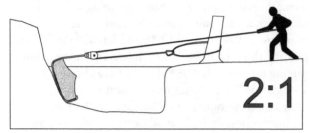

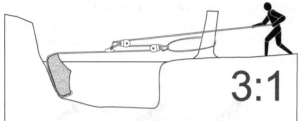

Fig. 5.16

The two most popular haul systems are the Z-drag and the pig rig. The Z-drag (known to climbers as the Z-rig) is simple to rig and gives an MA of 3:1. Stacking a 2:1 on it raises this to 6:1, or we can have two stacked Z-drags for 9:1. The pig rig, long a favorite of rescue squads, is two stacked 2:1 systems, giving a total MA of 4:1. The pig rig (short for piggyback rig) gets its name from the way it is set up: a haul system clamped to a haul/belay line, effectively separating the haul and belay systems. The Z-drag is somewhat simpler to set up and more familiar, while the pig rig gives a slightly higher MA with the same amount of equipment, and can be readily moved from one line to another.

Any haul system should have a ratcheting capability: it should hold the load under tension and release it when necessary. With haul systems like the Z-drag it is necessary to periodically reset the traveling Prusik as it is pulled closer to the anchor, which means that something has to hold the load while this is being done. This is usually a belay or some kind of *progress-capturing device* (PCD) so that the load doesn't fall if we let go of the rope. This is usually

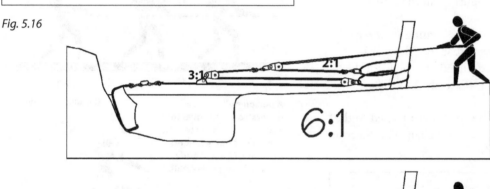

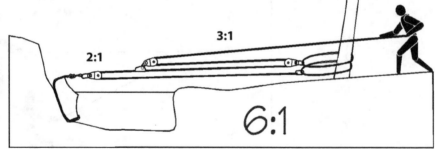

Fig. 5.17

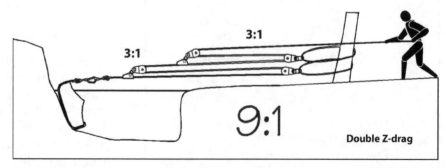

Fig. 5.18

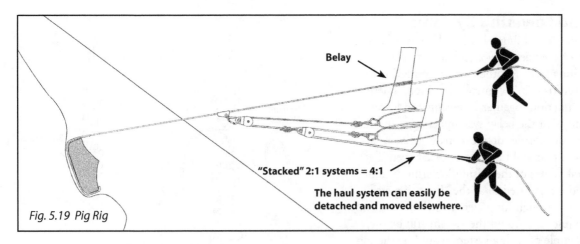

Belay

"Stacked" 2:1 systems = 4:1

The haul system can easily be
detached and moved elsewhere.

Fig. 5.19 Pig Rig

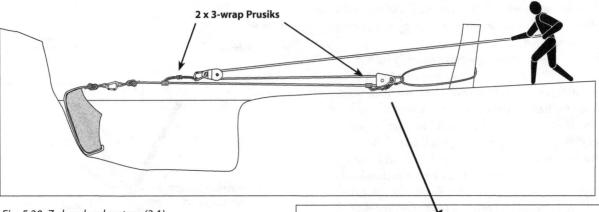

2 x 3-wrap Prusiks

Fig. 5.20 Z-drag haul system (3:1)

When setting up a Z-drag to haul a boat, it's a good idea to
have some sort of release for the progress capture device
(PCD) on the anchor in case the boat swings out into the
current.

TOP: Prusik-minding pulley with a load-releasing hitch

MIDDLE: Pulley set up with a "Bachmann Brake" – a
Bachmann knot and carabiner that "minds" the Prusik to keep
it out of the pulley. A mariner's hitch provides a release.

BOTTOM: An autoblock (French Prusik) functions as a PCD
and can be released under pressure with a sharp blow on the
top of the knot with a paddle (do not use your fingers!).

a cinching knot like a Prusik (sometimes called a
brake Prusik) that runs free when pulled but that
holds when released. If we choose to use a Prusik
here, there are some special considerations. During
hauling, the Prusik loop may be drawn up into the
pulley and become jammed. To avoid this we can
use a Prusik-minding pulley (PMP), whose squared-
off sideplates will keep the Prusik out of the pulley. If a PMP isn't available, an autoblock (French
Prusik) will function with a regular pulley. Since it
includes a carabiner, the Bachmann knot will also

work, although it is not quite as jam-proof as the
PMP. Yet another method is to use an entirely separate rope, or belay system, to hold the load.

If we want to be able to release the load under
tension, a load-releasing hitch or mariner's hitch
attached to the anchor, or a mariner's hitch attached
to the brake Prusik, works well. The French Prusik
can be released under load with a sharp blow with
a paddle on top of the knot, and for extreme cases
there is always the "rope wrench" —a sharp knife.

Belays and Lowering Systems

In addition to holding a load under tension, rescuers often have need to *belay*, that is, to catch or arrest a falling load. Another recurring need is to raise and lower loads in a controlled fashion. These two systems function together in raising and lowering operations with the belay system acting as a safety backup. Some authorities define a "true" belay as a separate system that will function by itself even if there is a total failure of the raising/lowering system. This is sometimes referred to as the "lightning strike test" i.e. if a bolt of lightning were to strike the scene and stun the rescuers, would the system still be on belay? An example of such a system would be the tandem Prusik rescue belay. A "conditional" belay, on the other hand, requires active intervention by the belayer (e.g., Münter hitch, Stitch plate, etc.).

Raising and lowering systems generally have a *main* or *working line* and a *belay line* to back them up. As always, the possibility of shock loading should be held to a minimum by keeping slack out of the system. A belay device, which is intended to catch a falling load, is not the same as a *progress capturing device* (PCD), which is meant only to keep the load from moving backward after the tension from the haul system is relaxed, as in a reset. Obviously the demands made on a belay device are much greater than on a PCD.

TIP: Use different color ropes for haul and belay lines. It makes giving commands easier.

Belays take many forms. The simplest are the *body belay* and *shoulder belay*, where the belayer uses his own body to provide friction. While these are simple, quick, and require no tools, they obviously have their limitations. In swiftwater use, they are used mainly for belaying individual swimmers with throw ropes or for scrambling up and down banks. Sport climbers and mountaineers sometimes use them, and even a limited form of rappelling can be done with a modified body belay (e.g. the Dulfersitz).

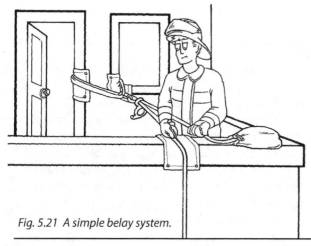

Fig. 5.21 A simple belay system.

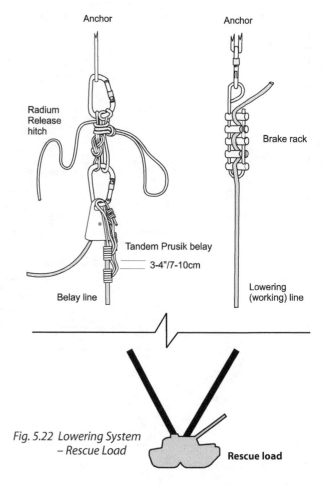

Anchor   Anchor

Radium Release hitch

Brake rack

Tandem Prusik belay
— 3-4"/7-10cm

Belay line   Lowering (working) line

Fig. 5.22 Lowering System – Rescue Load

Rescue load

Sport climbers who operate in the vertical realm, use dynamic (high stretch) ropes and deal with relatively light loads (a climbing load is typically defined as 176 lb/80kg), which allows them to use simple friction devices like Stitch plates, tubular devices and figure eights for belays.

Rescue loads, on the other hand, are considerably heavier—200kg/440lb for "light" mountain rescue use (these typically use 11mm/7/16" ropes) and 280kg/600 lb for "general" fire-rescue use (12.7mm/1/2" rope)—and require more substantial devices to arrest their movement. For flat and low angle moves a single Prusik on the haul line or a simple friction belay is adequate but as the angle becomes more vertical the belay requirements increase accordingly, as does the need for a separate haul and belay line. The ability of various belay systems to catch and hold vertically falling rescue loads has been a subject of much debate and testing in the past few years. Testing has shown that belay devices commonly used in sport climbing (e.g. Sticht plate, figure 8, Münter hitch) are not adequate, and that mechanical clamps like the Gibbs Ascender can damage the rope under a shock load. While the new generation of hard clamps such as the Rescuscender™ are much less likely to damage the rope than were older models, the current weight of opinion seems to favor using these devices as PCDs rather than as belay devices.

The tandem Prusik belay (e.g., dual 8 mm/5/16" three-wrap Prusiks of different lengths on a 13 mm/½" rope) sometimes called the *tandem rescue belay*, meets the requirements of the belay competence drop test.

> **The British Columbia Council of Technical Rescue Belay Competence Drop Test specifies a one meter drop onto three meters of kernmantle rescue rope with less than one meter of additional travel and less than 15 kN peak force. For a General Use rated device e.g. NFPA "G", a 280 kg/600 lb load and a 13 mm/1/2" rope are generally used. For a Light Use device, a 200 kg/440 lb load and an 11 mm/7/16" rope are used.**

Since it will slip somewhat under extreme load, it is less prone to sudden failure than systems using hard clamps. However, it requires some specialized equipment (i.e. Prusik-minding pulley) and must be tended during use. Lately some mechanical belay devices (e.g. 540 Rescue Belay, Petzl I'D) have been introduced that meet the standard as well.

River loads are difficult to define and are often quite unpredictable. As we saw in Chapter 2, water has great force (which increases exponentially) and weight, which often makes for a harsh working environment. However, compared to vertical rescue things happen much more slowly. Thus, as mentioned above, some rigging solutions like load-distributing anchor systems are still in use.

Although most rescuers use a separate belay and haul rope for steep angle and vertical applications, there are times when it is preferable to combine them.

Fig. 5.24 Traveling Brake Lowering

This is called *single rope technique* (SRT) and is often used in caving. An example of where this option might be better is on a lengthy rappel where there is a danger of the belay line becoming entangled with the working line. In most cases this is not as safe as a "true" belay, since the failure of any part of the system will result in a failure of it all. Single rope belays (e.g., bottom belay, Prusik safety) are often called *pseudo-belays* and are, in general, not recommended for rescue work except in rappelling.

A *lowering system* uses a friction device to control the descent of a person or load. It can either use a *fixed brake*, in which the friction device remains stationary and rope moves with the load, or it can use a *traveling brake*, where the rope is stationary and the load, attached to the friction device, moves. For both systems, a figure eight friction device is considered adequate only for a single person load and has the additional disadvantage of twisting the rope. For rescue loads, a brake rack or a specialized mechanical descender (e.g. Petzl I'D) rated for the job is a better solution.

- In the fixed brake system, the lowering is controlled by the brake operator at the friction device. The person(s) being lowered have no control over their descent, making good communications mandatory. By the same token, however, they are not distracted by the need to manage the lower.
- In the traveling brake system, the rescuer has more control, but must manage his own descent, complicating his task. This is the same thing as rappelling.

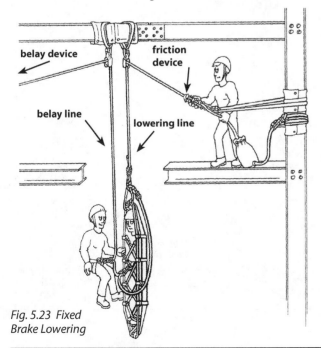

Fig. 5.23 Fixed Brake Lowering

Generally speaking, it's easier to lower than raise something. When lowering, gravity can be used as the principal force, and rock friction works for and not against you. However, most river rescue operations will involve raising.

Rappelling

Rappelling is a form of lowering where an individual uses a traveling friction brake to descend a fixed rope. Whole books have been written on this subject and we can do no more than review it here. Rappelling is not as commonly used in swiftwater rescue as in, say, caving, but a basic knowledge is still useful. Rappelling is generally agreed to be one of the more dangerous activities for rescuers. It is usually done unbelayed and there is always the chance of something going badly wrong, such as hair or clothing getting caught in the rappel device. This is especially true in a free rappel, where the rappeller does not touch the surrounding walls.

In swiftwater rescue, however, the two most common uses for rappelling are:

- going down a steep bank
- rappelling down from a bridge

In both cases, the exposure is usually considerably less than in climbing or caving.

The rescuer's own body can function as a simple friction device. The simplest is the *arm rappel* (guide's rappel), which is suitable only for very low-angle applications. The *body rappel* (Dülfer wrap, Dülfersitz) is quick but not secure—it is possible to "unwrap" and fall out of it. The rope is run from the anchor between the legs and wrapped around the hips, across the chest, and over the shoulder. The brake hand controls the descent. Use this system only for less exposed applications like going down the river bank—not for free rappels. Both these methods require no special equipment other than a rope. Rappellers should, however, wear protective clothing when doing either one.

The figure eight is the most commonly used rappel device. While it is normally used only for single-person loads, the rescue eight can be rigged with sufficient friction for a two-person load. Using two ropes in a double-rope rappel is another way to add friction. This can be two separate ropes or a single rope doubled at the anchor.

The brake rack is the preferred solution for longer rappels and heavier loads. It is fully capable of lowering rescue loads and will not twist the rope. However, if you choose to use a brake rack, *get hands-on instruction in its correct rigging and use.* It is possible to rig the bars incorrectly so that they come off the rope under load!

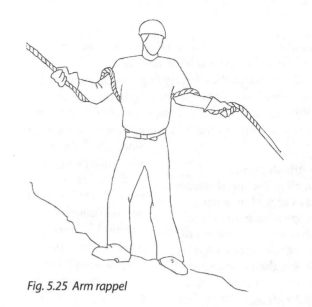

Fig. 5.25 Arm rappel

Stop & Hold Position

Fig. 5.26 Dülfer Wrap

Fig. 5.27 Basic Rapelling

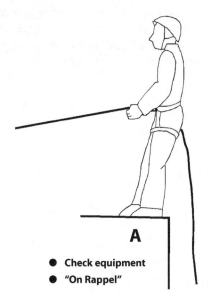

A

● **Check equipment**
● **"On Rappel"**

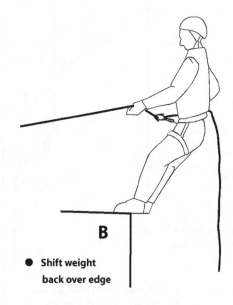

B

● **Shift weight**
  back over edge

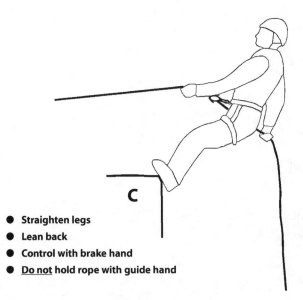

C

● **Straighten legs**
● **Lean back**
● **Control with brake hand**
● <u>**Do not**</u> **hold rope with guide hand**

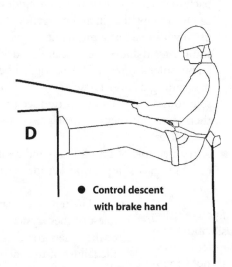

D

● **Control descent**
  with brake hand

The procedure for rappelling is the same no matter what device is being used. The rappeller attaches the rope to the descending device, then backs up to the edge. The rappeller's brake hand controls his descent by pulling more or less friction on the descender. Both the figure eight and the brake rack can be locked off during a rappel, allowing the rescuer to rest or to free his hands for other purposes.

There are several ways to improve the safety of rappels:

● Get adequate training and practice under controlled conditions.

● Use a separate backup belay. This may not always be possible, especially on long rappels where there is a danger of the belay rope twisting around the rappel rope.

● When using a figure eight, you can use a *bottom belay*. Here, a belayer at the bottom of the rope pulls down on it, increasing friction at the belay device and thus slowing or stopping the rappeller.

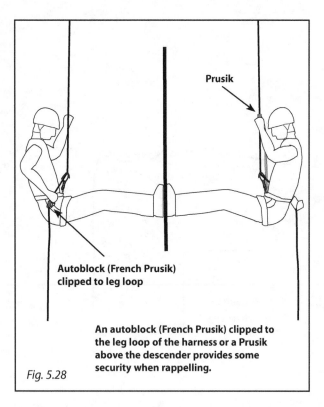

Prusik

Autoblock (French Prusik)
clipped to leg loop

An autoblock (French Prusik) clipped to
the leg loop of the harness or a Prusik
above the descender provides some
security when rappelling.

Fig. 5.28

Conclusion

The basic techniques of raising, lowering, and belaying loads are essential to any rope operation, as well as being the building blocks for the more complex operations described later.

- Another common safety measure is to use a backup cinching knot or camming device to catch the rappeller in an emergency. This is usually a Prusik or a mechanical device like a shunt rigged above the descending device or an autoblock (French Prusik) rigged below the descending device. These are not foolproof solutions and sometimes introduce problems of their own, but they are widely used and have saved lives.
- Carry a couple of Prusik loops and a safety knife or EMT shears in your pocket for emergencies.

TIP: If you're using a Prusik above the descender be sure that it's not too long i.e. that you can reach it if it engages; if you're using an autoblock below the descender, make sure that it can't be pulled into the descender when engaged.

The separate belay is the safest of these systems. None of the others guard against the failure of the anchor or the main rope. Rappelling is a complex and dangerous skill, requiring comprehensive training and constant practice to keep fresh. As such, it often requires a disproportionate amount of training time for the amount of actual use.

Rigging for River Rescue III
Advanced Techniques
Chapter 6

Fig. 6.0 A raft solidly pinned on Gunsight Rock at Troublemaker Rapid, South Fork of the American River, California.

It is now time to discuss the actual practice of rescue using the techniques covered in Chapters 4 and 5. Boats, whether rafts or hard boats, sometimes get pinned and need to be recovered. We will cover methods of doing this and appropriate safety considerations. Another common application is low-angle rescues to move rescuers from the river up the bank. Finally, we will take a brief look at high-angle techniques, specifically high lines.

Boat Rigging and Recovery

One of the most common river situations in which we will put our new rigging skills to use is in recovering pinned boats. By *pinned* we mean that the river pushes the boat against a rock, bridge piling, tree, or other obstruction in the river, holding it firmly in place. The force of the current on even a small pinned boat can easily exceed a ton. Nor is that the only force. Most pinned boats will be fully or partly submerged, adding the weight of the water to the force of the river. A very common pinning scenario is this: a boat heads toward a rock, and the crew, anticipating a collision, leans away from the rock (i.e., upstream). When the boat does hit the rock it stops momentarily and the water, helped by the weight of the crew, rolls it upstream. As the upstream gunnel goes underwater, the boat fills with water and is pushed against the rock. In the process, the boat often collapses and *wraps* around the obstacle, sometimes trapping crew members in or under it in the process.

Fig. 6.1 *To recover the raft on the previous page, rescuers set up a haul system to pull up one corner of the raft …*

In addition to the type of pin described above, where the center of the boat is pushed against the obstacle, there is also the *end pin*, where the ends of the boat are pushed against separate obstructions, and the *vertical pin*, where the bow of the boat gets pinned against the bottom of the river when running a drop. Vertical pins normally happen to sport kayakers running whitewater rivers.

The first step in recovering a boat is to correctly *assess* or size-up the situation. Consider:

● Are any people in danger? Is someone trapped in the boat or underneath it? If so, rescue efforts must begin immediately. If not, more time can be taken for a thorough assessment.

● Does the pinned craft present a hazard to other river traffic? This can be a significant factor on a river where there is heavy recreational boating traffic. If it does, the river or the accident site may have to be closed, or at least have the upstream traffic warned.

Once these two critical factors have been assessed, it's time to look at the pinned boat. As we have seen, there is a lot of force in that river. The trick is to make it work for and not against you. First, what kind of pin is it? Is the boat severely wrapped and underwater? Is one end higher than the other? If it is wrapped on a rock, the current will be pushing with more or less equal force on both ends of the boat. If you can lessen this force on one end, the water will tend to push harder on the other. If one end is higher than the

other, that's usually a good place to start. Normally, the easiest way to get a boat off a rock is to pivot it off, pulling or lifting on one end to lessen the pressure while letting the current push on the other. Trying to pull a boat directly against the current is extremely difficult.

Next, try to determine the method of extrication. Many boats can be freed without using mechanical haul systems. If you think that you're going to need to haul, you will need to find suitable anchor points on the boat and on shore. Of the two, the weakest are going to be the anchor points on the boat, whose designers never thought that someone would be pulling on it with this much force.

Consider also how you will access the pinned boat. In some cases you may be able to wade out to the accident site. However, because of the danger of foot entrapment, this should *not* be attempted by untrained personnel. Perhaps someone can swim out to it or catch the eddy behind it in a boat or rescue board. Again, they must have adequate training for the attempt. Can someone safely stand on or near the boat in order to rig it?

After that, you will need to get the rigging lines to the boat. Can they be thrown to the riggers on the boat, or shot over with a line gun? Or can someone ferry the rope over to the site with a boat?

Fig. 6.2 *… which was then pulled over to dump the water.*

Finally, take some thought about what will happen when the boat comes off. Many boats have been recovered from one pin only to take off downstream into another one because no one thought of this. In fact, it is a good idea to tie a line to the pinned boat *immediately* when you arrive on site. Often, boats are not as firmly pinned as they first appear, and some have washed off while the "rescuers" were still in the planning stage.

Although site organization and safety will be discussed in more detail later, it is worth mentioning here that *when you start stringing lines across the river, you must consider the safety of anyone upstream of you.* On a busy river, a rope across it can "clothesline" anyone coming downstream.

Once the rescuers have these things clearly in their minds, it is time to start recovery. Try simple, low-risk methods first and then progress to more complex systems if these don't work. Our goal is to *upset the equilibrium* of force that the current has imposed on the pinned boat, *dump the water* out if possible, and then *pull it to shore*. In many cases where the boat is not submerged too deeply and the forces finely balanced, just pushing or lifting on one end of the boat (the "armstrong" method) will allow the current to pivot the boat around the other side of the obstacle. A simple lever placed between the boat and the obstacle will multiply the strength of the rescuers' arms somewhat.

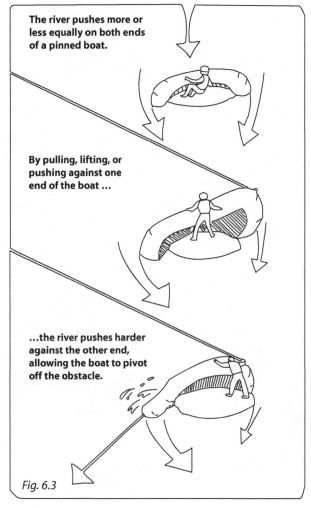

The river pushes more or less equally on both ends of a pinned boat.

By pulling, lifting, or pushing against one end of the boat ...

...the river pushes harder against the other end, allowing the boat to pivot off the obstacle.

Fig. 6.3

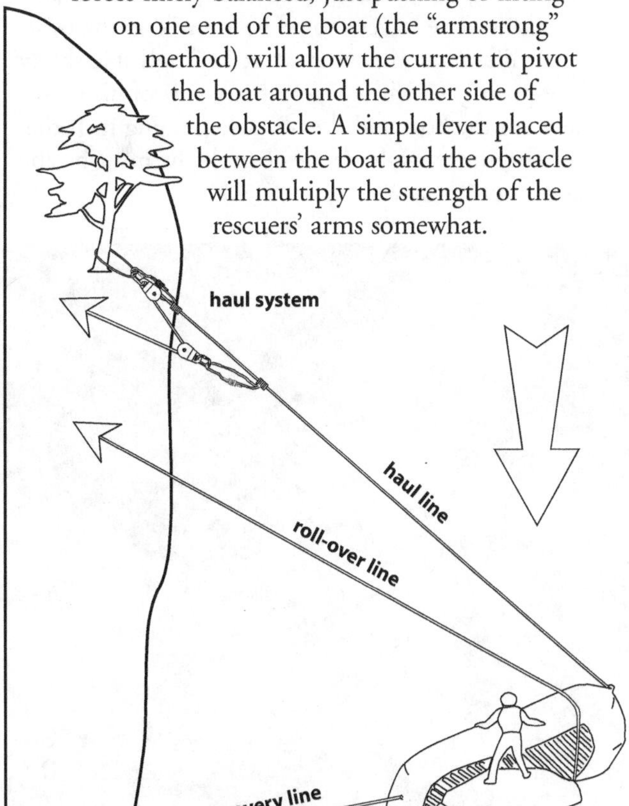

haul system

haul line

roll-over line

recovery line

Fig. 6.4 In some cases, three separate lines may be needed. In others, one line can serve all three functions.

If this fails, you may have to resort to a haul system. The most basic haul system (the tenboyscout system) is to simply attach a line to the boat and have a large number of people pull on it. If you have enough people, this may be all you need. If not, try a mechanical advantage system like the Z-drag or Pig rig.

Let's say that you have set up a Z-drag on a pinned boat. You have pulled with everything you have, but the boat remains stuck fast. What now? You can add pulling force and change the direction of pull slightly by using a *vector pull*. This technique makes use of a physical principle we have already met when discussing anchors—that pushing on the center of a taut line will greatly increase the force on the ends. Once the Z-drag has been pulled tight and locked off, attach a line to the center of the line going from the haul system to the boat. Then pull at right angles to the original direction of pull. The effectiveness of the vector pull falls off rapidly as the angle of the rope increases, but is quite powerful at first. It may provide just enough extra pull to get the boat off. If not, you can try stacking another pulley system on top of the first and pulling again, and then do another vector pull.

If the boat still won't come off, it's time to reassess the situation. Don't automatically assume that the problem is an inadequate hauling system. Most pinned boats will not require this much pull to get them off, and you're very likely to start pulling the boat itself apart. It's much more probable that something else is wrong—that you are pulling in the wrong direction or directly against the current, or that the boat is not rigged correctly. Before adding more pull, try changing some of these factors.

Consider also the direction of pull. Instead of pulling directly upstream, it usually works better to haul at approximately a 45° angle to the current. Although we will try to pull on the end of the boat with the least force on it, the location of a suitable anchor point on shore may effectively dictate where we set up our haul system.

Rigging a pinned boat can be a challenge. Part and often all of the boat will be underwater, with the current flowing swiftly over it. If you are the rigger, you will often not be able to see what you are doing, working against the force of the water will tire you quickly, your hands will become numb from the cold, and the rope will move around in the current like a live thing. The rigger should be aware that the boat may shift at any time and act accordingly. He should be especially careful not to get any part of his body between the boat and the obstacle pinning it. If the boat begins to wash off the obstacle, he must be able to get out of the way.

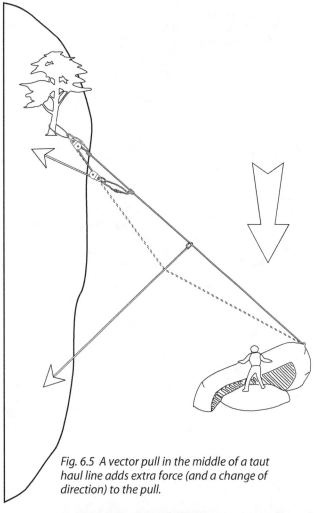

Fig. 6.5 A vector pull in the middle of a taut haul line adds extra force (and a change of direction) to the pull.

Fig. 6.6 Rigging a pinned boat can be a difficult and sometimes dangerous task.

Swiftwater Rescue

Fig. 6.7 The "Boatman's Eight"

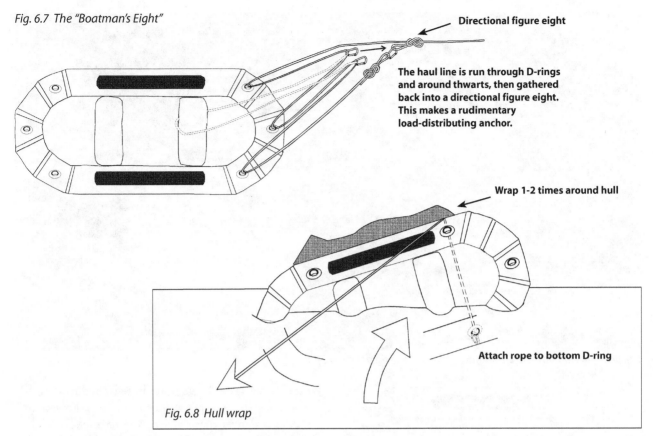

Directional figure eight

The haul line is run through D-rings and around thwarts, then gathered back into a directional figure eight. This makes a rudimentary load-distributing anchor.

Wrap 1-2 times around hull

Attach rope to bottom D-ring

Fig. 6.8 Hull wrap

Obviously, the rigger needs to tie the rope to something on the boat that will not break or come loose easily. If the boat is an inflatable, this might be an inflated thwart or one of the main tubes. Many watercraft also have D-rings, but these are intended only for light duty lashing, not heavy hauling. If you rig to a D-ring:

- A D-ring is generally stronger if pulled on from the side (i.e., in shear) rather than from directly above.
- Use a multipoint anchor system so that several D-rings (and maybe a thwart) share the load. The *boatman's eight* is a quick-and-dirty way of rigging a multipoint anchor through several D-rings. Tie an inline figure eight knot, leaving plenty of rope below it. Then run this through the D-rings on the boat, and tie a figure eight on a bight at the end. Clip this knot to the loop on the inline eight, then gather the rope between each D-ring and clip it into the loop also.

An alternative to hauling directly on the boat is to use a *hull wrap*, where the haul line is wrapped around the boat. This is the same principle as the no-knot, and has the effect of spreading the load over a much larger area, reducing the load on thwarts and D-rings. If rigged correctly, the haul line can also be used to roll the boat over to dump the water out, thus greatly simplifying the task of rigging. The best way to rig for this is usually to attach the haul line to the upstream side of the boat and then run it underneath and back over the hull to shore. If necessary, the line can be wrapped around the hull more than once. Getting the rope underneath the hull can be a problem. Sometimes it can be pushed underneath with a paddle or oar, or clipped into the upstream side and then a loop of rope slid over the ends of the boat. Yet another way of doing this is the *Steve Thomas rope trick*.

Fig. 6.9 Steve Thomas Rope Trick

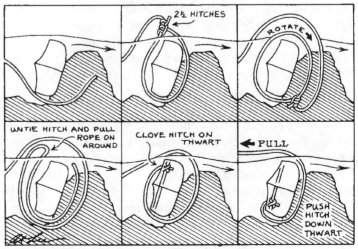

2½ HITCHES

ROTATE

UNTIE HITCH AND PULL ROPE ON AROUND

CLOVE HITCH ON THWART

PULL

PUSH HITCH DOWN THWART

Fig. 6.10 Big trouble on the Colorado – river rangers rig a pinned S-Rig.

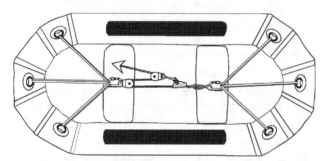

Fig. 6.11 The "taco" method: rig an anchor on one end of the boat to pull the other end up. With large craft pinned in a big river, this may be the only viable method.

If we are unable to find something to which to attach a line, we may have to use a *cradle rig*. Here, we use a cinching loop of rope around each end of the boat, tied together in the center. The haul line can then be attached at any point on the cradle, depending on what angle of pull we want. Another solution with an inflatable boat is to cut a *small* hole in the floor in order to tie an anchor loop around the main tube. Most modern boats are self-bailers, making this much easier—a line can be easily run through the gaps in the tubes and floor.

In some cases, we may want to use a separate line for each of the functions in unpinning a boat: one for hauling, one for rolling the boat over to dump the water, and one to pull it back to shore. In others, a single line may suffice for all functions.

Another method of reducing the force pushing on one or the other end of an inflatable boat is to deflate one or more of the air compartments. Deflating the

tube makes it catch less water but also reduces the boat's overall buoyancy. When doing this, however, there is a danger that water may enter the tube and add to our problems. Therefore, when bleeding air, the rigger needs to be sure that there is a good stream of bubbles and that he shuts the valve before the air pressure drops low enough to let water in the tube.

In really severe cases it may be necessary to cut the floor of an inflatable to lessen the pressure on the boat. Fortunately this is seldom necessary. Self-bailers with laced-in floors have an advantage here, since the lacing may be cut without doing permanent damage to the floor.

There are also cases where we can use the pinned craft itself as a haul anchor. With this method, sometimes called the "taco" method, we rig a haul system on one part of the boat to pull another part of the boat up out of the water. This works very well in cases where the boat is extremely large (it has been successfully used on the 39'/12 m "S-rigs" on the Colorado River) or where the boat is a long way from shore. We can rig the system end-to-end, so that it pulls the ends of the boat together, or, if the boat has a solid frame, to pull one side or an end of the boat up. Often we will be able to spill water from the boat this way and keep more from coming in.

Finally, we must mention the easiest method for unpinning a boat, one often overlooked. On many dam-controlled rivers the water can be turned off, whereupon the rescuers may simply walk out into the river bed and recover it. This is sometimes called the "Moses" method.

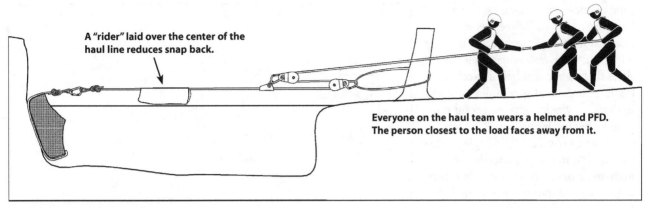

A "rider" laid over the center of the haul line reduces snap back.

Everyone on the haul team wears a helmet and PFD. The person closest to the load faces away from it.

Fig. 6.12 Boat hauling safety considerations

Swiftwater Site Safety Considerations

Unpinning a boat can be dangerous. Rescuers are in the water and close to a large, heavy boat that may move unpredictably at any time. Especially when no lives are at stake, caution is in order. Site management will be considered in more detail in the chapter on incident command, but these basic principles should be kept in mind:

- Establish an upstream spotter to warn the rescuers of river traffic, and downstream safety, (e.g., throw bags or boats) in case a rescuer gets washed off.
- Use the most experienced people for in-water work. Anyone going in the water should have proper training and the right equipment. *Untrained personnel are at extreme risk.*
- Be extremely careful when standing behind the boat. It's full of water and may shift or wash off at any time. Do not put your arms or legs where they might be caught between the boat and the obstacle.
- Consider tethering rescuers if the situation permits, and especially in situations where a downstream hazard exists. However, be sure that your tethering system is compatible with swiftwater use and that the personnel operating it have been trained in its use.

Once the boat is rigged, the haul system set up, and rope run across the river, other safety considerations come into play.

- It is quite possible that the anchor points on the boat may fail under tension and that the whole system may come flying back at you. Therefore, the personnel pulling on the haul system should wear a helmet and PFD for physical protection. The first person in the haul line should face away from the boat.

- To reduce the danger of the haul system snapping back at you, hang something near the center of the haul line, a *snubber* or *rider*, to reduce the snap-back. A small weight like a spare life jacket will do. You may also consider running the haul line through a directional pulley to keep the haul team out of the line of fire.
- The haul line may swing abruptly downstream when the boat comes off. Therefore, don't get downstream of a loaded rope, nor stand inside the bight of a rope that may become loaded, or astride it.
- Set up the haul and belay systems so that they can be quickly released if something goes wrong, (e.g., approaching downstream traffic, entanglement, etc.). Once you have used a rope for hauling, do not reuse it for a lifeline.
- Have a knife ready.

Low-Angle vs High-Angle

The traditional method of distinguishing between low and high-angle rescues has been this: *high-angle* is the vertical raising and lowering of rescuers or victims, and *low-angle* ("scree" or slope evacuation) is anything else. A more exact definition divides this into four categories:

- *flat* or *nontechnical*
- *low-angle* (15–40°)
- *steep-angle* (40–60°)
- *high-angle* (61–90°)

The vast majority of rope evolutions performed in swiftwater rescue operations will be low-angle/steep-angle. However, there are some situations, like bridge rescues, where conventional high-angle techniques are used. In addition, since many of the techniques normally associated with high-angle rescue, like high lines, have been adapted to river rescue, we have included them here. We will not spend a great deal of time on the traditional staple of high-angle rope rescue—the vertical raising and lowering of litters. While this is an essential high-angle technique, it is seldom used in swiftwater rescue, and since there are a number of excellent books that cover this complex subject in detail, it would serve no useful purpose to rewrite it all here.

Given a choice, a low-angle operation is almost always preferable to a high-angle one. The exposure is much less, as are the training requirements for rescuers and the stress on anchors, ropes, and other equipment. It often works better (and is safer) to pick a less steep part of the river bank for access, and then move personnel along the river banks, than to try to gain access directly to the accident site over a nearly vertical bank.

Low-Angle Operations

Since rivers have a habit of cutting themselves into gorges, swiftwater rescuers very frequently need to get up and down the banks from a road access point. Rescuers and equipment must move down, and rescuers, equipment, and evacuees must move back up. Although there are places where the walls of a river gorge may be more or less vertical, most are not. In low-angle operations, most of the motive power comes from the rescuers themselves, and not from the rope or haul system. The rope is used for security and as an assist. As the angle increases, the more the rescuers must rely on the rope for both functions.

If the bank is not too steep, a simple "scurry line" will suffice. Rescuers can move down, and uninjured victims up, holding the rope for added security. For steeper sections, experienced rescuers can use an arm belay or a body rappel. None of this requires any equipment other than the rope, and it is very quick to rig and use. For added security, rescuers can wear a sit harness (or simply a belt or webbing tied around the waist), and attach themselves to the rope with a camming device or Prusik loop. The rescuer holds the loop in one hand, and if he falls, the knot will cinch and catch him.

Fire ladders also provide a safe and convenient way of getting up and down the bank. If a litter needs to be brought up, two parallel ladders laid a litter's width apart work well.

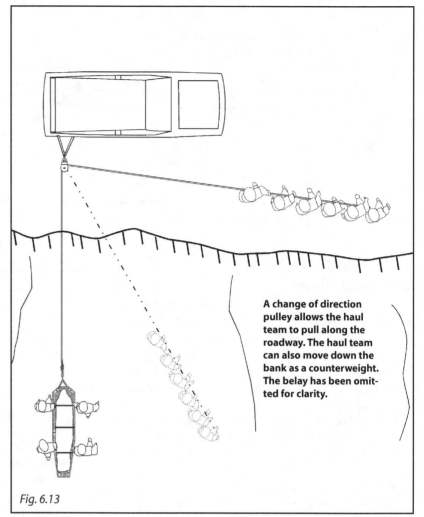

A change of direction pulley allows the haul team to pull along the roadway. The haul team can also move down the bank as a counterweight. The belay has been omitted for clarity.

Fig. 6.13

Rescuers can also use conventional rappel techniques to get down the bank. These are more secure, especially for steep sections, but require more training and take longer to rig (although a simple rappel can be set up and operated by one person). If the bank is steep enough to require a rappel, it is often better to go ahead and rig a simple haul system for both raising and lowering if there are enough personnel available to operate it. At some point during the rescue, you are probably going to have to rig a haul system anyway to get the rescuers and victims back up the bank. Since the rope is used primarily for security and assistance, these can be very basic.

Some common solutions are:
- A direct "tenboyscout" pull, often with a directional pulley so that the rescuers can pull in a more convenient direction, such as down the roadway.
- A "counterweight" system, where the haul team pulls downhill through a directional pulley.
- A 2:1 haul system. While simple to rig, it does require more rope, and the "diminishing V" may snag on brush or rocks as it moves uphill.

The victim, if uninjured, is usually put into a harness and assisted back up with a haul system. If the victim requires assistance, a rescuer can clip his harness and that of the victim together for the climb. Litter evacuations are similar and are covered in more detail in chapter 14. Vehicles are often used as anchor points on this type of rescue since it's easy to position them where you need them. Other common anchor points are trees and guard rails. When positioning access points, do not put them directly above the accident site, lest someone dislodge a rock into it.

Depending on the situation, the rescuers may choose to use a separate belay rope, a self-belay like a Prusik, or both. A belay Prusik or safety camming device like a Gibbs must be tended, and the tender careful to keep slack out of the system.

High-angle Operations

High-angle operations involve the more or less vertical raising and lowering of rescuers and victims. Here the entire force of raising, lowering, and belay is on the rope system, as is the safety of the victims and rescuers. Hence the exposure and the amount of training required is much greater than for lower angle operations. In swiftwater rescue, the most common high-angle operations are bridge rescues and cross-river high lines.

Fig. 6.14 A simple 2:1 haul system helps a rescuer assist a casualty up the grade.

High Lines

A high line is a fixed rope stretched across a river or gorge for the passage of personnel or equipment, as a rescue system to deliver rescuers to a point along it or to lift persons along the track line and transport them to the end points. An additional variation for river use is a high line where the load in the form of a tethered boat does not hang vertically but instead is pushed downstream by the current. Like some other specialized rope rescue subjects, the treatment of high lines in this chapter must necessarily be somewhat superficial. There are many variations, and only some of the simpler and more common methods are covered here. Rigging high lines requires hands-on experience and a thorough grounding in basic principles, things best learned in a reputable high ropes school.

The heart of the system, the main load-bearing rope, is called a *track line* or *main line*. The *span* is the horizontal distance between the *stations*, or anchor points, while the *chord* is the straight line distance between them. The *deflection* (sag, droop, or belly) is the amount that the rope hangs below the chord.

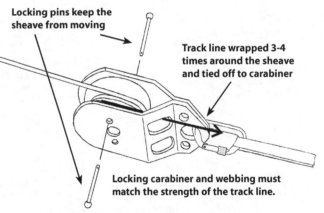

Locking pins keep the sheave from moving

Track line wrapped 3-4 times around the sheave and tied off to carabiner

Locking carabiner and webbing must match the strength of the track line.

Fig. 6.15 The Kootenay Carriage can be locked off and used as a high-strength tie-off for track lines.

Most high lines use a *carriage* made from pulleys to move the load across the track line. *Tag lines* attach to the carriage for horizontal movement of the load or belay. Often, however, it is necessary to raise and lower a rescuer (often with a victim, who is sometimes in a litter) at or near the midpoint of the track line. The simplest way to do this is to vary the tension on the track line to raise or lower the rescuer. Some situations, however, may require a fixed track line with a separate hoist system on the carriage to raise and lower a midpoint load.

Compared to the rope rescue systems discussed so far, high lines require much more time, training, and equipment to rig correctly. However, there are

Fig. 6.16 A drooping high line used for a recovery of a kayaker on the North Fork of the Nooksack River, Washington.

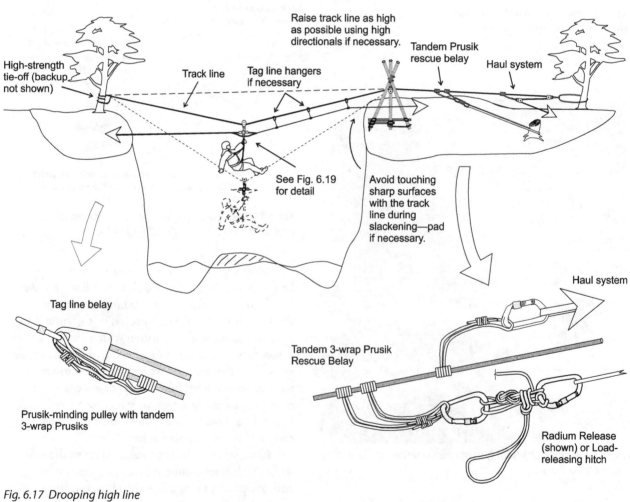

Raise track line as high as possible using high directionals if necessary.

High-strength tie-off (backup not shown)

Track line

Tag line hangers if necessary

Tandem Prusik rescue belay

Haul system

See Fig. 6.19 for detail

Avoid touching sharp surfaces with the track line during slackening—pad if necessary.

Haul system

Tag line belay

Prusik-minding pulley with tandem 3-wrap Prusiks

Tandem 3-wrap Prusik Rescue Belay

Radium Release (shown) or Load-releasing hitch

Fig. 6.17 Drooping high line

Fig. 6.18 After recovery, the kayaker's body was put in a screamer suit and transferred to shore. The operation took 18 people over 5 hours.

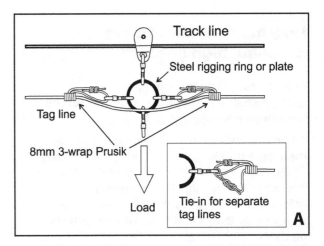

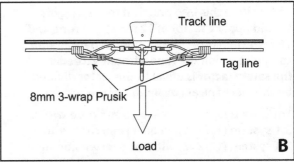

Fig. 6.19 Two variations of rigging the carriage for a drooping high line. A uses a steel rigging ring (a loop of webbing or cord can be substituted), while in B everything is clipped into the carriage pulley. In both cases the tag line doubles as a backup track line. If shorter tag lines must be used, they can be clipped directly into the rigging ring (inset A).

situations where they are appropriate, such as in very narrow river gorges where it isn't possible to use a helicopter, and the tethered boat variation is quite common.

A high line starts with a rope stretched between two anchor points. While this sounds easy, it is often quite a difficult and lengthy procedure in practice, especially if a river crossing is involved. Since high line anchors can be subjected to high stresses, they should be bombproof *and* backed up. If possible use a high strength tie-off (e.g. no-knot or tensionless hitch) that maintains the full strength of the rope. The Kootenay Carriage can be locked off and used as a high strength anchor in this fashion.

Drooping High Lines

In most cases, high lines for swiftwater rescue do not require long vertical lowers. Therefore, the *drooping* or *slack line high line*, which is simpler and quicker to rig than a high line with a mid-point drop, is often the best choice. The rescuer remains attached directly to the carriage and the track line is slackened to lower the rescuer to the accident site. Tag lines move the load from one end of the track line to the other, although the system can be simplified even further by eliminating the tag lines and having the rescuer pull himself out on the rope hand over hand (i.e. the Tyrolian traverse). Some systems, like the Kootenay Highline, ingeniously use the tag lines to back up the track line in case of a system failure. If you expect the tag lines to come under load, you must provide a suitable belay (e.g. tandem Prusik rescue belay) that

can be locked off. On longer high lines *tag line hangers* or *festoons* (a Prusik loop attached to the tag line every 25-30'/7.5-9 m or so and clipped over the track line with a lightweight non-locking carabiner) keeps the tag lines from drooping down too far. These are particularly useful for over-water high lines or boat lowers since they keep the tag lines from dipping into the water.

Once the track line has been run and suitable anchor points established on each end, the track line must be tensioned. This is the part of the operation that generally causes the most concern for the rescuers. As we have seen with both anchor systems and the vector pull, loading a taut line in the middle greatly increases the load on the anchors. Unlike the vector pull, however, in this case we do not want the anchors to move! In general riggers will want to leave as much slack as possible in the high line to reduce the load on the anchors. A quick look at Fig. 5.14 shows the relationships—an angle of 120° (sometimes called the "golden angle") equalizes the stresses so that each anchor bears the same weight as the load. A recommended technique to gain more slack and less tension, especially for drooping high lines, is to raise the

How hard should you pull a drooping high line?

Rope rescue instructor Kirk Mauthner suggests the following approximation for putting tension on a drooping high line:

An unknotted 11mm (7/16") low-stretch rescue rope has a breaking strength of approximately 30kN, and a 13mm (½") rope about 40kN. Therefore, for the high line to retain a 10:1 safety factor, the ropes should not have a tension greater than 3 and 4kNs, respectively, put on them.

To translate this into practical terms, riggers should apply a factor of 12 for 11mm rope and 18 for 13mm rope. The number of people who can pull tension the rope without exceeding the safety factor is equal to the factor divided by the mechanical advantage of the system.

Thus, six people can pull a 13mm rope with a 3:1 system (18/3=6), or an 11mm rope with a 2:1 system (12/2=6) without overtensioning it. These figures assume a steady pull by average adults using gloves with no jerking or "heave-ho's."

anchor points. If this isn't possible the riggers should use a high directional (see the section at the end of the chapter) to raise the leading edge of the track line to gain the same effect.

High Line With Mid-Point Drop

The mid-point drop option on high lines adds a separate raising and lowering system that can be moved along the span like a crane. Rescuers, with litters and patients if necessary, can be raised, lowered, and transported back and forth by means of tag lines. While high lines with fixed track lines and a mid-point drop are less often used in a vertical application for river use, they are used quite often in "horizontal" applications such as a boat on tether or telfer lower. Two of the most common options are:

- a directional lower (Fig. 6.20A) uses a simple change of direction pulley attached to a raising/lowering system located elsewhere.
- a 2:1 haul system (Fig. 6.20B) is somewhat simpler in that everything is located in one place and one of the tag lines can be eliminated. While being moved along the track line the rescuer clips himself into rigging ring, freeing the

haul line for use as a tag line. When it's time to descend he unclips, the rope support crew locks the other tag line in place, and he is lowered and raised via the 2:1 system. When hauled back up he clips in again, the tag lines are unlocked, and he is pulled to safety.

There are other, more sophisticated high line systems (e.g. Kootenay High Line) that are beyond the scope of this book to cover.

Fixed high lines should be tensioned initially with a simple 2:1 haul system pulled by a single person, otherwise the track line may be overtensioned when a load is placed on it. As mentioned above riggers should keep a safety factor of at least 10:1 on the track line when applying tension (15:1 in an NFPA-compliant system) and should attempt to raise the anchors where possible to reduce the stress on the anchors. Tension for high lines up 330'/100 m can determined by:

- **Calculation**. This has become easier in the age of hand-held calculators and portable computers, although reliable field measurement of small angles (15° or less) is difficult.
- **Estimation**. Including:

  The factor system mentioned above, which is applied *after* the line has been initially tensioned with a 2:1 system and has been loaded (preferred).

  Allowing enough slack to attain the "golden" 120° angle, which equalizes the tension on the anchors to that of the load.

  The *ten percent rule,* a quick and dirty method of estimating acceptable tension, which states that a load of 200 lbs over a 100-foot span should have a sag of ten percent (in that case, 10 feet). Use the total weight (including all gear) of the load and the entire span of the rope, and measure the sag *before* the rope is loaded. This widely used method, although not nearly as accurate as the others, will work in a pinch. It does not, however, take into account the variations in strength of different sizes of track lines and heavier loads.

- **Measurement**. The most accurate method of obtaining actual track line tension is to use a load-measuring device like a field dynamometer or force transducer for a direct measurement. If resources permit the system can be calculated or estimated during setup and then monitored during use.

DANGER: The risk of entanglement must always be considered when setting up a high line system over water where the rescuer may end up in the current. Once a person attached to a high line dips into the current, the forces on the system can increase dramatically, and unless the rescuer has a way to release himself, he is in great danger of drowning. When rigging the high line, try to set up a "run-free" system with no knots to catch as the rescuer washes out of the system. In addition, both the rescuer in the harness and those on shore should have a readily-accessible knife.

Once the track line has been tensioned, that tension must be maintained. The recommended method is to use tandem Prusiks, since these will slip before failure rather than damaging or cutting the rope like a hard clamp. This slippage also provides some warning of impending system overload.

River Applications

The high line with mid-point drop can be easily adapted to river use as a boat lower. Instead of a rescuer or litter lowered down vertically, the track line is stretched across a river, typically at a height of 10-15' (3-5 m), and the boat or raft is pulled downstream by the current, then moved back upstream by the haul system. Load calculation for a boat on tether can be tricky. Gravity plays a much smaller—indeed minor—role, and the load will not crash from height, but the force of the current can be difficult to judge and even more difficult to "engineer." The load will vary with the design of the watercraft and the speed of the current, and can change considerably across the river. If the watercraft fills with water or otherwise gets caught in the current or a feature such as a hole, the forces on the system can jump abruptly to very high levels. For this reason riggers should attempt to set up a system that can be released quickly if need be. The actual mechanics of rigging a boat lower are covered in chapter 8.

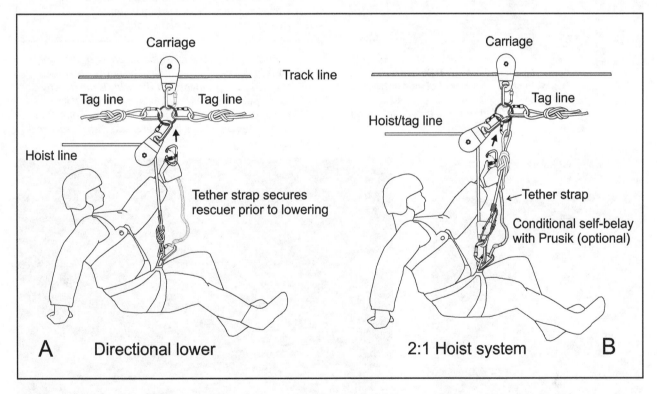

Fig. 6.20 Two methods of attaching a hoist system to a high line carriage. The variation shown on the right uses the hoist line as a tag line. A Prusik on the lower pulley adds a measure of safety.

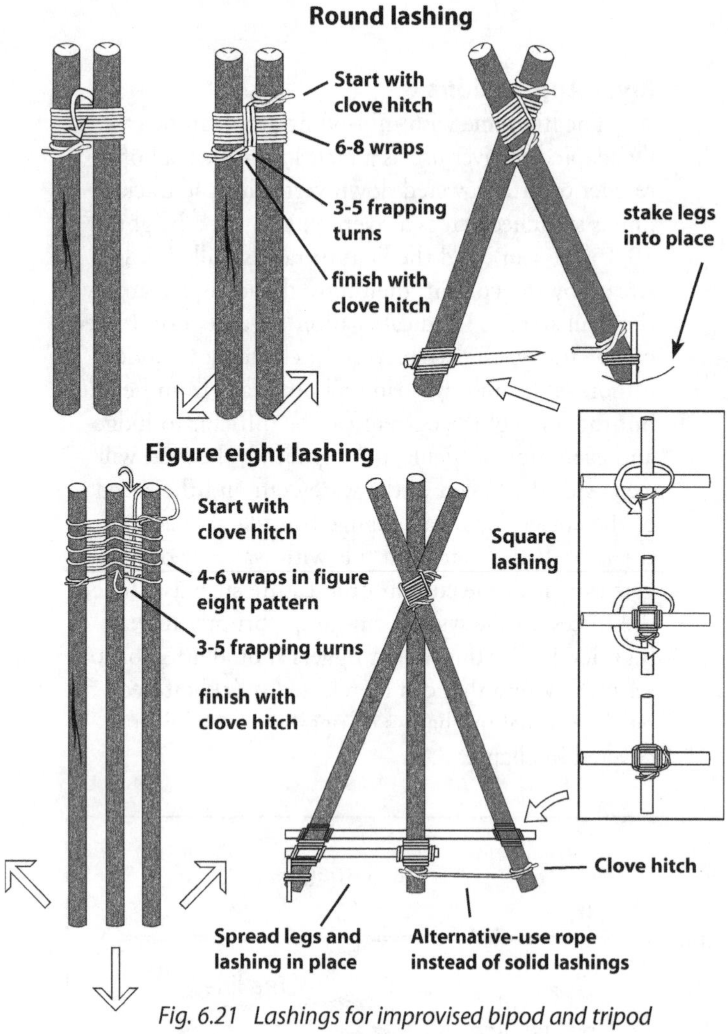

Round lashing

Start with clove hitch

6-8 wraps

3-5 frapping turns

finish with clove hitch

stake legs into place

Figure eight lashing

Start with clove hitch

4-6 wraps in figure eight pattern

3-5 frapping turns

finish with clove hitch

Square lashing

Clove hitch

Spread legs and lashing in place

Alternative-use rope instead of solid lashings

Fig. 6.21 Lashings for improvised bipod and tripod

A-Frames, Tripods, and High Directionals

As mentioned above riggers often find it necessary to raise or redirect track lines or haul lines to reduce tension, clear obstacles, keep rescuers above the water, avoid abrasion or clear edges. Raising an anchor point in this manner is generally referred to as a *high directional anchor* (HDA), although it's actually a directional pulley, or as an

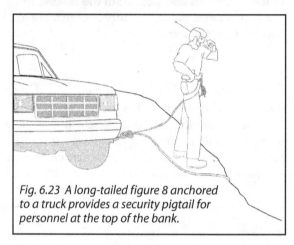

Fig. 6.23 A long-tailed figure 8 anchored to a truck provides a security pigtail for personnel at the top of the bank.

Fig. 6.24 A commercial tripod set up for a steep angle bank lower/raising operation. The haul line is raised to avoid edge problems but the belay is run directly over a pad on the edge to avoid a shock load if the tripod fails. A haul system with a change of direction is mounted on a truck bumper along with a brake rack for lowering.

Fig. 6.22 Instructors Jim Segerstrom and Phil Poirier fabricated a very serviceable A-frame from sections of 2" steel tubing bolted together braced by an aluminum radio mast tube.

Swiftwater Rescue

artificial high directional (AHD). Any object in the right spot capable of holding the anticipated load can be used, including bridges and other artificial structures, trees, and the like. However in some cases these things are not present and riggers have to provide their own. These can be improvised frames such as A-frames, bipods or tripods fabricated from local materials such as poles or beams lashed together or commercial versions such as the SMC Terradaptor or the Arizona Vortex Multipod, which can be rigged in a number of configurations. Riggers should factor in any additional loads on the anchor caused by the haul or track line's change of direction through the directional (see Fig. 5.14).

While riggers may need to run a track or haul line through a high directional, they should avoid doing so with the belay. If the high directional fails and the belay is run up with it, the resulting drop causes a major shock load to the system. Better to keep the belay low and run it as directly as possible over an edge so as to have it move as little as possible in the event of a main line failure.

Fig. 6.25 A typical high-angle site.

High-Angle Site Organization & Safety

High-angle rescue sites have their own safety considerations apart from river safety measures. The two must sometimes be combined, often uncomfortably. Like river safety, high-angle safety depends in large part on adequate training and proper equipment. The first job of a safety officer at a high-angle site, then, should be to

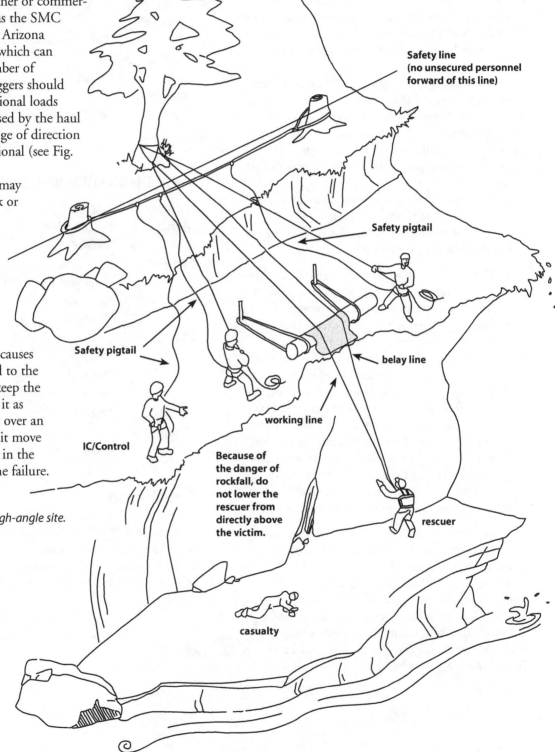

Safety line
(no unsecured personnel forward of this line)

Safety pigtail

Safety pigtail

IC/Control

working line

belay line

Because of the danger of rockfall, do not lower the rescuer from directly above the victim.

rescuer

casualty

exclude from the site all personnel who do not have these. Adequate equipment for high-angle purposes includes a rock climbing helmet and proper footwear.

A key concept in high-angle site safety is the establishment of a safety line. This should be far enough back from the edge that a person standing behind it is in no danger of injury from falling or rockfall. Forward of this line, all personnel *must* be clipped into some form of protection with a pigtail. In most cases, it will be expedient to run a rope along the safety line for rescuers to clip into. In other cases, a belay may have to be run from a distant anchor point.

When setting up a high-angle rescue site, consider the position of the victim. In most cases you *do not* want to have the rescuers come down from directly above, since any dislodged object may hit the victim. If at all possible, it is better to lower the rescuer off to one side in order to minimize this possibility.
Depending on the situation, these safety measures may be modified for low-angle sites.

When organizing a high-angle rescue, consider filling the following positions:

- The *operations officer* is not necessarily the incident commander, but has the final say on all matters affecting the high-angle rescue.
- The *safety officer* is a knowledgeable person without other duties whose responsibility it is to observe the rescue site for anything that might affect the safety of casualties and rescuers.
- The *control* actually coordinates the rescue. All communication comes through him, and he should be positioned with a view of the entire operation. This frees the leader from the minute-to-minute responsibility of the operation and allows him to see it as a whole. Anything control cannot see should be covered by a:
- *Spotter*, who relays information to control.

The more complex the system, the more important is communication between members of the rescue team. A standardized system of signals, rehearsed in advance, is absolutely necessary. The standard climbing signals are a good place to start.

In addition to these, rescuers need to develop standard signals for common operations like raising and lowering. Some commonly used ones are shown in the following two charts.

STANDARD CLIMBING VOICE SIGNALS

BELAYER	CLIMBER/RESCUER	MEANING
	"On Belay"	Are you ready?
"Belay On"		Belay is ready
	"Climbing"	I am climbing
"Climb On"		Understood
	"Slack" or "Give"	Give me rope
"Thank You"		Understood
	"Up rope," "Tension" or "Take"	Take up slack
"Thank You"		Understood
	"Off Belay"	I am secure
"Belay Off"		Understood
"ROCK!"	"ROCK!"	Look out for rock
		— Echoed by all

RAISING/LOWERING VOICE SIGNALS

BELAYER/ HAULER	RESCUE/ LITTER CAPTAIN	MEANING
	"On Belay"	Are you ready?
"Belay On"		Ready
	"Down Slow"	Lower slowly
	"Down Fast"	Lower faster
	"STOP"	Stop the operation
"Two-oh"		20 feet of rope left
	"Haul Slow"	Haul slowly
	"Haul Fast"	Haul faster
	"Set"	Set the safety grab
	"Off Belay"	We are secure
"Belay Off"		Understood
	"Off Rope"	We took the rope off

Conclusion

All swiftwater rescuers will probably have occasion to use the boat rigging and low-angle rescue techniques described in this chapter. While high-angle techniques are used less often, they are the basis of some of the boat-based lowers covered in later chapters.

Shore-Based Rescue
Chapter 7

Shore-based rescues account for the majority of "saves" in swiftwater. While they do have definite limitations, they keep rescuers out of the water, thus reducing their chance of injury. Some shore-based rescues, like simple reaching ones, can be performed by people who have had little or no training. In general, shore-based rescues require less training and equipment than those involving boats or other complex equipment like helicopters. Hence, they are one of the first rescues to consider. However, anyone contemplating performing any water rescue should be able to rescue themselves *if* they somehow fall into the water, as discussed in Chapter 2. This is always a possibility, even for the simplest and most straightforward reaching rescue.

Reaching Rescues

The most basic method of rescuing someone in the water is to simply extend something to them and pull them to shore. This can be just about anything: a jacket, jumper cables, a tree branch, a paddle or oar, a ladder. In short, just about anything can be used that both victim and rescuer can grab. Obviously, reaching rescues are limited to conscious victims relatively close to the banks of the river, although the technique can also be used effectively from boats.

Fire companies can use inflated fire hoses (which are discussed in more detail later in this chapter), pike poles, extension ladders, and other fire equipment for reaching rescues. Firefighters can also attach floats (e.g., ring buoy, inner tube) to the end of a ladder and push this out into the current to a victim.

Truck-mounted aerial fire ladders have also been used successfully for reaching rescues. There have been a number of rescues made by parking an aerial ladder truck near an accident site like a low-head dam or a stranded vehicle and extending the ladder to the victim. A rescuer was then able to go out on the ladder and lead or carry the victim back to safety. If the design of an aerial ladder does not permit this, it may still be possible to use it like a crane for a pickoff. This technique is described in Chapter 11 (victim stranded in midstream).

Wading Rescues

Rescuers can also go into the river by wading, and this can be a very effective rescue technique in shallow rivers. Wading may also get rescuers near enough to victims for a reaching or throwing rescue without using a boat. Wading obviously has its limitations: it can only be used in relatively shallow water and slower currents, and wading rescuers do put themselves in more danger than if they had remained on shore. Still, it is a very fast method, requires only basic swiftwater training, and the rescuers need little equipment other than basic river gear. Since a wading failure usually means a swim, however, self-rescue training is a must.

There is admittedly a contradiction here—to first tell you that when swimming you should keep your feet up and not try to stand, and then say that wading rescues are OK. Wading in swiftwater *does* entail some risk of foot entrapment. However, with proper training, it is a manageable risk. Most foot entrapments occur when a swimmer, who is moving at the speed of the current, jams his feet down onto the riverbed to stop himself. When wading, however, there is time to pick your footing before putting your entire weight down.

Before wading across any stretch of swiftwater, consider what would happen if you either slip or begin to float and have to swim. You must leave yourself a reasonable amount of distance before the next hazard to get to shore. Consider also what kind of bottom the river has. The safest bottoms are those made of sand or gravel since these give good footing and pose a minimal danger of foot entrapment. A smooth, hard, very slick bottom (a situation often found on mountain streams) may make it difficult or impossible to regain your footing once it is lost, and difficult for you to get to shore when swimming. The most dangerous bottom is one with irregular crevices or potholes on the bottom, into which a foot might slip.

Fig. 7.0 Swiftwater Wading

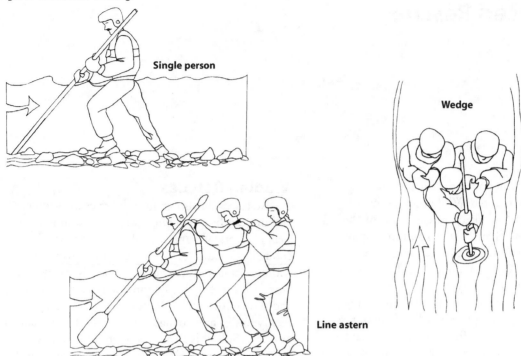

Single person

Line astern

Wedge

One-Person Wading. A single person can wade through moving water, although it is more difficult than doing it with a group. To do it effectively, you will need a staff of some kind on which to lean (e.g., a canoe paddle with the handle down). It should reach at least to your chin. Face upstream and lean forward, letting the current take your weight. The staff and your two legs form a tripod, which is more stable than your legs alone. Move slowly across the current, facing upstream, moving only one of the three points (your two feet and the staff) at a time. You will have to lift the staff swiftly and jam it back down on the bottom or it will be swept between your legs. As the water deepens you will have to lean more and more into the current. Eventually your feet will float off and you'll have to swim for it.

Two-Person Wading. Adding another person makes wading considerably easier because we now have four legs instead of three. Two people face and grab each other by the shoulders of their life jackets. They then wade across the current, facing either up and downstream or sideways to the current.

Wedge. A single person with a staff can be expanded into a wedge by adding more people. Additional pairs of people grab the shoulders of those in front to form a wedge facing upstream. The wedge, while quite stable, depends for its integrity on the front person. If he fails, the whole wedge comes apart. The people behind the leader can help by pushing down on his shoulders to improve his footing. The line astern, in which the rescuers stand in a single line, is similar.

People Pivot. The two-person method can also be expanded for more people, with a resultant increase in stability. Just add more people, each holding the life jacket of the person next to them. The assembled multitude moves across the current in helical fashion. The circle pivots on either the upstream or downstream person (opinions vary as to which method works better—my personal preference is to pivot on the upstream person). The pivot person stands in place while the others move. As the next person comes into line, he then becomes the pivot, and so on. The people pivot works well for people of different heights: it will often remain intact even if shorter people's feet lose contact with the bottom.

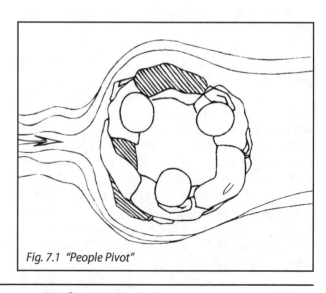

Fig. 7.1 "People Pivot"

Line Rescues. A staff can also be used for wading with three or more people holding it, facing upstream in a line abreast.

Throw Bags

More people are probably rescued with throw bags than with any other single method. The throw bag is cheap, simple, and lightweight, and while practice is essential to maintain proficiency, the training is simple. It is not, however, a perfect method: rescuers are subject to skill decay and there is some risk of entangling a victim in the rope meant to save him.

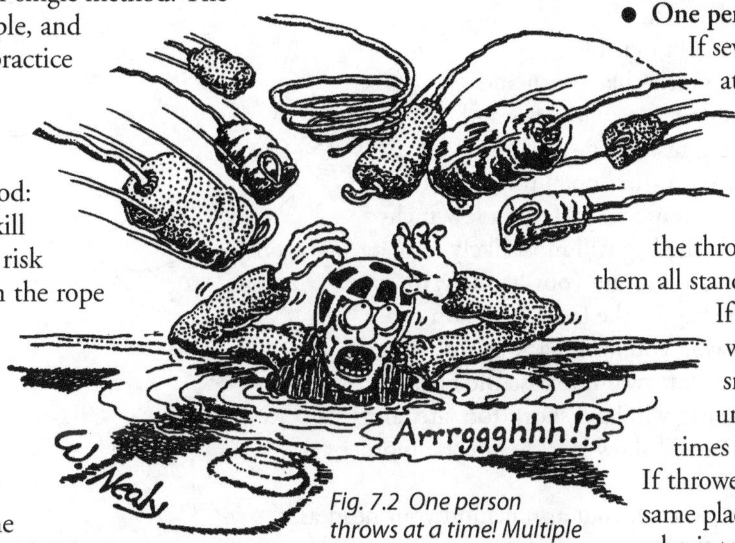

Safety Considerations

- **Carry a knife.** Anyone working around ropes in the water should carry a knife. Entanglement is always a possibility, and rescuers must always be ready to cut themselves or others free.

- **Do not tie yourself to the rope.** To do so is to invite drowning, since if you are pulled into the water and the rope tangles with an obstacle, the current will most likely hold you under.

- **Practice, practice, practice.** Like any other physical skill, throw bag skills decay after a surprisingly short period of time. Proficient use is safe use.

- **One person throws at a time.** If several rescuers throw at the same time, the danger of entanglement is greatly increased. It is usually better to space out the throwers rather than have them all stand (and throw) together. If several lines entangle when thrown, they may snare the victim, with unpredictable and sometimes adverse consequences. If throwers must stand at the same place, they should agree who is to throw first.

- **Hang on to the rope when you throw.** If you don't, the whole rope will go into the river, creating an instant hazard. This may seem obvious, but it is a surprisingly common mistake.

Fig. 7.2 One person throws at a time! Multiple ropes risk entanglement.

Arrrgggghhh!?

Fig. 7.3 Underhand Throw

There are two type of throws: *underhand* and *overhand*. Each has advantages and disadvantages. The underhand throw is easiest for most people to learn and use but requires a clear space in front of the thrower. The overhand throw is somewhat more difficult, but enables a rescuer to throw over obstacles like brush and other people; it works best with smaller bags and is the preferred throw from inside a raft.

To make the underhand throw, open the mouth of the throw bag about halfway and pull out a few inches of rope. The end inside the bag will most likely have a loop on it—do not wrap this around your hand. With your throwing hand, grasp the bag by the loose fabric on top. Select your target, and swing your arm back behind you. When your arm comes back forward, release the bag. If you release too soon, the bag will land short; too late, and it will fly upward. Practice will show you the precise moment of release to get the range you want.

For the overhand throw, cock your arm behind your head as if throwing a football after taking out a few inches of rope to hold in the other hand. Then throw the bag forward in a slight upward arc. The release point is not nearly so critical here as with the underhand throw.

When throwing the throw bag, use the following procedure:

- **Establish contact with the victim**, preferably eye contact, if possible before throwing. Yell "ROPE" or whistle to get the victim's attention. A throw bag against the side of the head is a last resort.

- **"Lead" the victim** by aiming slightly downstream of him. Wait until the victim is closest to you, but still slightly upstream, before you throw. There is some disagreement of whether, if you have to err, it is better to miss upstream or downstream of the victim. If the rope lands upstream of the victim, he can't see it. However, it will probably float faster on the surface than a person in the water, especially if he is backstroking against the current. Hence it will be easier for him to catch. On the other hand, if the rope is thrown downstream of the victim, he can see it but, in order to catch it, he may have to swim forward to get it, and thus come out of the safe swimming position.

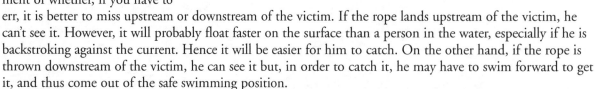

Fig. 7.4 Overhand Throw

- **Throw *over* the victim if possible.** This minimizes the possibility of a range error. Especially with the underhand throw, it is very easy to throw short if you aim directly at the victim.

- **Consider what will happen when the victim grabs hold.** The amount of force on the rope will increase suddenly and dramatically. The rescuer will then have to assume some sort of belay (discussed below) to accommodate the increased force.

TIP: Wet the throw bag before using it. The water adds weight to the bag and gives better range.

Throwing the Throwbag

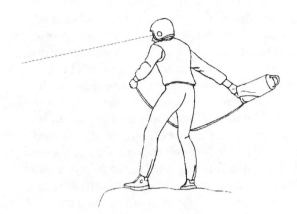

1. Pull out end of rope
2. Establish contact

3. Aim

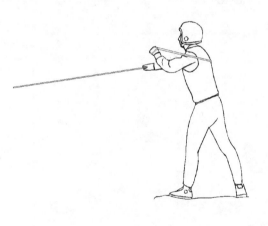

4. Throw

5. Belay

6. Restuff

Fig. 7.5

Swiftwater Rescue   105

Belaying a Swimmer

Throw rope belays need not be elaborate. A simple body or shoulder belay works well as long as you are dealing with a single victim in slow current. For swifter water or multiple victims, more secure belay methods may be required, such as those described in Chapter 5. One of the simplest and most often used is a friction belay around a tree. A belay for a swimmer can be either *static* or *dynamic*. A static belay does not move, i.e., a rescuer assumes a belay and holds it in place. While secure, this stops the swimmer with a jerk when the rope comes taut, and may cause him to let go. The stretchy poly ropes used in most throw bags absorb some of the shock, but they do not eliminate it. A dynamic belay, on the other hand, arranges the belay so that the force is fed to the swimmer more gradually, reducing the shock load on both swimmer and rescuer. The faster the current, the more critical it is to reduce shock loading.

The rescuer may do a dynamic belay by:

- **Running along the river bank**, if the situation permits, while at the same time pulling the swimmer toward the bank.

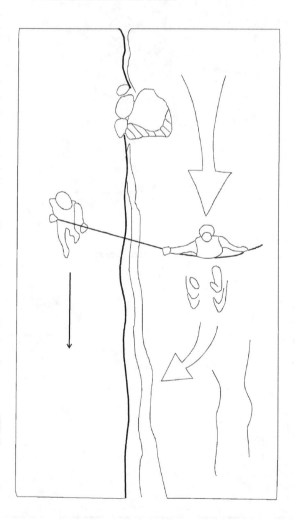

- **Allowing the rope to run over his body** (or through a belay device or around a tree) enough to reduce the shock loading. Unless great care is taken, however, this is a great way to get friction burns. The chance of rope burns be reduced by wearing a wet life jacket when belaying.

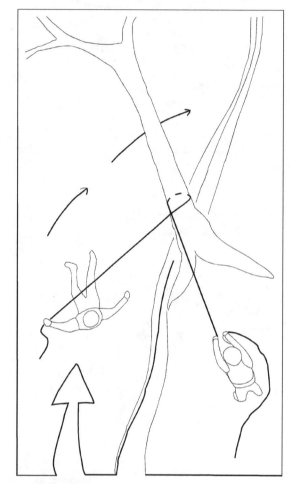

Fig. 7.7 Static Belay

- **Using a "two-stage" belay**, where one person makes the initial throw and belay to slow the swimmer, then hands off to another rescuer who pulls them in. This works well for bridge and flood channel rescues.

Fig. 7.8 *A static body belay with the rescuer tethered by means of a rescue PFD. The belay is set to open downstream to avoid entanglement.*

The rescuer must assume the belay quickly, and when belaying, he should take care that the bight of rope on his body *opens* as the victim goes downstream. This avoids having the rope wrap around him. The swimmer can help get himself back to shore quickly by angling his body to the current with his head toward shore. This creates a ferry angle that pushes him toward shore.

TIP: To get the victim to shore even faster, use a vector pull on the throw rope. After the rescuer is on belay and the rope taut, another rescuer wades out and pulls on the center of the throw rope, thus applying extra force. If the situation does not permit wading, a line can be clipped over the throw line to do the same thing from a distance.

Setting Ropes

The act of positioning throw ropes on shore is usually called "setting rope." The knowledge of where to set the ropes is as important as how to use them. It is embarrassing to be standing there with a throw bag in your hand, only to realize that the person you want to rescue is downstream of you and out of range of your rope. To avoid this, consider the following factors:

- **Where is an accident most likely to happen?** Often a swimmer submerges briefly after falling in the water. In most cases, after an accident it takes some time for a victim to reorient himself. Thus, we often see ropes set too far upstream to be effective.
- **What will happen** *after* **the accident?** Where would the river's currents carry a victim? How swift is the current?

- **What is downstream?** Where are the hazards? Is there a calm pool downstream or something nasty like a strainer or a big rapid?
- **Where will the swimmer land?** Once the swimmer has the rope, he should be pulled quickly into a convenient eddy. Allow some extra room in case the victim grabs the bag instead of rope. If at all possible we want to avoid setting a rope just above a major hazard, especially if it would require a swimmer to approach above it to get the rope. However, in some cases we may want to position a rope just above a hazard if it appears that a swimmer may be carried into it.

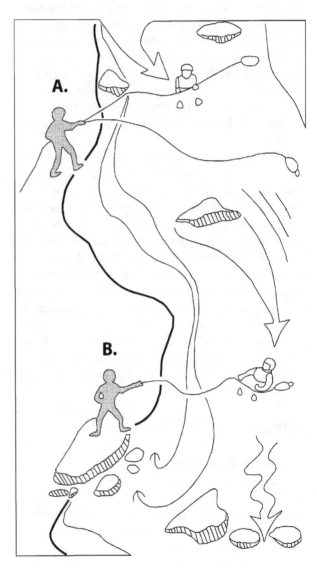

Fig. 7.9 *The rescuer at (A) throws with the swimmer slightly upstream, allowing him enough time for a second throw. He is then able to swing the victim into an eddy.*

The site at (B) catches anyone the first rescuer misses, but is less suitable because the rescuer must pull the victim in quickly to avoid having him swept into a boulder sieve.

If so, the rope should be far enough upstream of the hazard so that the rope, even at full extension, will not allow the swimmer to enter it.

- **Set up throw ropes in depth.** Do not allow the swimmer to get downstream of the last rope, unless there is another rescue system below it.
- **Integrate throw bags with other systems**, including rescue boats. Throw bags can be effectively used from boats as well as from shore.

The Second Throw

An important part of rope throwing technique is the ability make a second and subsequent throws. This may be either because you missed with the first throw (yes, it does happen) or because there are multiple victims. Whatever the case, any rescuer needs to be able to make any subsequent throws as quickly and accurately as the first. Rescue 3's training standard is that you be able to make an effective second throw within 20 seconds of the first. Obviously, this does not leave much time for a leisurely restuffing of the throw bag.

The fastest way to make the second throw is to coil the rope rather than to stuff it back into the bag. This can literally be done on the run. However, the range and accuracy of a coiled rope is not as good as a bagged one. Therefore, if time allows, you should restuff the bag. There are two recommended methods. First, hold the bag at waist level and run the rope over your shoulder; open the mouth of the bag all the way and stuff handfuls of rope into it. Second, kneel, open the bag, place it on the ground, and stuff handfuls of rope into it. With practice either can be done quickly.

Yet another method for a quick throw is to fill the bag with water and toss it. However, you should be careful with this method since a full bag is heavy and can cause serious injury if it hits someone in the head. For the same reason, do not attach anything hard (like a carabiner) to the throw bag.

Tag Line Rescues

Tag line rescues utilize a line extended across the river. They are simple, quick and easy to rig, and can be used for a variety of purposes, including foot entrapment rescues, low-head dam rescues, and moving rescuers across the river.

Line Ferries

Descriptions of rescue systems often start with the innocuous phrase "first, get a line across the river." While this may sound simple, it is not. In fact, it is usually the most difficult and time-consuming part of the rescue. Rescue 3 founder/instructor Jim

Segerstrom, in a magazine article, listed no less than 14 different methods of getting a line across a river. In general these break down into:

- **Walk the rope across.** Sometimes there is a convenient bridge (this is often overlooked), a catwalk, or perhaps a shallow spot upstream or downstream of the rescue site where the rescuers may wade across.
- **Throw the rope across.** A throw bag works for narrow rivers up to 60–70' (18–20 m) across. Various commercial "line tossers" (discussed more fully in Chapter 3) can extend the throwing range to about twice that distance.
- **Shoot the line across**, using a slingshot, bow, line gun, or cannon.
- **Ferry the line across**, using a boat, a swimmer, or even a helicopter. We will defer a detailed discussion of line ferries to Chapter 8.

Cross-river tag lines are necessarily limited to fairly narrow rivers. While this may sound like a major limitation, the vast majority of watercourses fall into this category: small rivers, creeks, ditches, and flood and irrigation channels. The wider the river, the harder it is to get the line across, and the more difficult it is to manage it once there. As the length of the rope increases, so does its weight, adding to handling problems. The two other major factors affecting line management are the speed of the current and the size of the line. We have already seen how the force of the current increases exponentially with its speed, and this also applies to a rope in the water. It is also true that the more rope we put in the water, the greater the force there will be on it. Segerstrom notes in his article that the load on a rope may be "more than 600 pounds in a 100-foot-wide channel moving at 15 feet per second."

One way to reduce this force is to use a *messenger line*, a smaller line (e.g., parachute cord) sent across the river first and then used to pull heavier lines across. In extreme cases, we may have to use several different weights of lines, progressively working our way up in size. Another method is to hold the line up out of the water, although obviously this will only work on narrow rivers. Some systems, such as river high lines, will have the line clear of the water when fully set up, while others, like snag tags, will not, and will consequently put more force on the rope and rescuers.

> **TIP:** Use rope bags for on-site rope management. By this we mean not only a throw bag but also larger bags for bigger and longer ropes. Bags reduce the entanglement potential and make ropes easier to control when paying out in a cross-river operation.

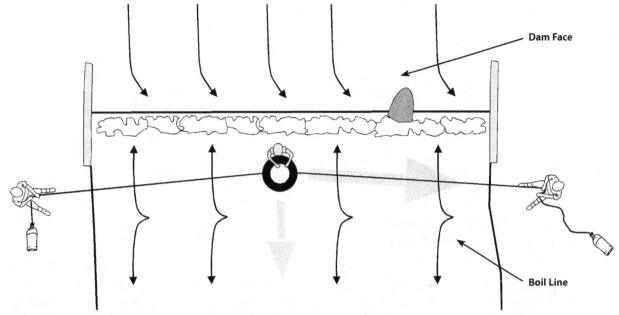

Dam Face

Boil Line

Fig. 7.10 A floating tag line. If conditions permit, it is much easier to pull the victim sideways than downstream.

The tag line can be either a continuous line with the float attached to it (the usual method is with a Prusik) or from two lines, each attached to the float. Once the tag line is across the river, the float can be maneuvered to the victim by shore-based teams. Since the float stays on top of the water, this is relatively easy. Effective communications between the teams on either bank are a must. Once the victim has hold of the float, he may be pulled sideways to the banks or one side may be released, converting it into a pendulum rescue.

> **TIP:** A quick way to make a floating tag line is to clip two throw bag ropes to a PFD. First zip up the PFD and tie or clip a throw rope through the arm hole. Throw the first bag across, then pull the PFD out into the river.

Fig. 7.11 A pre-rigged floating tag line using a large inner tube.

Floating Tag Line. The most basic type of tag line rescue is the floating tag line. This consists of a line across the river with some sort of float attached to it. It is normally used to get a flotation device to a victim who is fixed in position, whether in a hydraulic, a low-head dam, or on an obstacle. The float gives the victim something to grasp and provides some flotation. The float can be just about anything: a ring buoy, an inner tube, a PFD, or even an empty milk jug. The larger the float, the easier it is for the victim to grab, but the more resistance it will offer to the current.

Fig. 7.12 A floating tag line in action.

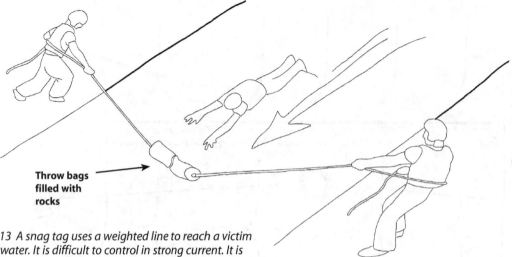

Throw bags filled with rocks

Fig. 7.13 A snag tag uses a weighted line to reach a victim underwater. It is difficult to control in strong current. It is also very hard on the throw bags.

Snag Tag/Weighted Line. A variation on the floating tag line is the snag tag, or weighted line. This is the same as the floating tag line except that a weight instead of a float is used. The weight gets the rope under the surface of the water, a necessity in some instances, but it also puts a tremendous amount of force on both the rope and rescuers. The idea of the snag tag is that the weight will sink the rope underwater to get to or underneath a trapped victim or pinned boat. One might, for example, sink the line down far enough to get it underneath the leg of a foot entrapment victim. Much of this is theoretical, however, since in practice the snag tag is very difficult to control in fast current. Since the water pushes the weight to the surface and bounces it around, the line must be "sunk" quickly to get it underwater. Effective use of a snag tag requires practice, coordination, and timing on the part of teams on both banks, and in any case it is difficult to get it exactly where you want it. For these reasons the snag tag is best used as an ancillary technique rather than a primary one.

Bare Line. A totally bare line can also be used to good effect. It is easier to handle than either the floating tag line or the snag tag and may be forced underwater for brief periods. The bare line should not be used to try to intercept swimming victims. Rather, it is used to stabilize a foot entrapment victim or a pinned boat, and can be used in conjunction with the cinching rescues described below. In a foot entrapment rescue situation, the line is worked up to the victim, who will hopefully be able to grab it and hold his head up. This technique is simple, requires no equipment other than a rope, and poses little danger to the rescuers.

More than one tag line can be used at a time. For example, one could be used for stabilizing a foot entrapment victim, while another might be used for pulling him upstream.

Variations of the bare line can also be used to get rescuers to an accident site and ferry them across the river. These are discussed in more detail in Chapter 11.

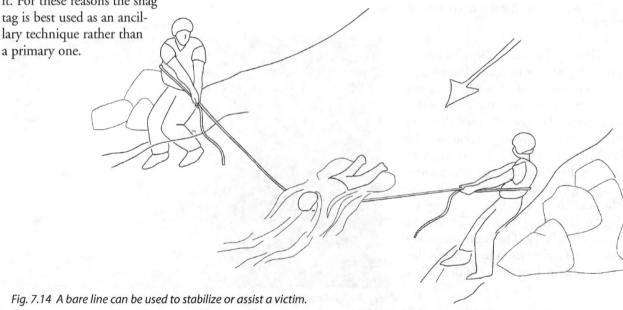

Fig. 7.14 A bare line can be used to stabilize or assist a victim.

This happened at a known sieve below a rapid called "Marbles." The rapid has since changed a bit and now the sieve is much more open but at the time it was a fairly ugly spot. There isn't much of an eddy between the exit move of Marbles and the sieve itself, and the sieve has caught several folks (one went in backwards and had a hard extraction, I went in there once too, but forwards, and self-rescued—at low water).

On the day these pics were taken I had run first and eddied immediately below the sieve, but I couldn't see the sieve from the eddy I was in as long as I was in my boat. The next guy was coming through behind me but I heard his boat and paddle hit the rocks in such a way that I just KNEW he must be going into the sieve. I scrambled out of my boat and started running/scrambling up the rocks to the sieve while I opened my throwbag. Harris (white helmet) was eddied out upstream and he had seen the victim going into the sieve and he was in action too. I was able to throw my rope to Harris while I was still in motion and he caught it and we had

the tag line ready to go to the victim really fast. We both got the rope around our waist on belay and tight and the victim got one or both arms over the rope before his boat had gone too far into the sieve. As he was popping his skirt the boat shot into the sieve but we had the tagline pretty tight and it stabilized him nicely. It all worked out like it's supposed to on that one day and it felt pretty good. They don't all go that well of course but this one did and I was glad I hadn't hesitated when I head the "funny" sounds that made me think (correctly) that he was going sieve-ward. – Gordon Dalton

Fig. 7.15   Fig. 7.16

Fig. 7.17 The "Carlson Cinch"

First, set up a stabilization line to keep the victim's head up.

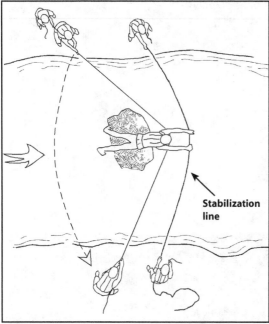

Stabilization line

Then, a cinch line is passed under the victim and passed back to the other shore behind him.

Cinching Rescues. The bare line can also be used for a cinching rescue: that is, the line can be deliberately wrapped around or cinched down on the victim so he can be pulled to shore. In a foot entrapment rescue, rescuers must often have some sort of purchase on the victim in order to rescue him, since he almost always must be pulled upstream against the current. Although cinching rescues work best on narrow rivers they are relatively simple and require little special equipment. However, they do present a contradiction, since we have already cautioned rescuers several times not to wrap the rope around themselves. Cinching is justified only in situations like foot entrapment (or body recoveries) where the victim is in relatively shallow water and will probably drown anyway (or has already drowned) if something is not done.

Even so, rescuers should not tie off any of the ropes on the system described below. Since cinching rescues are appropriate for foot entrapments, we will consider them in more detail in Chapter 11.

A cinching rescue requires a practiced, knowledgeable rescue team on both sides of the river.

It proceeds as follows:
- The rescuers first establish a *stabilization tag line* to help the victim hold his head up while the rescue proceeds.
- The rescuers then establish another tag line, the *cinch line*, across the river in front of the victim, then bring it up underneath him. The line must be at least twice as long as the width of the river.
- The rescue team on one side clips a short *control line* onto the cinch line, then passes the cinch line back across the river *behind* the victim.
- The team receiving the cinch line clips it back onto itself to form a loop. They also clip a short control line to the cinch line.
- The rescuers then cinch the loop down around the victim. The two control lines on opposite banks control the size and position of the loop. In order to keep the loop from cinching the victim too tightly, the rescuers should put a stopper knot 30–40" (75–100 cm) from the carabiner that completes the cinching loop.
- Using a combination of cinch and control lines, the victim is pulled upstream until free, then swung over to shore.
- A variation of this technique is to form the loop on shore, move it over the victim, lower it down over him, and then cinch it down.

Fig. 7.18

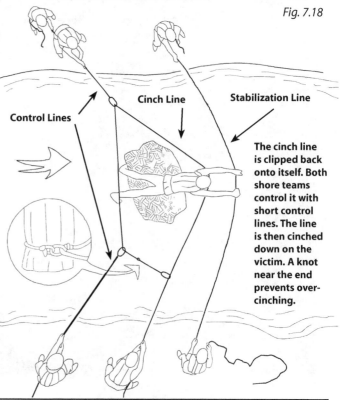

Control Lines

Cinch Line

Stabilization Line

The cinch line is clipped back onto itself. Both shore teams control it with short control lines. The line is then cinched down on the victim. A knot near the end prevents over-cinching.

Swiftwater Rescue

Mid-River Cinches (Kiwi Method).
While the cinching techniques described above work well on narrow rivers, they are much harder to rig on wider rivers. One method first used by whitewater guides in New Zealand (and hence often referred to as "the Kiwi method") is this: if a victim is pinned close enough to shore for a line to reach, it can be thrown upstream of them and allowed the float down around them to trail in the current. Rescuers on shore can then retrieve the line either with a boat, by swimming or with a line capture device such as the Crossline Reach System or the Snag Plate (see Chapter 3) to create a cinch. If the line is first tied off and a line capture device used, a simple cinch can be made with very few people.

Fig. 7.20 A near-shore rescue on the Gauley River in West Virginia. A group of kayakers have made an impromptu cinch and wading rescue. The pinned kayaker has a line attached to the cow's tail on his rescue jacket.

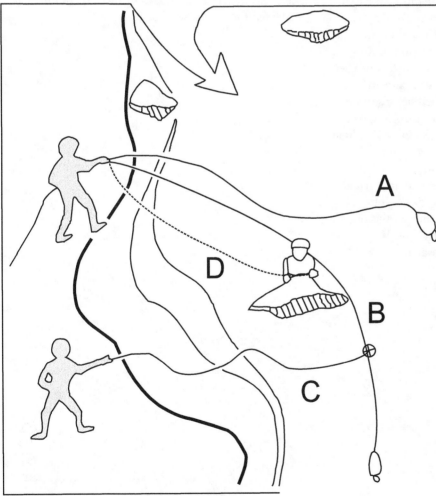

Fig. 7.19 For near-shore rescues on wide rivers, throw a line upstream of the victim (A) and allow it to drift down past him (B). Retrieve the line with either a line capture device (e.g. Reach or Snag Plate) or by swimming (C), then pull it into a cinch (D).

Inflated Fire Hose Rescues

Most fire trucks carry an excellent tool for shore-based rescues—the fire hose. If the hose is sealed and inflated with gas instead of water, it will float on the surface, offering a substantial flotation device for a swimmer. Although this technique has been in use for many years (it also has obvious applications for ice and lake rescue), many departments still seem to be unaware of it. Once inflated, a fire hose can be extended from shore to the victim who is either out in the current or in a fixed location like a low-head dam. It is also quite useful in bridge rescues. These advanced techniques are discussed in more detail in Chapter 11. The larger the hose diameter, the more flotation it will provide. Standard fire hose in the U.S. comes in 50' (15 m) lengths with a diameter of 2" (6.5 cm). This is adequate for most situations, although larger diameter supply pipes will provide more flotation. The hose is normally inflated to a pressure of 25–30 pounds per square inch (170–210 kPa). The hose can be inflated with air from a scuba bottle, a nitrogen cylinder, or even CO_2 from a fire extinguisher. Fire companies may purchase commercially-made hose inflation devices, or they may fabricate their own locally. All that is needed is an airtight cap on the ends of the hose and an inflation fitting. A field-expedient method of sealing the hose is to fold over the ends and tie them down. Folding over the end toward the victim to make a loop is often a good practice anyway, since a loop on the end makes the hose easier to grab and hold.

The simplest hose rescue is to extend an inflated fire hose to a swimmer. While this works well for a victim in a hydraulic, it is not as flexible a solution as using a throw bag. Another disadvantage is that a hose, when put into moving water, is quickly swept downstream. However, it may be possible to suspend the hose over the water, either from a bridge or by attaching it to a rope held on the other side of the river, then dropping it down to the water's surface just before the victim arrives. As with fixed lines, however, the hose should not be tied off on both ends with the victim holding on to the middle while still in the current. One or the other end of the hose must be released so that the victim can swing into shore.

A field-expedient float or even a small raft can also be made from fire hose. For example, a 50' (15 m) length of hose can be folded several times, tied together, and then inflated. Another use for inflated hoses is to lace them through the rungs of a ladder. This makes a floating ladder that can be used as a float or adds flotation for soft surfaces like mud or ice.

> **TIP:** A hand fire extinguisher can serve as an emergency fire hose inflator. Blow CO_2 into the hose, then fold the ends of the hose over and tie them. This is not as good as an inflation kit, but will work in a pinch. Short sections of hose can even be inflated by mouth!

Conclusion

Shore-based rescues are quick, simple, and relatively safe ways of rescuing victims in swiftwater. *Reach* and *throw* rescues fall into the low-risk end of rescue options and should be considered first if conditions permit. Although shore-based rescue techniques are less effective on wider rivers, many of them can be adapted for boat-based rescues as well.

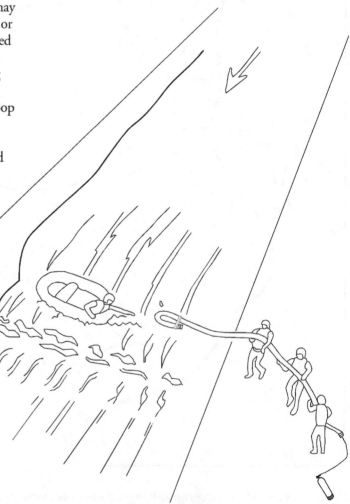

Fig. 7.21 An inflated fire hose makes a simple and effective shore-based reaching rescue. A tag line from the opposite shore is helpful to guide the hose into position and stabilize it in the current.

Boat-Based Rescues
Chapter 8

Fig. 8.0 Barry Miller shows off a homemade flip line holder made from a section of bicycle inner tube.

Shore-based rescues, while safer for the rescuers, have definite limitations. Most shore-based rescue devices only work within a short distance of shore, typically less than 75 feet (23 m). Rescue boats overcome the distance limitations of shore-based rescue by going into the river. By doing so, however, we greatly increase the danger to the rescuers. This danger can be reduced by proper training of rescue boat crews in self-rescue and swiftwater boat handling. We can also reduce risk by using boats to assist rather than perform rescues. An example of *boat-assisted* rescue would be using the boat to perform tasks like ferrying lines and supplies across the river, or to evacuate casualties, rather than actually picking up a victim in the middle of a rapid. A *boat-based* rescue, on the other hand, uses the boat either as part of a rescue system (e.g., some form of tethered boat rescue), or as a free rescue boat in the river.

This chapter is primarily concerned with the techniques of swiftwater boat handling. Any person who chooses to put themselves in swiftwater in a boat must first master swimming self-rescue. In Chapter 3 we discussed the different types of boats available for swiftwater rescue, and there is no need to go over it again here, except to differentiate where different types of boats require different techniques. This chapter cannot, in the space available, do more than summarize the technique of whitewater boat handling.

Self-Rescue

Ideally, anyone who gets into a boat should have appropriate self-rescue training and personal protective equipment. However, this will not always be possible, particularly for those who are being rescued. At the minimum, nevertheless, each person in the boat should have a PFD and some instruction, however minimal, in self-rescue. This means carrying extra PFDs for potential victims and the ability to give a short, concise, safety briefing.

On narrow rivers the best rescue technique is usually for boats and boaters to get to shore as quickly as possible. However, on bigger rivers, particularly during periods of high water, it may not be only more convenient but absolutely necessary that the self-rescue be done in the river. It may be also in some cases safer than swimming to shore, especially if the water is flowing through the trees. A capsized boat need not be a disaster provided the boaters know how to deal with the situation in moving water. Whitewater kayaks, for example, are squirrely little craft that flip in a heartbeat. However, kayakers quickly learn to execute an Eskimo roll to turn the boat back upright again. Some open canoes can be rolled as well.

Inflatables can also be righted while still in the current, provided they are properly rigged and the crew knows what to do. First, the boat must be rigged securely; that is, there must be no trailing lines or loose equipment. Second, the boat must be rigged with flip lines. These are short lines mounted on the sides of the boat that reach over the hull to the other side. They allow crew members to get on top of the boat and flip even a fully-loaded boat back upright. Flip lines must be packaged so that they are secured during normal operation, yet can be pulled free quickly when needed. Some boats have another line rigged permanently across the underside of the hull to assist the crew in getting on top of the boat, but this may cause snagging in shallow or debris-laden water.

Fig. 8.1 Rigging flip lines.

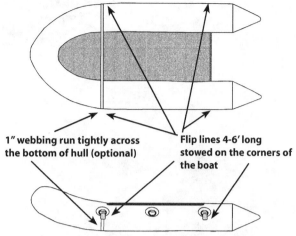

1" webbing run tightly across the bottom of hull (optional)

Flip lines 4-6' long stowed on the corners of the boat

All loose lines and throw bags must be secured.

Yet another option is for boatmen to wear the flip lines on their person, usually in a pouch on a belt (always being sure they can be released if need be). Boatmen climb to the top of the raft, clip into a D-ring, and pull the boat over.

Fig. 8.2

The flip sequence is as follows:
- The boat flips.
- The crew assembles on one side of the raft after counting heads and checking for injuries.
- One crew member gets on top of the boat, assisted by those still in the water.

Fig. 8.3

- He then helps the others up. At this point, depending on the situation, the crew may opt to paddle the upside-down raft to shore and reright it there.

Fig. 8.4

Otherwise:
- The crew digs out the flip lines and stands on one side of the boat. The boat flips back over and they fall into the water.

Fig. 8.5

- The crew reenters the boat. In order to get someone in the boat more quickly, one crew member often goes underneath the boat while the rest are climbing on top and wraps his arms and legs around the thwart. When the boat rights, this leaves him inside the boat where he can assist the rest of the crew in re-entering the boat.

Obviously, these rerighting techniques will not work on larger boats. However, some of these, including some of the larger ones (e.g., some IRBs and 44' USCG boats), are designed to be self-righting.

Some special self-rescue considerations apply to swimming with an overturned or swamped boat. Depending on its design, preparation, and the situation, the boat may be either a help or a major hazard. A water-filled boat is extremely heavy, and there have been several accidents when swimming boaters were caught between a swamped boat and a rock in the river. If you swim, then find the boat immediately and *get upstream of it*. This does not necessarily mean that you should abandon the boat altogether.

Inflatables and boats that have been equipped with extra flotation, even when overturned or swamped, may provide welcome extra buoyancy in a rapid—as long as you stay upstream of them. You may be able to get on top of the boat and get out of the water altogether—an important consideration in extremely cold water.

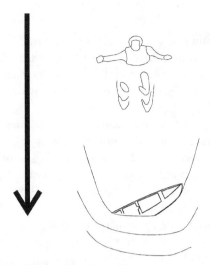

Fig. 8.6 When swimming with a swamped boat, stay upstream of it.

Power boaters must also keep in mind the potentially lethal consequences of hitting someone, either a victim or a crew member, with the propeller. If someone falls overboard or otherwise gets ahead of the motor, an operator *must* be alert to instantly kill the engine. A cut-off (dead-man) switch attached to the motor operator's clothing will ensure that the motor shuts off if the boat capsizes or the operator goes overboard. The danger of propeller strikes can be reduced by using prop shrouds and can be eliminated altogether with jet propulsion units.

Boaters may also combine boat rescue with self-rescue. Swimmers can hold on to a boat in a rapid, then swim it to shore in a calmer stretch. They may also be able to swim to shore with a line or painter, then pull the boat in behind them. Once the boat is on shore, it can then be righted and dumped.

Boat Preparation

Any boat intended for use in swiftwater may benefit from the following modifications:

- **Extra flotation.** If a boat swamps or capsizes in the current, the amount of flotation may determine what happens next. The higher the boat floats, the less likely it is to be pinned on a rock in the current, and the easier it is to tow. Flotation may be either solid foam or air-filled bags.
- **Reinforcement of critical points.** Handles, D-rings, eyes, and other attachment points may come under more stress than they were originally designed to handle, particularly if the boat is being used in a lowering system.
- **Extra lashing points** for gear.
- **Safety modifications.** These should always include a "deadman's switch" attached to the operator to shut the motor off if the boat capsizes or the operator goes overboard. A prop

Fig. 8.7 The front grab handle on this Jon boat has been reinforced for swiftwater use with a backing plate.

shroud or jet drive reduces the danger of someone getting cut up by the propeller. Rescuers may also want to fit the boat with a perimeter line ("chicken line") around the outside, as well as lines ("painters") on the bow and stern. Foot cups on the floor will help the crew stay in place in rough water.

Rigging for Swiftwater

Whitewater boatmen say "always rig for a flip." It is something any swiftwater boat operator should also keep in mind. At any time during a rescue the boat may swamp or capsize. Although obviously a serious problem, the effects of such a mishap can be minimized by proper rigging.

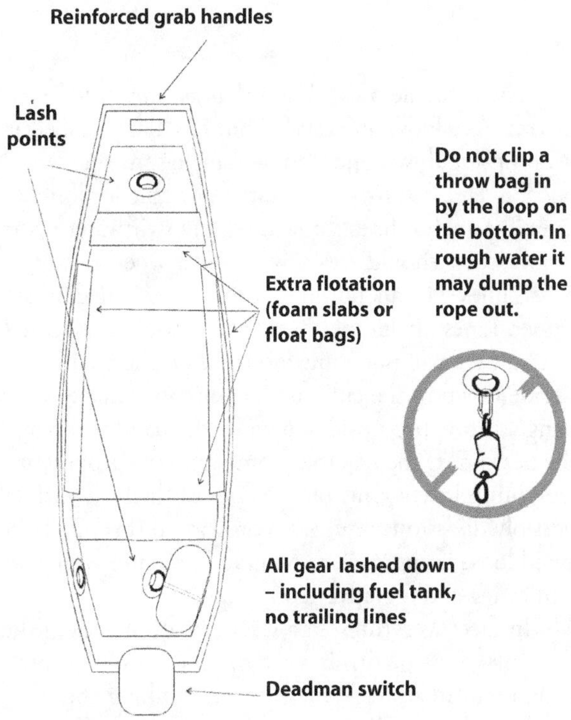

Reinforced grab handles

Lash points

Do not clip a throw bag in by the loop on the bottom. In rough water it may dump the rope out.

Extra flotation (foam slabs or float bags)

All gear lashed down – including fuel tank, no trailing lines

Deadman switch

Fig. 8.8 Boat modification for swiftwater operation.

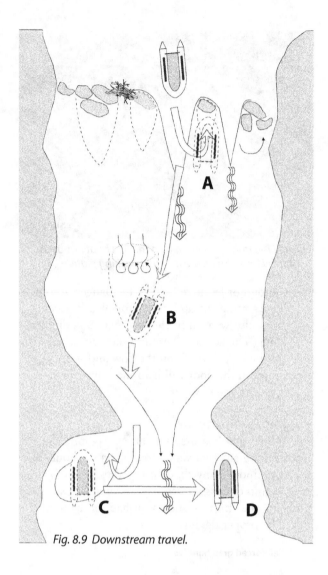

Fig. 8.9 Downstream travel.

The second rule is to *pad any sharp edges*. Rescuers have been injured by hitting the sharp edges of metal boxes, gunwales, and the like in a capsize. Glue foam over exposed edges or pad them with something like a PFD.

Third, take spares and backups. This includes paddles or oars in case the motor fails, and a spare paddle or oar, too. There should be extra PFDs for victims, and spare parts (e.g., spark plug, shear pin) for the motor. On extended missions, you might even consider an extra motor.

Additional rescue equipment might include a boat hook, throw bags, state and federally required boating equipment (e.g., fire extinguisher), a bailer if needed, and communications equipment such as a radio and a bullhorn.

> **TIP:** A whitewater paddle hook converts a paddle or oar to a boat hook.

Boat Handling

Swiftwater boat handling, as we saw in Chapter 2, is mostly a combination of route finding and the three basic river maneuvers: eddy turns, peel-outs, and ferries. These techniques, as well as weight shifts, leans, and high sides, must be mastered before proceeding to more advanced topics like upriver and downriver travel and boat-based rescue.

Downstream Travel

One of the dangers of long or continuous rapids is that a boat will get too much downstream momentum—that is, the boat will be carried from one rapid to the next, unable to stop or scout. To prevent this, we break the rapid into smaller, more manageable segments by *eddy-hopping*; that is, by going from one eddy to another. This technique is also useful in rescue work, where we want to be able to get as close as possible to an accident site. Let's briefly review the example rapid we looked at in Chapter 2. Last time, we ran straight through it, but this time we'll do some things differently.

Instead of running straight down the tongue in the top center of the rapid (Fig. 8.9), we'll catch the eddy behind the rock to the left (A). This allows us to stop in the middle of the rapid. When we are ready to leave, we peel out back into the current and head downstream. However, about halfway down the rapid, we cut in behind a hole (B). This does two things: it slows our downstream progress a bit, and pulls the boat over toward the right shore. This sets us up to catch another large eddy against the river right bank (C). We can also ferry across the current to go from one eddy to another, or from one bank to another, without going downstream (D).

The first rule of rigging is that *all lines must be secured*. Any loose line can fall out of a boat—especially an upside-down one—and entangle someone with possibly fatal results. This is especially true for long lines like anchor lines. In fact, on any swiftwater operation rescuers should seriously consider whether any excess lines, like anchor lines, can be eliminated altogether. Lines, including throw bags, should be secured so that they will not come loose even if the boat is upside-down in the current. A common mistake is to hang a throw bag upside-down by the handy loop at the bottom. If the bag then somehow comes open, it will dump its contents on the floor of the boat with possibly disastrous results. A corollary to this rule is to avoid loose loops, such as on grab lines, into which an arm or leg could inadvertently slip.

In fact, everything in a rescue boat, including the gas tank, must be firmly secured. Do not just throw equipment in a boat—tie it down, thinking about what will happen if the boat is upside-down. Then go flip it in a safe place to check.

For effective boat control when running downstream, boaters need to be moving either faster or slower than the current. Otherwise the boat will go where the current goes, which may not be what you want. Oar-powered boats typically row upstream to slow themselves, back ferrying across the current to position themselves. Most other boats travel faster than the current. However, as we have seen, it's easy to get going too fast downstream, particularly with a motor. Motor boats may reduce their speed in several ways, including using eddies and slack water as already mentioned, running the motor in reverse to slow the boat, or by "backing" down rapids with the boat pointing upstream. Another very effective method, especially for smaller boats and inflatables, is to use paddles or oars for downstream travel, and reserve the motor for upstream travel and cross-current ferries.

Upstream Travel

So far we have discussed only downstream travel. However, motors allow us to travel upstream as well. Jet boats, in fact, routinely travel upstream on large whitewater rivers like the Snake and Salmon. Let's take a look at a rapid and see how we'd get up it. From the eddy at the bottom of the rapid (Fig. 8.10), we power back up the first chute (A). In general, we want to stay close enough to shore to stay out of the main current, but far enough out to avoid shallow water and obstacles like boulders. If we want to stop, we can power back up into an eddy from downstream. The motor also gives power boats the ability to "stall" against the current. In this case, we continue up into the slower-moving water of the hole at the top, and from here we can either continue upstream or ferry across the river into the eddy on the right bank (B).

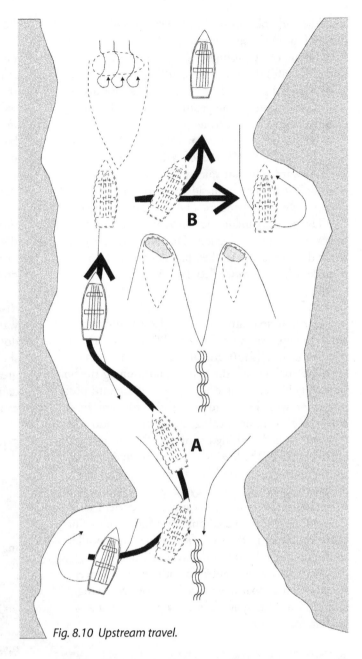

Fig. 8.10 Upstream travel.

Paddle Rafts

Paddle rafts are cheap and simple to operate, which is why they are used recreationally every day of a busy summer on rivers all over the world. The most useful size for most rescue organizations is a four to six-person raft. Motor inflatables, including smaller IRBs, can also be maneuvered downstream using paddle raft techniques.

Paddle rafts travel forward downstream, propelled by the paddles of the crew. Normally, the crew sits in a balanced configuration; that is, they sit in pairs on either side of the raft. A guide or paddle captain sits in the stern and calls commands. Each crew member has a paddle and holds himself in the raft by either pushing his foot into a soft cup made of raft material glued to the floor, or by pushing his foot (*not* his ankle) under the raft's thwarts. The paddle commands are:

- **Forward paddle.** Everyone paddles forward. Stroking in unison greatly improves efficiency.
- **Back paddle.** Everyone paddles backwards.
- **Right turn.** The crew members on the right side backpaddle while the left side members paddle forward. The raft spins to the right.
- **Left turn.** The right side paddles forward; the left side backpaddles. The raft spins to the left.

The paddle captain may also make steering strokes to correct the raft's direction. The raft crew uses turns to set the correct angle, then paddles forward (or back) to make whatever move they have in mind.

Oar Rafts

Oar-powered rafts, usually called oar boats, operate in the opposite manner of paddle rafts. Where a paddle raft travels forward, an oar boatman usually rows the boat against the current, positioning the boat by setting his ferry angle and back ferrying into position. By rowing backwards and facing forward, he is able to see where he is going. Boatmen can also row forwards (called pushing or "portegee"), but this is less efficient. While the leverage of the oars makes it easy for the boatman to spin the raft quickly, oar boats are relatively slow. Thus, they have severe limitations in rescues that require power like cross-current ferries and upstream travel. Some rafts have the rowing frame mounted in the stern of the boat and paddlers up front, which increases the available power.

Rafts are easily adapted for other types of river rescues such as boat lowers, are relatively forgiving, and can carry heavy loads. Oars are often used as a backup for motors.

Fig. 8.11 An oar boat.

Power Boats

Most small power boats (including motorized rafts) have an outboard motor mounted on a transom at the stern of the boat. The operator sits in front of the motor, steering it by physically turning the motor with a tiller bar. This can cause some confusion since, in order to turn the boat to the left, the motor must be turned to the right. Larger boats often have a steering wheel mounted forward, which makes steering somewhat more instinctive. Power boats differ from the nonpowered craft mentioned so far in that the power comes from the end rather than the center, so that the operator must push the stern around so that the boat assumes the correct angle. Otherwise, the principles of boat handling are much the same as nonpowered craft.

Obviously, the power boater must also take care that the propeller does not slice up someone in the water, that it is not damaged on rocks, and that it does not hang up in the shallows. Power boaters should also keep in mind that flood waters carry a great deal of debris that can clog intakes and foul or damage propellers and that loose rescue lines can get caught in a propeller.

Boat Rescues

The simplest and most direct method of boat rescue is to bring the boat next to the victim and haul him in like a hooked tuna. It is also one of the most dangerous. Smaller boats like canoes and kayaks can be overturned by panicked victims, and attempting to chase someone down in a rapid is considerably more difficult than running the rapid alone. Before boat crews venture into the water to rescue others, they should be thoroughly familiar with boat handling *and* the stretch of river in which they intend to operate. Boat crews should also have practiced methods of getting someone back in the boat, since it could be one of their number.

Rescuers, especially those in smaller craft, should make some attempt to ascertain the victim's mental state before attempting a rescue. Although we will discuss victim psychology in more detail later, it is enough to say for now that rescuers should hail the victim, establish eye contact if possible, and try to decide whether the victim is capable of assisting in his own rescue or if he is a potential threat. Panicked victims obviously require extra caution.

Larger craft, like rafts and power boats, have the capacity to take a victim on board, and this should certainly be done if possible. Swimmers, unless the river has few obstacles, should be approached from downstream if possible, so that if the boat is swept into a rock, the swimmer will not be caught between

Fig. 8.12 To get someone back in a raft, first grab them by the shoulder straps of their PFD…

Fig. 8.13 … then pull them up and back over the tube.

the boat and the rock.

The rescue boat operator's first task is to get his boat close enough to the victim to rescue him. In swift current, the best solution is often to get downstream of the victim and use the motor to stall against the current. As the current brings the victim close to the boat, the operator must then match his speed to that of the victim. If he does not, the victim will either drift away—or worse, go underneath the boat. For an effective rescue, both the victim and rescue boat should be moving at the same speed as the current. If the boat has a propeller, the operator must kill the engine, or at least put it in neutral, as the victim comes alongside.

After bringing the boat next to the victim, the rescuers must get him on board. In most cases, this can be done by simply lifting the victim up over the gunwales into the boat. To do this, the rescuers normally grab the victim either under the armpits or by the shoulder straps of his life jacket, then lift. Attempting to drag someone in by grabbing their hands or wrists does not work very well. In an inflatable, a single rescuer can grab a victim as described above and then fall backwards into the boat, pulling the victim in with them. There is some controversy about which way the victim should be facing when pulled in. The "West Coast" solution is to have the victim facing outward, away from the tube, as he is pulled in. Proponents point out that this keeps the victim's PFD from snagging on the boat's chicken lines or D-rings. The "East Coast" school, however, maintains that the victim should face the tube,

since the human body bends that way more readily when pulled over it. In practice, both methods seem to work equally well.

Rigid-hulled power boats present a more difficult challenge, since they typically have more freeboard than inflatables, and the hull often flares outward, especially near the bow. Smaller hard-hulls (e.g., Jon boats) are also often more prone to capsizing while taking a victim on board than are inflatables. One solution is to take the victim in at the stern, where the hull is generally lower and the danger of capsize less. Some boats even have a small platform there expressly for this purpose. However, the motor must be shut off when doing this. If the victim is capable of assisting himself, a boarding ladder—either solid or one fabricated from rope loops—will work. Injured or unconscious victims can be retrieved with

Fig. 8.14 A boarding net can ease the task of getting a heavy victim over the gunnels of a rigid-hulled boat..

a boarding net. The net is attached to the gunwales and brought under the victim's body. The rescuers then pull the outside of the net up, rolling the victim into the boat.

Any victim, upon being pulled into the boat, should immediately be given a medical survey and a PFD. In many cases, victims will require treatment for hypothermia as well.

Boats can also be used as rescue platforms. Many of the techniques of shore-based rescue can also be used successfully from boats. Rescuers can reach to victims with boat hooks, oars or paddles, or they can throw ropes to them. Larger boats, including IRBs, inflatables and rigid-hulled power boats, are very stable and make excellent rescue platforms. Throw bags can get to victims, such as ones trapped in a low-head dam, that a boat crew might not want to approach directly for a rescue.

There may be times when rescuers are unable to get very large victims into the boat. For example, a 240-pound (109 kg) man with an additional 60 pounds (27 kg) of soaked clothing (not an unusual scenario) equals a total weight of 300 pounds (136 kg). He might then have to be lifted vertically three feet (1 m) to clear the gunwale, which might well be beyond the abilities of some rescuers. In this, the rescuers might opt to tow the victim to safety. If so, the tow should be kept short—only as far as needed to get the victim to safety or to a place where he can be gotten into the boat (e.g., a shallow spot, an island, or shore). Towing greatly increases the incidence of hypothermia, since there is now water flowing rapidly past the victim.

Smaller boats like kayaks will have to tow victims to safety; something that requires considerable skill and strength on the part of the kayaker. On smaller rivers, the victim can hold on to the grab loop of the kayak, kicking with his feet to assist. On larger rivers, the swimmer can crawl up onto the back deck of the kayak and hold on to a broach loop, security bar, or the cockpit rim. Rescuing an unconscious victim with a kayak is difficult; however, a cow's tail attached to a rescue PFD can be used in an emergency.

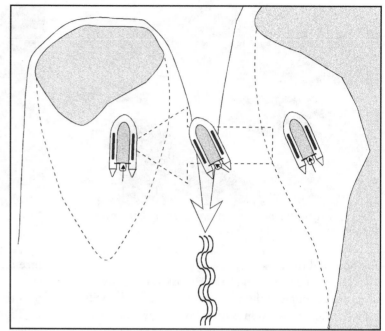

Fig. 8.15 *The rescue ferry. The ability to move across a jet of current from one eddy to another without losing position is an important rescue boat-handling technique.*

Rescue Ferries

A very common rescue situation is to have a victim stuck on an obstacle in the river. Typically, this is an island where someone has ended up after a swim or a capsize, or a car that has run off the road or been caught in rising water. A scenario like this recently played out in a small town through which a river ran. An inner tuber had gotten stranded on a midstream island. Although the current was swift, there were no real rapids, and there were large eddies on either side of the river. The local rescue squad turned out and began trying to get to the island by powering their boats directly across the current from the eddy to the island. The results were predictable: as the rescue boats entered the current, they were instantly blown downstream, almost capsizing in the process. After watching the progress of the rescue attempts for awhile, the tuber eventually jumped in the river and swam to shore.

The correct solution to this problem would have been simple:

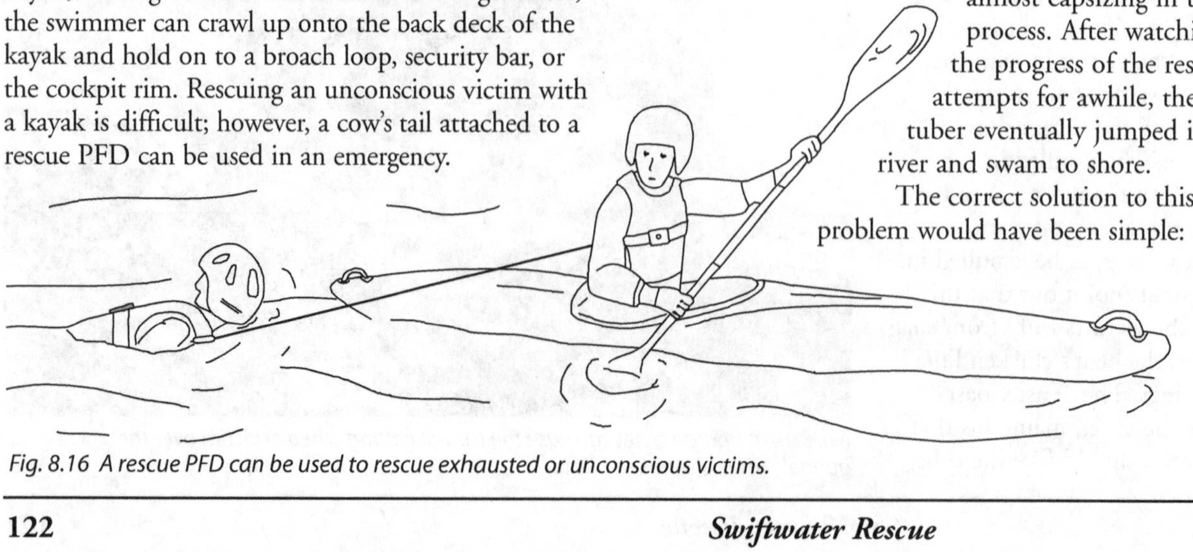

Fig. 8.16 *A rescue PFD can be used to rescue exhausted or unconscious victims.*

Swiftwater Rescue

ferry from the eddy on shore to the one behind the island, pick up the tuber, and ferry back. The problem lay not with the squad's equipment, but rather that they did not know how to do a swiftwater ferry. Any obstacle in the river creates an eddy behind it, and although all eddies may not be usable (some are extremely turbulent), rescuers should always consider them when planning a boating rescue. Even submerged obstacles often create slack water that will aid a rescue boat. The general rule for rescuing someone from a midstream obstacle with a boat, then, is to try to get the boat into the eddy *behind* the obstacle.

If a cross-current ferry isn't possible, the rescuers may have to catch the eddy from upstream, which is somewhat more difficult. If the eddy is small they may have to use their imaginations. This is illustrated by another incident, this time on the flooded Nantahala River. Four teenage boys had rented a raft and gotten on the river. Shortly after putting in, their raft hit a tree and wrapped. The raft ended up underwater, with the boys standing on a barely protruding tube, holding the tree. The rescuers, a group of local raft guides, approached from upstream, hoping to catch the tiny eddy behind the tree. It looked hopeless, but they had one of their number stationed in the bow of the raft with a small throw bag. As the raft passed the tree and began to make the eddy turn, he threw the rope to one of the boys, who caught it and helped the raft into the eddy. The boys got into the raft and were ferried to shore, completing the rescue.

Line Ferries

In the chapter on shore-based rescues, we discussed some of the various ways of getting a line across the river. Here we will look at another common method—ferrying the line across with a boat. Line ferries are simple in concept but often quite difficult in practice. The method is exactly the same as the rescue ferry described above, except that the boaters now have a line trailing behind them. If this line gets into the water, which it almost always will eventually, the drag will begin to pull on the boat, upsetting the ferry angle and eventually pulling the boat back downstream. The wider the river, the more of a problem this becomes.

There are several ways to reduce the current's drag. One we have already discussed is to send a smaller messenger line over on the initial ferry, then use it to pull over a heavier line. Another is to hold the line as far out of the water as possible. This is easier on larger boats that allow a rescuer to stand during the ferry. The shore team can also help by getting their end of the line as high as possible, either by going farther up the bank or by climbing a tree. A third method is to simply dump a lot of rope in the river and hope that the boat will get to the other side before it goes taut in the current. In any case, the rope must be paid out rapidly to minimize drag on the boat. This may be done either from the shore or from the boat, depending on the situation. A rope bag will minimize the dangers of entanglement. Because the line will inevitably be carried downstream, the general rule is that the length of the line to be ferried should be at least twice the distance to be crossed.

Power boats have a definite advantage for line ferries, since a motor can overcome much more drag than a muscle-powered craft. Still, on wide rivers in swift current, even big motors will be overmatched. Speed is critical on a line ferry, since every second the line stays in the water and every additional foot of line that goes into the water, means more drag on the boat. A fast ferry will often succeed where a slow one will not.

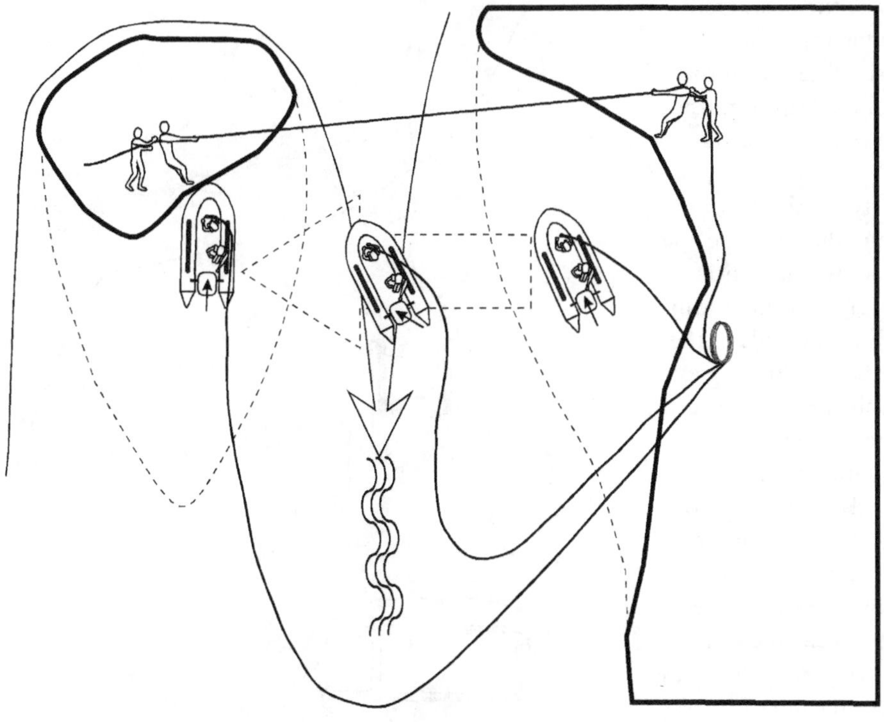

Fig. 8.17 A line ferry is another important rescue boat-handling technique. The rope should be about twice as long as the distance to be crossed.

Fig. 8.18 A line ferry. One crewman stands to hold the rope as far as possible out of the water.

A critical point in any line ferry is the transfer of the line from the boat to the opposite shore. The amount of drag on the rope usually peaks just as the boat reaches the opposite shore. Many boats have been abruptly jerked back into the current at this point. As soon as the boat reaches shore, the line must be belayed and, if possible, lifted out of the water. The best way to do this is to organize three teams: the on-shore team, who manage and pay out the rope; the ferry team, who operate the boat; and the opposite-shore team, who receive the rope from the boat team and then manage it on the other side of the river. The opposite-shore rope team must meet the boat, grab the boat and the rope, and secure the rope. Sometimes the opposite-shore rope team must be carried in the ferry boat. If so, they must be ready to leap from the boat and assume control of the rope as soon as the boat touches shore.

In addition to the normal safety considerations for ferries, the boatmen must ensure that they are able to release the rope if something goes wrong. Since the drag of the rope will progressively increase during the ferry, it should not be attached to the ends of the boat, since this may throw off the boat's ferry angle. The best attachment point, if possible, is near the center of the boat, free to swing through a wide angle.

Rope ferries can also be done with smaller boats like canoes, kayaks, and rescue boards, and even with swimmers. Although these lack the power of motor boats, a skilled paddler or swimmer can do a surprisingly good job, especially on narrow rivers. However, these craft usually carry only one paddler, who must use both arms to paddle, steer, or swim. A rescue vest is a big help for line ferries here, since attaching a line to the vest's D-ring keeps it out of the paddler or swimmer's way and allows him to release it if necessary.

Swimmers and small boats can gain an advantage by starting considerably upstream of the ferry site. This allows them some extra time to get across the river before the rope comes taut. If the situation allows they may even choose to start as far upstream as the river is wide.

Tethered Boat

So far we have been talking about free boats. However, there are times when we want a boat to stay more or less in one place, such as when a victim is fixed in position. There may also be a time when the boat is approaching a dangerous hazard like a low-head dam and we want a method of reeling the boat back if the crew gets too close. For example, in the San Marcos incident recounted in the prolog, the rescuers tied a rope to the back of the raft for exactly this reason. However, rescuers should keep in mind that attaching any rope to a raft will add a great deal of drag if the rope gets into the current, and this may prevent the raft from maneuvering.

Single-point tether. In many cases a raft can simply be tethered with a line to the bow and lowered down to a trouble spot. If the current will take the raft to the right place, this is a very fast and simple technique.

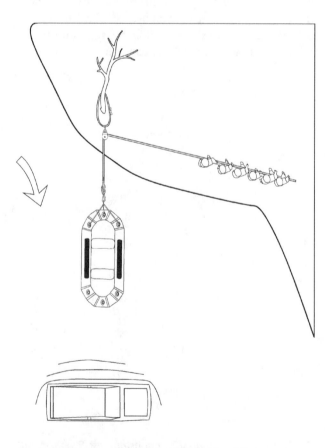

Fig. 8.19 Single-point tether with directional change.

Swiftwater Rescue

Fig. 8.20 *A single-point lower used to rescue a stranded paddler.*

2- and 4-point tethers. If we tether a raft from each bank, we can hold it in position and move it around in the river where we want. This is the same idea as the floating tag line, except that we are now using a raft. The 2-point tether has one line going from the raft to each bank, while the 4-point has two. The 2-point is quicker and easier to rig, but the

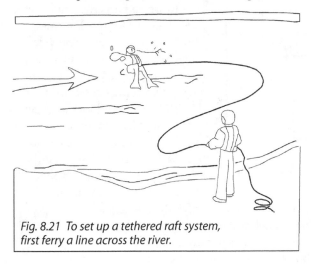

Fig. 8.21 *To set up a tethered raft system, first ferry a line across the river.*

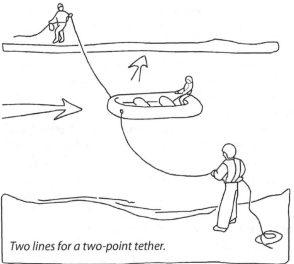

Two lines for a two-point tether.

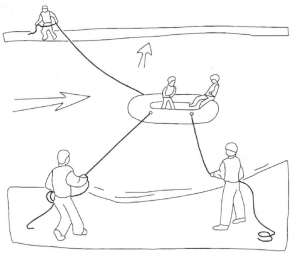

Fig. 8.22 *Ferry an additional line across...*

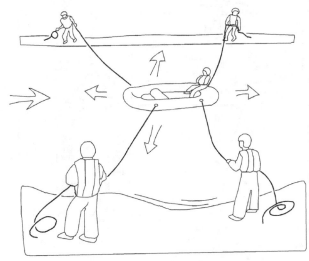

... for a four-point boat tether.

4-point tether gives better control. Both systems are limited to relatively narrow, slow-moving rivers. They do not work well on rivers over 200 feet (61 m) or on river bends. For the system to work properly, the control lines must be held out of the water, and this becomes progressively more difficult as the river gets wider. Nevertheless, these systems are quick to set up and require little equipment.

On a raft, the control lines are normally attached to the D-rings on the outside of the tubes. For the 2-point, the lines are attached near the bow, while in the 4-point, the lines go to the bow and stern. Setup is relatively simple: after attaching the near-shore lines to the D-rings, the rescuers ferry the upstream control line to the opposite shore. This can be done with any of the methods described above, including ferrying the line across with the raft. For the 4-point, the rescuers can either ferry both lines across at the same time, or pull the raft across with the upper control line, then set the lower control line when on the opposite bank.

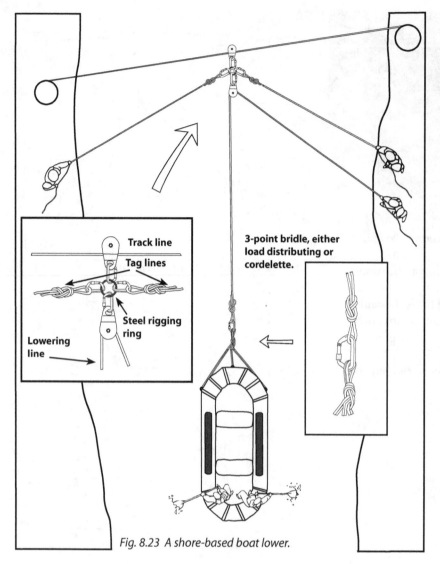

Track line

Tag lines

Steel rigging ring

Lowering line

3-point bridle, either load distributing or cordelette.

Fig. 8.23 A shore-based boat lower.

raft is operating primarily under its own power or with another system, and the tether acts not as a control but as a safety.

Boat on Tether / Telfer Lower / River Tyrolean / Movable Control Point. For really precise control on fast water, rescuers should consider setting up a boat on tether, also called a Telfer lower, a river Tyrolean or highline, or movable control point system. If it seems familiar, it is the same high line with midpoint lower system discussed in Chapter 6, adapted for river use. Here, a line goes across the river, but instead of lowering a rescuer vertically down from the midpoint, a rescue boat (typically a raft) is lowered downstream. The boat on tether requires quite a bit of equipment, time, and experience to set up properly, making it a less desirable for quick rescues. However, once set up, it provides a stable working platform that can be precisely positioned in the current. The system is usually set up above the accident site and lowered down to it. If the situation permits, it is usually better to lower the rescue boat past the accident site and then bring it up into the eddy behind. However, there are situations, like low-head dams, in which the rescue craft will be positioned below the site and pulled up to it. It is normally used to approach fixed accident sites, such as pinned boats and stranded victims. Since it is easy to move the rescue craft in a regular, repeatable grid pattern, the boat on tether works very well for body searches.

To set up the boat on tether, we must once again get a line across the river. This time, however, the anchor line must meet the same standards for construction and anchoring as any high line (e.g., ½"/13 mm or better static nylon rescue rope). Since the anchor points may be subjected to considerable force, and since the rope should be 10-15' (3-5 m) off the water, the ideal anchor point is a large tree (BFT). If no tree is available, rescuers may have to use tripods or A-frames to get the rope up off the water. If possible, the anchor line should be angled slightly downstream, since this will aid in launching.

Almost any watercraft can be used for the lower. Rafts are normally used, but if the rescuers opt for a

Each control line should be at least twice as long as the river is wide.

Operation of the system requires at least one person on each control line. They hold the control lines out of the water, thus reducing the drag and the need for a belay. The line holders should simply hold the ropes and not belay them. If the force is too great for them to do this, then another system should be considered. Once the system is established, the boat can be moved up and downstream as well as back and forth.

For best results on the river, the raft should be loaded towards the stern to keep water from building up on the bow and swamping the raft. The crew should also have knives in case they need to cut free, and paddles or oars for backups. Communication between the shore teams and the raft is critical, and should be run through a control.

A variation of the 2-point tether is the "V" tether. This is similar to the 2-point system, except that the lines are attached near the stern of the raft. Here the hard boat, they should ensure that the attachment points are strong enough to take the pressure. If there is any doubt, they should be reinforced. On rafts, the pressure can be reduced on individual D-rings by using a load-sharing anchor for a bridle.

There are two basic variations of the boat on tether, one where the rescue craft is lowered and raised by a shore-based team, and the other where it is lowered by the crew of the rescue boat. If it is a shore-based lower, it is generally set up with a simple directional pulley on the anchor line carriage run to a shore-based haul team. If necessary, it is relatively easy to set it up for greater mechanical advantage. On a system controlled by the boat crew, however, the raising/lowering line is either run directly to the rescue craft or from a bridle on the front of the boat through a directional pulley on the carriage and back to the boat.

In both types of lowers the rescue craft is moved back and forth across the river by means of tag lines. If the downstream

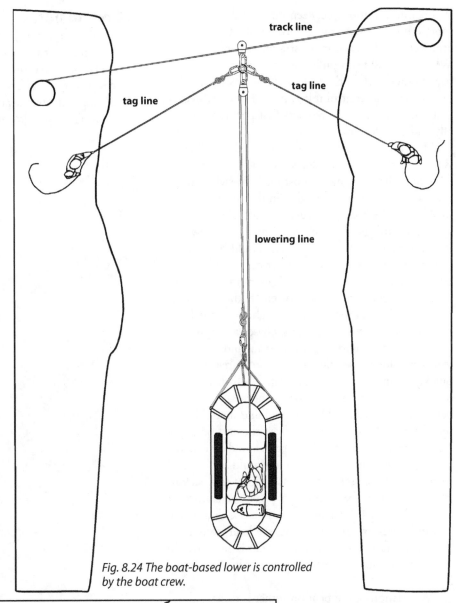

Fig. 8.24 The boat-based lower is controlled by the boat crew.

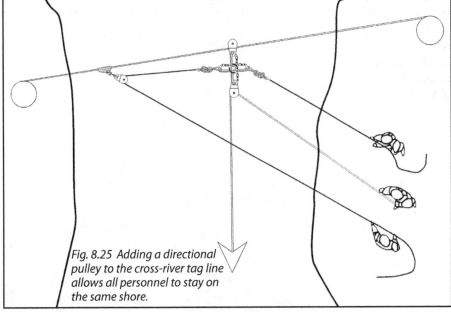

Fig. 8.25 Adding a directional pulley to the cross-river tag line allows all personnel to stay on the same shore.

angle of the anchor line is large enough, it may be possible to do away with the lower tag line. Tag lines normally go to each shore, but the far side tag can also be run through a pulley on the anchor line, so that both tag lines come back to the same shore. While this adds some complication, it does make coordination between the tag line operators easier.

The advantage of a shore-based system is that it frees the rescuers to concentrate on the task at hand. The down side of this arrangement is that communications between raft and shore become more critical. Rescuers will find that the boat in this arrangement does not move straight up and down. When the haul team pulls the boat up, it also moves toward them; when they release it, it moves away. The boat is easier to control if it is moved in one direction at a time; that is, if the rescuers first position it laterally and then lower or raise it.

The most critical part of using a tethered boat is the launch. It is here, especially when crossing the eddy line, that the boat is most likely to swing sideways and swamp or get tangled in the control and haul lines. The boat (particularly a raft) should be loaded toward the stern to avoid having the bow dig in and ship water. Crew members should be positioned on either side of the stern with a paddle as the boat is launched to ensure that it does not swing sideways in the current. As the boat enters the current, the force will load the system, straightening the boat out, after which the positioning phase can begin.

Whatever system is used, the haul and tag line teams should ensure that all lines can be released quickly in an emergency. Since there is little lateral force on the boat, the tag lines are normally not belayed, but simply held. The haul line should be set up without a knot or throw bag on the end, so that it can run free if necessary. The crew should be experienced in handling a free boat and should have knives to cut free if necessary, as well as paddles as a backup.

With a shore-based boat on tether the control or incident commander is on shore, in a position where he can be seen by tag line teams, the haul team, and the rescue craft commander. In a boat-based lower his position is optional, either on shore or in the rescue boat.

A variation of the boat on tether is to have the rescue craft pulled *upstream* into the backwash of a low-head dam. Here the anchor line is set over the dam and the rescue boat launched below the backwash. To ensure that the rescue boat does not get caught in the backwash, it has a couple of tag lines attached to the stern, tended from downstream of the boil line.

Conclusion

There are many variations of boat rescues. Some use a free boat maneuvering in the current, while others use a boat or shore-based tether to control the boat or act as a safety. As with other types of rescue, training—in river awareness, self-rescue, and boat handling—is the key to successful boat rescues.

Fig. 9.0 Firefighter Don Lopez struggles to rescue a 15-year-old girl from a swollen creek near Santa Rosa, CA.

**Pope Gregory I (A.D. 590–604) gives us a
description of an early in-water contact
rescue. A monk, Placidus, fell into the
water, whereupon another monk, Maurus,
walked out on the water and grabbed
Placidus by his hair, pulling him to shore.
While this system has much to recommend
it in terms of speed and simplicity, it is not
a practical option for most of us.**

Any rescuer entering the water to perform a rescue, especially one that requires direct contact with a
victim, must have a high level of skill, training, and
experience. Of the rescues covered so far, these are the
most dangerous. A swimming rescuer lacks the power
and security of a boat, and must rely heavily on his
own knowledge and skill for survival. He may also
have to grapple directly with panicked victims. Yet
swimming rescues have undeniable advantages—if
they are dangerous, they are also fast and simple. By
relying on the rescuer, we can cut out a lot of complex
equipment and its potential for failure. There is no
waiting for helicopters to arrive, boats to be deployed,
or lines to be ferried across the river. Time is almost
always at a premium in any river rescue, and in-water
rescues maximize whatever time is available. In some
cases, such as when dealing with an unconscious victim, there may be no other way to rescue the victim
other than by hands-on contact. Training and experience can reduce the danger of swimming rescues to
acceptable levels.

River Swimming

Before attempting a swimming rescue, swimmers should be thoroughly familiar and comfortable with self-rescue swimming, including both feet-first and head-first swimming, catching throw ropes, and how to deal with hazards like strainers and holes. Any rescue swimmer must be fully equipped, including a PFD, thermal protection, and helmet. In shallow, rocky rivers, rescue swimmers may want to consider physical protection as well. Drysuits, while warm, do not offer much physical protection from bumps and bruises. Wetsuits are much better in this regard, but in any case a swimmer may want extra padding for his shins and knees. The rescuer's dress and PFD should not inhibit swimming, and should be carefully checked to ensure that they are snag-free. Swim fins greatly increase a swimmer's efficiency, but these must be smaller fins (like the Churchill) that are suitable for river use rather than those intended for scuba divers. Hand fins also increase the swimmer's speed, but inhibit his ability to grasp a victim or rope. Some hand fins with open fingers and minimal webbing will work, however.

Rescue swimming is done in an aggressive forward crawl stroke in order to get to the victim as quickly as possible. When swimming with fins in shallow water, the swimmer is well-advised to use a flutter kick with his legs straight in order to avoid banging his knees. He should also be aware that crosscurrents may catch the fins more than his bare legs. It is not unusual for novice swimmers, swimming downstream, to be rolled over backwards as the current catches their fins beneath them. Swimmers with fins should also be careful around the upstream side of boulders, since water pressure may pin the fin against them.

Fig. 9.1 Practice swimming on the Colorado River. These swimmers are crossing from the wave train to an eddy.

Compared to a boat, a swimmer is much slower and has very poor visibility at water level, which makes route finding much more difficult. Large waves restrict visibility even more. A rescue swimmer should think of himself as a small boat, and use boat-handling techniques like eddy turns, peel-outs, and ferries to move around in the river. However, since the swimmer lacks the power of a boat, he must allow more time for maneuvers, and use river features like eddies as much as possible to help him. Swimmers must also be alert to change position to feet downstream when the situation calls for it, such as when going over a drop.

Because of this lack of speed and power, crossing eddy lines may prove difficult. Since the swimmer's full body hits the eddy line, he may find himself surfed off downstream. One way to prevent this is to swim full speed towards the eddy line, then roll over it when you hit it.

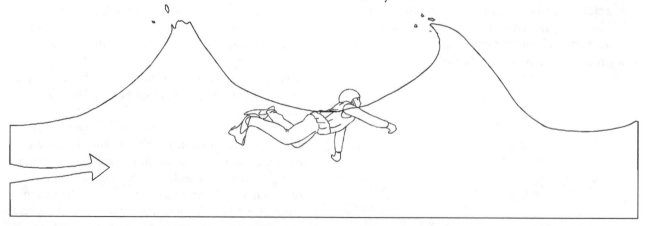

Fig. 9.2 A forward crawl works best for rescue swimming in deep water. Swim fins greatly increase the swimmer's speed and power, but require experience.

Interception Strategies

Since a swimmer lacks the speed of a rescue boat, he must consider appropriate strategies for getting to the victim. If both rescuer and victim are in the current, and the rescuer must chase the victim downstream, the resulting chase may be a long one. Moving downstream may move both rescuer and victim toward greater danger away from the support of teammates and lengthen their stay in the water. In addition, since most victims are looking downstream to see where they are going, they may not see the rescuer. Therefore, in most cases the best position for the rescuer to intercept the victim is from downstream. If the rescuer is downstream of the victim, he can "stall" against the current by swimming upstream. The current will then rapidly bring the victim to him, and it will keep him closer to any shore-based support.

If the rescuers have a boat, they may choose to approach the victim in the boat, and then deploy the rescue swimmer as they close. Nonpowered boats will find that, like swimmers, the best strategy for interception is to get downstream of a victim and stall against the current. On wide rivers the rescue swimmer will be, of necessity, based out of a boat rather than on the shore.

Rescue Boards

A rescue board can be a useful accessory to any swimming rescue. It provides extra flotation for the swimmer and acts as a small boat hull, reducing drag and allowing him to swim faster. It is normally used with swim fins, and lower leg protection is strongly recommended. Rescue boards come in different versions: a small size originally made for ocean surf as the Boogie Board™, a larger river board made by Carlson capable of holding two people, and lately an inflatable one.. Any of them provides something to put a victim on, making it easier to swim. The smaller version, which comes with a "surf leash," can also be used as a rescue flotation device (RFD) for tow rescues.

While still suffering from many of the disadvantages of a swimmer (lack of visibility, slow speed), a rescue board fits well into cramped spaces like the trunk of a police car or the equipment compartment of a fire truck. It is also very quick to deploy and, since it is relatively stable, it is much faster on which to train novices in basic river maneuvers than with a boat. It does, however, require a high level of fitness to use effectively.

> **TIP:** A swimmer's lower legs and ankles will take a pummeling when using a rescue board. Some form of protection is highly recommended. This can be as simple as taping a slab of ensolite foam to your shins and ankles, to wearing more elaborate defenses like hockey or baseball shin guards.

In practice, the rescue board operates much like a small boat. For contact rescues, the rescuer approaches the victim and places him on the board. Then he moves behind the victim, holding him between himself and the board.

Victim Psychology

Since in-water rescues often require the rescuer to make contact with a victim under rather hazardous conditions, a rescuer needs to be able to make an assessment of the victim's mental state. The success of the rescue and the safety of the rescuer depend on it. Rescuers must often vary their tactics and techniques based on the victim's mental state as much as his physical characteristics. In general, rescuers may expect to find a victim in one of four states:

- **Normal survival behavior.** This does not necessarily mean that the victim is calm (he may be highly agitated), or that his responses are appropriate for the situation. It does mean that he is still capable of rational thought and making some sort of purposeful movement in an attempt to stay alive. He is usually able to follow simple instructions and may be able to assist rescuers.

Fig. 9.3 A rescue board offers a swimmer extra flotation and protection for himself and a victim.

- **Panic.** The victim "loses it," that is, loses the capacity for rational thought and appropriate survival action, and exhibits panic instead: counterproductive, random, nonpurposeful movement such as screaming and thrashing in the water.
- **Counterpanic.** The victim withdraws ("derealizes") from a threatening environment. He becomes totally unresponsive and simply floats along, oblivious of attempts to rescue him. A counterpanicked victim may ignore a throw rope literally dropped into his hands. However, a totally passive victim may also be injured, exhausted, or hypothermic.
- **Instinctive drowning response (IDR).** IDR is also called Drowning Non-Swimmer (DNS), since it usually involves a non-swimmer. The signs are: 1) head tilted back, 2) mouth open, 3) arms flailing in unison, and 4) head bobbing up and down. This behavior is instinctive and cannot be controlled by the victim. It may even appear that the victim is playing. However, *anyone exhibiting IDR is drowning and must be rescued immediately*—once they sink below the surface they will be much harder to rescue or recover.

Of these, the most difficult victims to deal with are the panicked or drowning ones. A counterpanicked victim presents less of a threat, but is also unable to assist. However, *any* victim will do whatever he feels is necessary to survive, including grabbing or climbing on top of a prospective rescuer.

The principal causes of panic are:
- Sensory overload. There is too much sensory input about the dangers and too little time to process it.
- Lack of knowledge about the situation the victim suddenly finds himself in, and of appropriate responses and defenses.
- The perception that death or serious injury is imminent and that escape has been cut off or that rescue is not possible.

Obviously, the victim's perception is all-important. How they react will be determined by this and their individual psychological makeup. A person may panic when they are in no danger because they perceive that they are, or they may blunder into a life-threatening situation through ignorance of the dangers. Panic can often be controlled by giving the victim more information. This might be a public education program, a pretrip safety briefing, or even giving instructions to

a victim in the water. If so, the instructions must be kept clear and simple. Another aid to panic prevention is to reassure victims that they are not in imminent danger of dying and rescue is on the way.

Probably nothing gives a victim in the water more reassurance, however, than being able to grab something—anything—that floats. Since that anything could be you, rescuers in the water should approach victims with caution if there is any doubt as to their mental status. When just out of reach, try to make an initial assessment; if possible, make verbal and eye contact and evaluate the response. Realistically, however, there will not be time in most rescues to conduct an in-depth interview with a victim. The rescuer's actions and options are covered below in the section on combat swimming.

Fig. 9.4 A towing rescue, using a Boogie Board™ as a rescue flotation device.

Tow Rescues

Lifeguards often use tow rescues, and these work equally well on the river. Here the rescuer passes an object to the victim and then uses it to tow him to safety. By putting something between himself and a victim, the rescuer reduces the chance that the victim will grab him. While a tow can be made with just about anything (e.g., a belt, a rolled towel, or a short loop of rope), it is generally better to use some kind of floating object. This gives the victim extra flotation (vital if he has no PFD) and something more substantial to hang on to. It may also help the victim avoid or control his panic. Lifeguards use a variety of rescue flotation devices (RFDs), including rescue tubes, cans, floats and buoys. For the river rescuer, a small rescue board like the original Boogie Board™ works well. It comes with a short "surf leash," giving the rescuer the option of shoving the board to the victim and hanging on to the leash. He can then tow the victim to safety on the board.

Fig. 9.5 "Reverse and Ready"

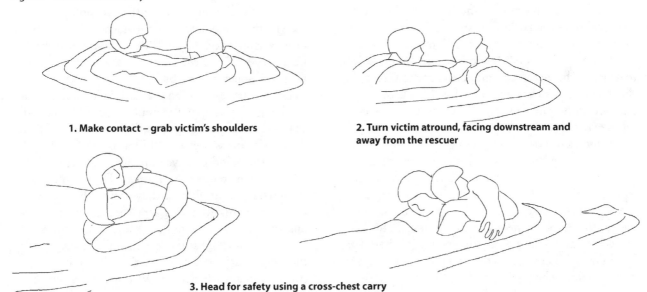

1. Make contact – grab victim's shoulders

2. Turn victim atround, facing downstream and away from the rescuer

3. Head for safety using a cross-chest carry

Contact Rescues

A contact rescue is any rescue in which the rescuer makes direct contact with the victim. Although the rescuer contacts the victim in many boat and board rescues, the term "contact rescue" generally refers to an in-water rescue where the rescuer physically grabs the victim and swims or drags him to safety.

Combat swimming means securing the victim, including grappling with him if necessary. As we have seen, while many victims will be cooperative, almost any victim is capable of causing a rescuer a great deal of trouble. A rescuer can approach a victim from any direction. If he is swimming aggressively, especially with fins, he should be able to choose the time and place of the rescue. Tactically, the best place to do this is usually a calm spot below a rapid. Swim fins are a great help when swimming with a victim, since one or both arms may be engaged by the victim.

The rescuer's goal is to get the victim into the proper position for rescue swimming: facing away from the rescuer and on his downstream side. The rescuer approaches the victim, grabs the opposite shoulder of his PFD, and spins him around so that he is facing away from him. From there he can use a cross-chest carry and begin swimming towards shore with a side stroke. The process

of doing this is called "reverse and ready." The rescuer should also be making verbal contact with the victim, giving reassurance and explaining how the victim can assist. With a cooperative or counterpanicked victim, this process will often be quick and smooth. During the rescue, the victim always stays on the downstream side of the rescuer, even if going into an obstacle. This fits in with the general philosophy that the safety of the rescuer comes first.

A panicked or uncooperative victim is much more of a challenge. If possible, the rescuer should approach from behind and quickly clamp the victim to him while offering calming words. Even if the victim sees the rescuer, he may still be able to reach

Fig. 9.6 This rescuer has approached a struggling victim from behind. This allows him to establish control quickly.

out and grab his PFD and quickly spin him around. If the victim sees the rescuer and lunges for him, the rescuer should roll backwards and attempt to stay out of reach. He should continue to talk to the victim and try to calm him down and instruct him on how to assist. If the victim continues to come toward him, the rescuer can kick water in his face (this is very effective with swim fins) or, if this fails, put his feet on the victim's chest and push him away.

If the victim does succeed in grabbing the rescuer, the rescuer may have to break his hold. One method is to grab the victim's thumb and push it backwards to loosen his grip. If the victim's hand does release, the rescuer may be able to quickly spin him around backwards. Another method is for the rescuer to tuck in his chin and grab the victim's elbows. He then digs his thumbs as hard as possible into the interior aspect of the victim's elbow joints and pushes up. While doing this, the rescuer should be attempting to calm the victim and get him to cooperate.

Rescuers may sometimes be presented with a victim without a life jacket. If the situation permits, the victim should always be provided with some type of flotation device. A good PFD is best, but is difficult to get on someone while in the water. A rescue tube, developed for use by lifeguards, may be the best solution. The rescue tube is a flexible foam float that can be quickly wrapped around a victim's trunk and snapped into place, even in turbulent water. While it does not work as well as a PFD, it is certainly better than nothing and greatly simplifies a rescuer's task.

Live bait (tethered swimmer). Free-swimming rescues have some obvious drawbacks. The biggest is that any victim, no matter how cooperative, is a serious hindrance to a swimmer. A rescuer must use at least one arm to assist the victim, and there is always the danger that rescuer and victim will go downstream into a worse danger. We can control these problems to some extent by tethering the swimmer, either from shore, from a boat, or even a helicopter.

Fig. 9.7 Rescue swimming

A conscious victim can often be towed by his PFD or clothing, especially if the rescuer is wearing fins.

An effective technique for discouraging a panicked victim from grabbing you is to roll back and kick water in his face. Failing this, you can push him away with your feet on his chest.

Fig. 9.8 Live bait rescue.

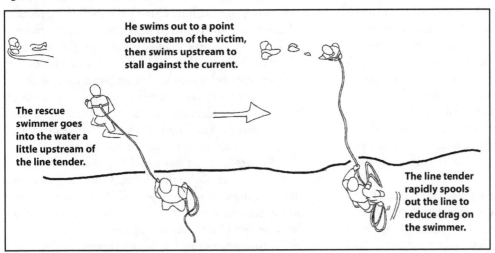

He swims out to a point downstream of the victim, then swims upstream to stall against the current.

The rescue swimmer goes into the water a little upstream of the line tender.

The line tender rapidly spools out the line to reduce drag on the swimmer.

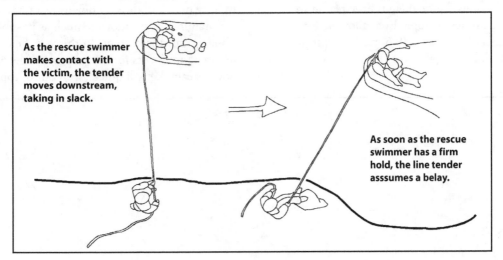

As the rescue swimmer makes contact with the victim, the tender moves downstream, taking in slack.

As soon as the rescue swimmer has a firm hold, the line tender asssumes a belay.

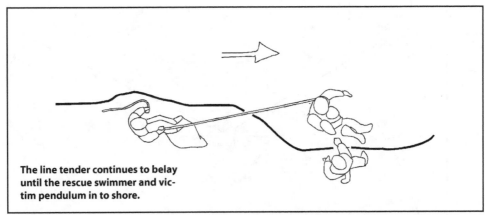

The line tender continues to belay until the rescue swimmer and victim pendulum in to shore.

A tether gives us firm control of the swimmer and, when the swimmer makes contact with the victim, of both. "Live bait" (a term borrowed from our Russian friends) is one of the few systems that can effectively rescue unconscious, drowning, exhausted, or hypothermic victims. On these and panicked victims, the rescuer can use both hands. Once in contact, he need not swim, but is reeled rapidly in to shore by his tenders.

However, live bait has disadvantages, too. The tethering line can snag on an obstacle like a midstream rock. Any line in the water is subject to the same limitations we saw in line ferries—it puts a drag on the swimmer and has a finite length, making it suitable only for close-range rescues. For planning purposes, live bait should be considered the equivalent of a throw bag, except that a live rescuer is on the end of the rope.

A live bait swimmer should have the same qualifications and basic equipment as any other rescue swimmer. However, they *must* be equipped with a quick-release rescue harness (either a separate harness or one built into the rescue jacket) as described in Chapter 2. The rescuer clips his harness to the tether line, either directly or by means of a cow's tail. In any case, the rescuer must be able to get free of the tether quickly if something goes wrong. If the swimmer clips directly to the harness, he should use a locking carabiner to avoid inadvertently clipping himself to one of the harness loops.

To reduce drag, the line should be the smallest size that can reasonably be expected to do the job. Since the release system is only rated to 730 lbf/3.25 kN and the rescuer's body will only stand about half that without serious injury, there is little reason to use large diameter rescue rope. Even allowing for a substantial safety factor, a ⅜" (9.5 mm) polypropylene rope will suffice.

Any live bait swimmer must have one or more *tenders* to manage the line. During the swimming phase of the rescue, the tender(s) flake the rope out rapidly so as to cause as little drag as possible on the swimmer. Once the swimmer has the victim (or if he runs out of line), they belay him so that he pendulums back to shore. In order to give the swimmer as much line as possible, the tenders may wade out in the water if conditions permit. The length of the line will depend on conditions, but will normally be 70–150' (21–46 m).

The rescue swimmer should be initially positioned downstream of the victim. Once he gets a visual fix on the victim, his object will be to intercept him in the current. Since he has only a limited amount of line, he should start slightly upstream of the tenders and swim aggressively out in the current, attempting to ferry out until directly downstream of the victim, maintaining his position as much as possible by swimming against the current. In this way, he can gain a few extra moments for interception. If he gets very far downstream, he will quickly run out of rope.

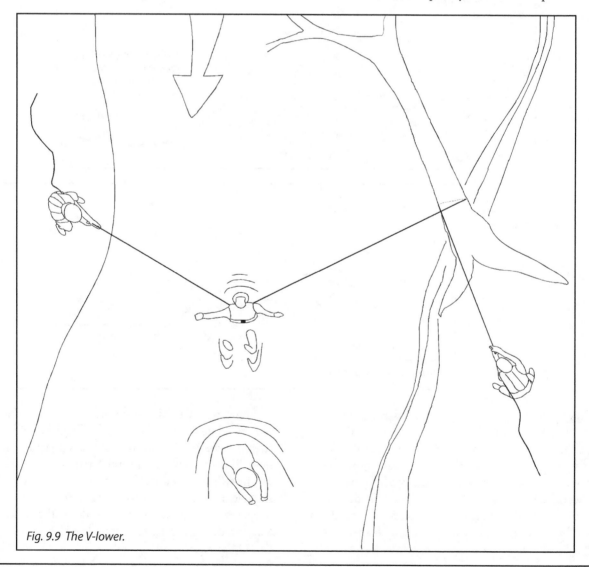

Fig. 9.9 The V-lower.

When selecting the site for a live bait rescue, rescuers should look for an open site with no obstacles (trees, boulders, etc.) that might snag the tether. They should also look for a site where the current might bring the victim closer to shore, or one where the rescuers might get farther out in the river, such as on an island. Live bait can also be used from a boat, although care must be taken with the tether to avoid fouling on the boat, especially in the propeller. Just as important, they should consider what is below. While a tether can be used to get the swimmer back in quickly, the rescuers must also consider what will happen if the swimmer has to release. Don't set this system up just above a major hazard, where the rescuer would have no chance of swimming back in.

V-lower. Live bait is a dynamic system meant to intercept a swimmer in the current. For a victim fixed in position, however, the V-lower provides a way to lower a rescuer down to him. It is a variation of the floating tag line, except that the float has been replaced with a rescuer. The rescuers attach tag lines from the D-ring of the rescuer's PFD (which has a quick-release system as described previously) to each shore. A tender team on shore belays each line. The type of belay required will depend on the swiftness of the current, but a rescuer causes a lot more resistance than a float.

In practice, this system works best in shallow water, where it helps the rescuer wade slowly downstream. It is a good way to get a rescuer down to a foot entrapment victim. Even getting the rescuer upstream of him may break the force of the current enough to allow the victim to be freed. Rescuers should be very cautious when lowering a rescuer down to the upstream side of any accident site that might pin him if he has to release the tethering rope. While the V-lower can be used in water too deep for the rescuer to stand, it becomes harder to operate and less effective as the speed of the current increases. Eventually, the force on the tenders and especially the rescuer will become too great.

A variation of the V-lower is the *one-line system*, which can be done with no special equipment other than a rope. First, the rescuers stretch a line directly across the river at the surface (this is an exception to the rule of always putting a line across at an

angle). The rope, which should be roughly twice the length of the river's width, must be firmly belayed on either side with a method that allows a controlled release (e.g., Münter hitch, figure 8). Next, a rescuer slides down to midcurrent on the *upstream* side of the rope, holding it at belly level. The pressure of the current will keep him in place against the rope, but he can escape at any time by simply somersaulting under the rope. Then, the belayers on shore gradually slacken the rope, allowing the rescuer to move downstream. As he does so, the "V" shape of the rope makes him more secure. While this method is quick and simple, it places great stress on the rescuer's body and, consequently, can only be used in shallow water or slower currents (see figure 11.10).

Conclusion

In-water contact rescues offer a quick and simple way to make a rescue, if the rescuer has the training, skill, and equipment to do it. However, these rescues are dangerous to the rescuer and require an exceptionally high level of training and fitness.

"Now, Grog! Throw! . . . Throoooooow! . . . Throw throw throw throw throw throw! . . .

Fig. 9.10 Live bait rescue - original version.

Helicopters
Chapter 10

Fig. 10.0 US Coast Guard rescue swimmer Shawn Lesko fits a rescue strop on a driver trapped on his car in the Smith River, CA. He was hoisted to safety by a Coast Guard HH-65C helicopter.

General Considerations

Why helicopters? It's a question that should probably be asked more often than it is. As rescue tools they have many advantages and some disadvantages as well.

The advantages are:

- Speed. Helicopters can drastically cut the time required for both deployment of rescuers and patient evacuation. Under central control they allow speedy concentration of resources to trouble spots.
- Mobility. Helicopters are much less constrained by terrain than boats or ground vehicles. They can access victims on wide rivers, low-head dams, and other places that conventional vehicles cannot. They can be removed from danger areas for disasters like hurricanes and then speedily redeployed after it, even over flooded areas.
- Visibility. A helicopter can dramatically expand the vision of a rescue team, especially when trying to locate rapidly moving victims on a river, or managing diverse resources in a complex incident.

Against these very real advantages must be weighed the disadvantages:

- Physical limitations. Both temperature and altitude reduce a helicopter's efficiency and lift capacity. Mechanical failure, either outright during flight or as a limiting factor in aircraft availability, is also a factor.
- Lack of visibility during darkness and bad weather. While FLIR, illumination devices (e.g. NightSun), night vision goggles and other technological advances have made darkness less of an issue, bad weather is still a limiting factor for any aircraft.
- Communication problems with the on-scene rescuers, who may not be able to talk directly to the helicopter. Aircraft from different agencies and services may not be able to talk to each other.

In addition, though a helicopter may be fast in flight, it may be slow to scramble, the rotor wash over a water rescue can make hypothermia problems dramatically worse, and a crash on the accident site would be catastrophic. Yet there are a lot of helicopters out there.

In 2011, public safety agencies in the U.S. had over 600, with the Los Angeles County Fire Department alone having 8 helicopters and the city fire department 5.

Pilot training is vital. A European study of helicopter accidents in alpine rescue accidents (IKAR, 1991) indicated that only a relatively small percentage of mishaps (2.7% engine problems, 17.3% not engine related) were due to mechanical failure. The vast majority (80%) were due to pilot error, often related to inadequate training. Among the special issues that need to be addressed in swiftwater pilot training are vertical reference flying for short-hauls, vertigo problems caused by hovering over moving water, and the effect of moving water on loads and hoist equipment.

Another issue for both pilots and crew is preparation for ditching in the water, a skill not always considered important in the continental interior. Yet ditching is a distinct possibility for any helicopter engaged in swiftwater rescue operations. Crews who expect to engage in it should have appropriate water personal protective equipment (see section on PPE below) *and* "dunker" training, since helicopters have a nasty habit of inverting when they hit the water.

In the last edition we noted that few helicopter crews were trained for swiftwater rescue. This has changed very much for the better, and there are now large areas of the US (e.g. Texas, Southern California, North Carolina) and other countries that have helicopter rescue units well trained and equipped for

Fig. 10.1 "Dunker" training is essential for aircrew members expecting to undertake swiftwater rescues. This PIG (Portable Immersion Gadget) is made from PVC pipe.

swiftwater rescue, and some federal agencies like the US Coast Guard have added it to their skill sets as well. This has led to a rethinking of the priority of helicopter employment as a "high risk" option, particularly after the rescue of several "lower risk" boat crews by helicopters. Incident commanders need to consider the relative qualifications of each rescue unit, whether they are using boats or helicopters, and assess the risk accordingly. A well-trained helicopter crew can make lower risk rescues than a less well trained boat crew (and vice-versa). Time, however, is always a factor in swiftwater rescues, and if they can be begun safely they should not be put off "until the chopper (or the boat) gets here."

Still, the best practice remains helicopter conservatism: it is just another piece of gear with advantages and disadvantages like anything else. Tim Setnicka cautioned many years ago in *Wilderness Search and Rescue* (Appalachian Mountain Club 1980) that "helicopter use is not the panacea for SAR problems. We must guard against becoming 'helicopter junkies,' always depending on a helicopter to perform and support our operations."

- **Helicopter-*supported* rescues.** The heli-copter moves personnel and equipment to the rescue site; assists at the rescue site by performing tasks like line or equipment ferries, sling loading, medical evacuation; or acts as a mobile command post or spotter. The helicopter operates out of a conventional landing zone (LZ) or helispot, observing normal aviation safety procedures. Equipment and rescuers go from the LZ to the rescue site and evacuate casualties from the accident site to the LZ. In good weather conditions this poses the least risk for both aircraft and rescuers.

- **Helicopter-*based* rescues.** Here, the helicopter either deploys rescuers or evacuates casualties directly at the accident site, or in any case operates from a hover or without a prepared LZ. This includes techniques like tethered rescuers, single-ski landings, helicopter rappeling, helicasting, short-hauls, hoist operations, and the like. These greatly increase the risk level—from pilot error, mechanical malfunction, improper rigging, and other causes.

Air ambulance services routinely transport thousands of patients every year without problems. However, in SAR operations, the benefit to the patient must be weighed against the risks of air transport. A helicopter evacuation from a prepared LZ in good weather may present little more risk than a ground evacuation. On the other hand, risk factors

Swiftwater Rescue

in remote areas and in bad weather can accumulate that might limit or rule out the use of helicopters. Unfortunately, there have been instances of lightly injured patients who have been injured or even killed during attempted on-site helicopter evacuations or rescues.

When considering an air evacuation or rescue, especially using a high-risk option, first consider the following:

- **Weather and visibility.** Darkness, high winds (>30 knots), a ceiling of less than 500' (150 m), and a visibility of less than one-half mile (.8 km) all greatly increase the risk factor in helicopter operations.
- **Patient condition.** Is this a rescue, an evacuation, or a body recovery? A recovery or a nonserious injury does not normally justify a high-risk rescue or evacuation. A "serious" injury is usually considered to be a patient with one or more of the following:
  - ▲ Significantly reduced Glascow Coma score
  - ▲ Airway problems
  - ▲ Respiratory distress
  - ▲ Circulatory problems (low blood pressure, symptoms of shock, no radial pulse, etc.)
- **Rescue/evacuation options.** Rescuers should first consider options other than helicopters. Often even a seriously injured patient can be evacuated a short distance to a safer LZ. Even a long and arduous conventional evacuation might be preferable to a high-risk one. Will a longer duration evacuation compromise the patient's condition?
- **Rescuer skills.** Have the aircrew and the ground rescuers been trained in the types of rescue being contemplated? Signals? Rigging? Safety procedures? On-the-spot improvisations greatly increase the risk of a mishap.

In any case, the final decision rests with the incident commander, based on the totality of circumstances. Unfortunately, there have been instances when helicopters have been employed as the rescue method of last resort, under great pressure, without adequate communications or crew training.

Personnel Safety

Any rescuers whose duties will take them either on or around helicopters should be aware of basic aviation safety procedures. The most basic rule of helicopter safety is to *stay away from the rotors*. Most common helicopters

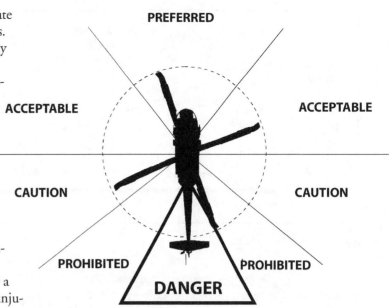

Fig. 10.2 When approaching a helicopter, keep the following zones in mind. Ground personnel should wait until the aircrew indicates that it is safe to approach.

have two rotors: a large one on top of the fuselage and smaller one at the tail. Some newer models ("NOTARs") have eliminated the tail rotor by directing the turbine's exhaust through the tail boom. When operating, the rotors can be difficult to see, and getting hit by either will cause death or serious injury. *Do not approach a helicopter from the rear*; the best direction of approach is from the front where the pilot(s) can see you. If possible, do not approach from the uphill side, since this puts you closer to the blade arc. Be aware also that, as the pilot cuts the power after landing, the main blade will dip lower. In any case, it is necessary to duck as you pass the blade arc on any operating helicopter.

Incident commanders and safety officers should ensure personnel stay clear during approach and don't go rushing to the helicopter immediately after landing. Instead, ground personnel should follow the

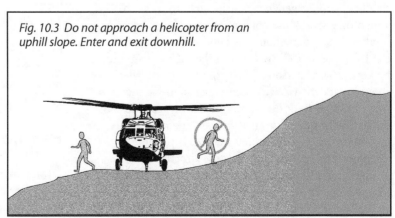

Fig. 10.3 Do not approach a helicopter from an uphill slope. Enter and exit downhill.

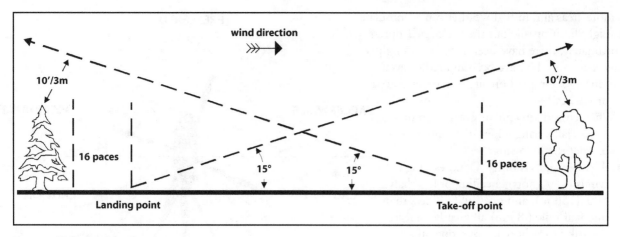

Fig. 10.4 Helispot clearances

instructions of the crew—cautiously approach the aircraft from the front, and do not come within the blade arc until directed to do so. If the situation permits, the safest way of all is to wait until the engine shuts down altogether and the blades are still.

Landing Zones

Although helicopters can take off and land vertically, it is much easier and safer for them to take off and land into the wind like an airplane. A vertical (maximum power) takeoff or landing will get a helicopter in and out of an abbreviated LZ. However, this significantly reduces the safety margin and should be reserved for emergencies.

The best location for an LZ is usually on top of a ridge or hill. For normal operations, the long axis of the *landing zone* (LZ, also called a *helispot* or *helipad*) should be oriented up and downwind. The exact clearance dimensions are dictated by the size and type of helicopter, but, in general, the end of the LZ should be at least 150' (46 m) from the nearest 50' (15 m) high object in the flight path. There should be nothing projecting on the LZ more than 12" (30 cm) from the ground.

Secure any loose items on or near the LZ. Even small helicopters generate considerable rotor wash, which may send loose objects into the blades or jet intakes. Bystanders should be kept well away from the LZ (at least 200'/61 m), and the LZ itself should be kept clear of personnel and equipment. Rescuers should mark the LZ with something solid—either heavy enough to not blow away or firmly staked down. Night operations may require chemical lights or other methods of marking the LZ. If possible, establish radio contact between the aircraft and ground crews. If visibility is marginal or the rescuer's location not known with precision, the ground personnel may have to signal the aircraft with mirrors

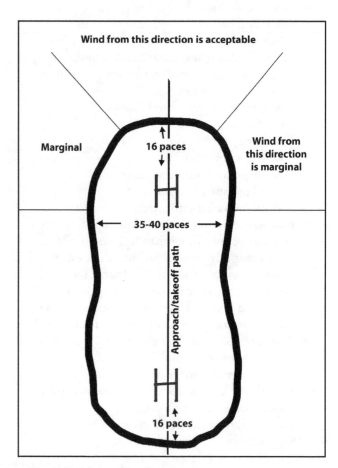

Fig. 10.5 Helicopter landing zone

or pyrotechnics. Ground crews should also mark the wind direction for incoming aircraft. This can be done by radio or hand signals (see appendix), smoke, ground signal panels, or even brightly colored clothing.

Helicopter Performance

Like any other aircraft, helicopters are sensitive to certain environmental factors, the most important of which are those that affect air density: temperature, atmospheric pressure, and humidity. On hot days or at higher altitudes, the air is less dense and the helicopter's performance suffers accordingly. Some helos perform better under these conditions than others. All helicopters must perform load calculations for different density altitude conditions.

One of the essential characteristics of a helicopter is its ability to hover. If it is close (one-half its rotor diameter or less) to the ground, a cushion of air forms beneath it that aids the hover. This is called *ground effect* and the hover is called HIGE, or Hover In Ground Effect. This effect reduces power requirements (and increases load capacity) for a hover. Large obstacles like cliffs, rough terrain, and moving water can upset air circulation and reduce the ground effect cushion. As the helo gains altitude, this effect dissipates and results in HOGE, or Hover Out of Ground Effect. The helicopter is now totally power-dependent for a hover.

In case of engine failure, a helicopter can in some cases *autorotate* back to the ground by allowing the rotor to freewheel. The pressure of the air will continue to turn the rotor blades, allowing for a reasonably safe landing. However, for an effective autorotation most helicopters need an altitude of at least 300' (91 m) and some forward speed. Unfortunately, most rescue operations described in this section take place at altitudes lower than that, which means that in case of mechanical failure, the landing is apt to be rather abrupt.

Helicopter Types

While the rescuer on the ground need not have detailed knowledge of helicopters, it is important to have a general working knowledge of their size and capacity. While it is far beyond the scope of this book to examine the many models of helicopters in detail, we can, following U.S. Forest Service and Department of Interior practice, divide them into several broad categories and show examples of each. These categories (called types) are based on size, seating capacity, and payload, and provide a rough but convenient planning yardstick. The load figures listed here are for *planning and illustration purposes only* and are not intended to substitute for required load calculations. There are significant differences in the actual sizes and capabilities of helicopters within each type. Anyone contemplating working with a certain type of helicopter should thoroughly check its characteristics before doing so.

Some commonly used helicopters for search and rescue are described in Appendix 1.

TWO-WAY RIDGE HELISPOT

Fig. 10.6

Prevailing wind

Safety circle

300'/100m

300'/100m

No tree or other object should project above this line

Pad

20%

20%

Type 1 Helicopters
Safety circle 110 ft.
Touchdown pad 30 x 30 ft.

Type 2 Helicopters
Safety circle 90 ft.
Touchdown pad 20 x 30 ft.

Type 3 Helicopters
Safety circle 75 ft.
Touchdown pad 15 x 15 ft.

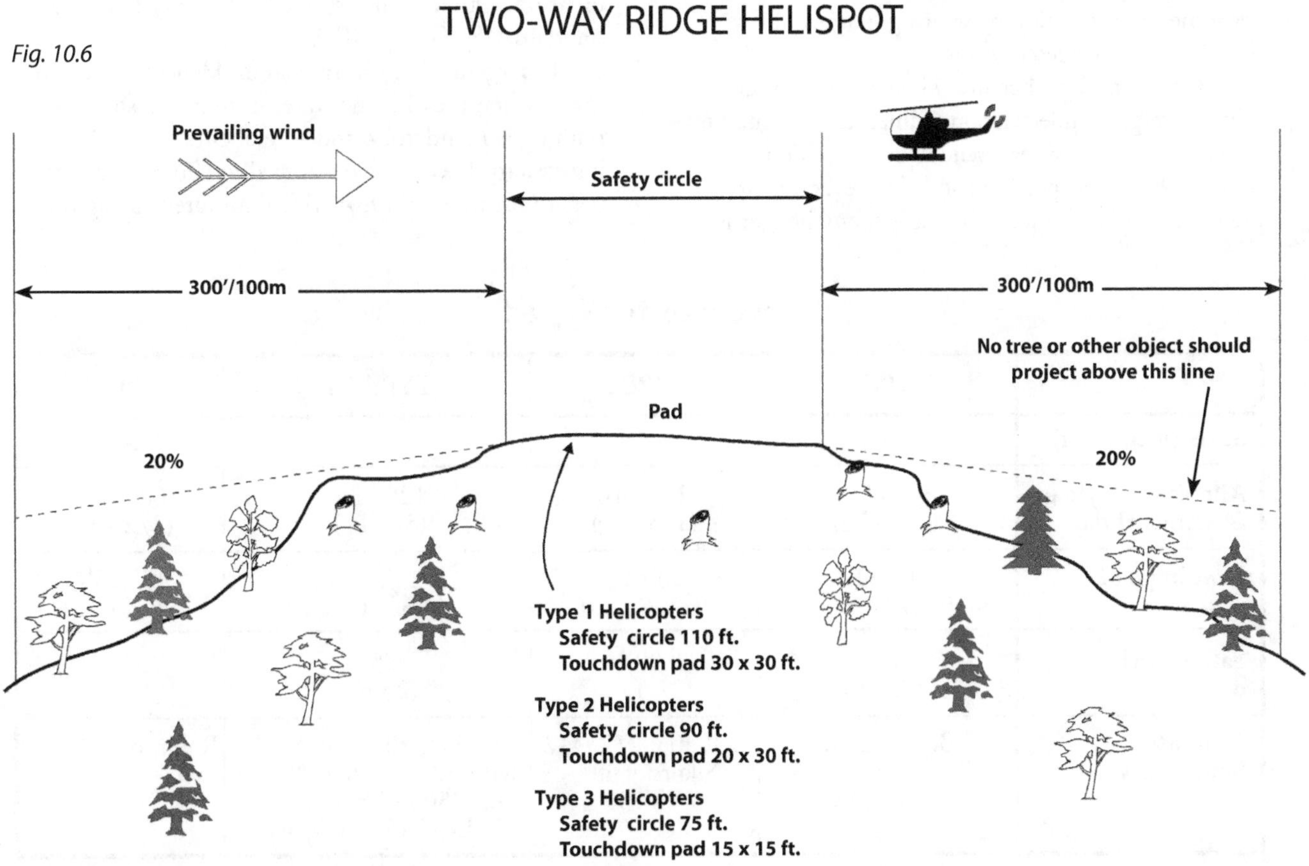

Aviation Personal Protective Equipment

The requirements for aviation and swiftwater PPE differ substantially and sometimes conflict. Aviation PPE seeks to minimize the effects of fire and high-speed impact, both of which are of less concern in swiftwater situations. However, water rescuers whose duties take them into helicopters ignore these dangers, particularly fire, at their great peril. Earlier we recommended synthetic pile garments of nylon or polypropylene for water rescuers. However, if exposed to flame or heat of over 300° F, petroleum-based synthetic fibers (nylon, polyester, polypropylene, plastic, etc.) will burn and melt, sticking to the skin and causing severe burns, even if they are worn as undergarments. Swiftwater rescuers, then, must sometimes make an uncomfortable choice as to which environment they wish to be prepared for. The risk of burns may be reduced *somewhat* by substituting wool for synthetic pile. Wool will burn if left in contact with an open flame, but will not melt and stick to the skin, and its insulative ability is comparable to synthetics. Recently woven garments made from Nomex® have been developed that offer the insulating benefits of synthetics while being fireproof, although they remain expensive. Similar considerations apply for foot protection. Aviation authorities recommend leather boots but this will not always be possible for water rescuers, who may get some fire protection by wearing Nomex® or wool socks under water rescue footgear.

Helmets are another area where the two disciplines diverge. While aviation helmets are designed to take up to six Gs, water rescue helmets are designed only to blunt low-speed impacts. However, due to the prevalence of head injuries in crashes, any helmet is beneficial. As a rule, all personnel engaged in flight operations should wear a helmet of some kind.

Normal aviation PPE consists of:

Aviator's protective helmet (flight helmet). Provides protection for the head, ears, and temples. A floating suspension ensures protection from high speed impacts. It also contains communications gear for the pilot and crew members. Examples are the Gentex SPH-5 and HGU-56/P, and similar models.

Fire-resistant clothing. The most common name for this polyamid or aramid clothing is Nomex®. Unlike other synthetic fibers, Nomex® does not melt and stick to the skin or support combustion. It will resist temperatures of up to 700° F (371° C). The most common form of garment is a coverall or flight suit. A Nomex® hood adds protection for the head.

Leather boots. These should extend above the ankles so that the Nomex® flight suit can be drawn down over them.

Gloves. These should be either Nomex®, leather, or a combination of the two.

Personal flotation device. Current flight regulations require that an FAA or U.S. Coast Guard-approved personal flotation device be worn for all flights over water beyond power-off gliding distance to shore. Aircrews sometimes prefer inflatable PFDs, since they are less bulky while inside and, in the event of a crash, easier to get outside of the fuselage before inflation. Other agencies, like LA County Fire, use standard USCG Type III/Vs.

Emergency breathing system. Helicopters often invert when they hit the water during a crash. Dunker training is mandatory, and several emergency breathing systems have been developed to allow crew members to breathe for a few critical minutes during the

Helicopter Types

	TYPE I	TYPE II	TYPE III	TYPE IV
Seats (incl. pilot)	15+	9-14	4-8	2-3
Allowable payload @ standard day	5000 lbs (2268 kg)	2500 lbs (1134 kg)	1200 lbs (544 kg)	600 lbs (272 kg)
Landing pad	30'x30' (9.1x9.1 m)	20'x20' (6.1x6.1 m)	15'x15' (4.6x4.6 m)	15'x15' (4.6x4.6 m)
Safety circle diameter	100' (30.5 m)	90' (27.5 m)	75' (23 m)	75' (23 m)
Example helicopters	Boeing 234, Sikorsky 61 & 64, UH-60, S92, EC-225	Bell 205, 212, 412 Sikorsky 58T AW139	Bell 206L, 407 MD-500 ser., 600 & 900 AS-350 series, EC-135 EC-145, HH-65	Robinson 22, 44

egress of a ditched aircraft. These include basic systems like Spare Air® and HEED™ to more elaborate ones like the HABD and SBA systems. All require prior training and presence of mind for effective use.

Extrication Devices

Since over-water operations, especially on swiftwater, usually prohibit landing to load casualties, various solutions have been devised to either lift victims into the helicopter or to fly them along underneath it on a fixed line to a safer location. While the methods of doing this will be discussed in more detail later in the chapter, some of the equipment is worth looking at here.

Fig. 10.7 Horsecollar Sling

Sling (horsecollar). The horsecollar is a simple padded sling. The individual being hoisted slides it over his head and shoulders and fits it under his armpits. He then crosses his arms over the sling and is lifted to safety. While some horsecollars have an "idiot strap" to keep the victim from falling out, this method is best used for experienced,

uninjured people. It is the least secure method for lifting and should only be used if other, safer alternatives are not available.

Water basket. The water basket is a tubular steel basket with a side rail and flotation. In practice, the basket is lowered into the water, the victim climbs into it and is hoisted to safety. It is used mainly by the Coast Guard for open-water rescues. This method works well for inexperienced and lightly injured personnel, but those with severe injuries should be evacuated in a litter. In moving water large objects like baskets risk being caught by the current.

Forest or jungle penetrator. This device was developed to get a rescue device down to pilots through the dense jungle canopies of Southeast Asia. It has a narrow, pointed steel body about five feet (1.5 m) long that is heavy enough to smash down through forest canopies. To use it, the person being rescued fastens a sling around himself under his armpits, and folds down a seat, sits on it and is lifted to safety. MAST (Military Assistance to Safety and Traffic) helicopters often use this device, which is sometimes deployed on a hoist with a rescuer already on it. As the San Marcos incident detailed at the beginning of this book shows, it is not the most secure device available and should not be used for swiftwater operations.

Fig. 10.8 Helicopter extrication devices. Quick Strop (front left), Cinch Collar (front right), Jungle Penetrator (left center), Water Basket (rear).

Fig. 10.10 Screamer Suit

Fig. 10.9 Billy Pugh Net

Fig. 10.11 The Bauman Bag (right) is a simple, secure method of securing a patient for aerial evacuation. Spinal injury patients require the addition of a backboard or other rigid stabilization device.

Billy Pugh Net. The Billy Pugh Company has developed helicopter nets of various sizes (2–10 persons) for personnel transfer, light cargo use, and rescue of mobile and immobile persons in the water, from oil rigs, and from the tops of buildings. These nets are made in several sizes, with and without floats. All have one side open, allowing the helicopter to scoop up the victim. However, the open side and the lack of a positive security device for the victim also raises the possibility that he may fall out. While these nets seem to work well on open water, they are not suitable for swiftwater use.

Screamer suits. This device is an body suit with the arms and legs removed and a lifting harness incorporated. It is secure, quick to don and fits almost anyone. Once donned, it requires no knowledge, participation, or experience on the part of the victim, and is a very secure way to lift an uninjured or lightly injured victim. The disadvantages are that it requires a knowledgeable person to put it on, is difficult to put on while in the water, and it is difficult to get over a PFD.

Evacuation harnesses. Unlike a rescuer's harness, an evacuation harness serves only to safely cradle a casualty for evacuation. Most, like the Petzl Evacuation Triangle, operate like an oversize diaper and are thus easy to secure over a victim wearing a PFD. They are secure and have the advantage of being relatively victim (and rescuer) proof, but are not suitable for spinal injury patients.

Soft bags. The *Bauman Bag* is a soft stretcher bag supported by a single overhead rigging point. It is quite secure for patients and is used primarily for evacuating injured personnel. Compared to the traditional Stokes litter, the Bauman Bag is lighter and easier to store and rig. It is also less likely that rescuers will make mistakes in patient packaging when using it (there have been cases of patients falling out of Stokes litters during flight operations because the rescuers incorrectly packaged them). While the Bauman Bag does not, by itself, have the necessary rigidity for spinal protection, this can be added by first putting the patient on a rigid backboard such as the Ferno-Washington scoop stretcher, then putting him in the Bauman Bag.

Fig. 10.12 The Petzl Evacuation Triangle fits the casualty like a large diaper.

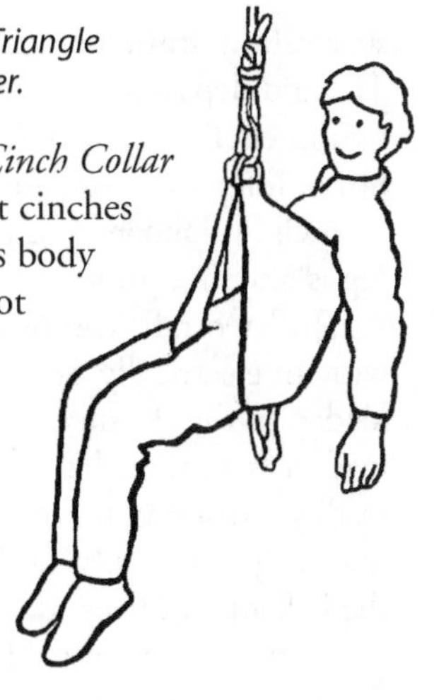

Cinching devices. The *Cinch Collar* resembles the horse collar but cinches securely around the casualty's body when loaded. Since it does not have a hard ring float, it is suitable for situations where less flotation is needed or where a hard float might injure the victim. The Cinch Collar is secure but uncomfortable and suitable only for short distances. It is easy to store and transport, and is very widely used.

Sky Hook. The *Sky Hook Rescue System*™ is a large, locking aluminum hook attached by a short tether to a rescuer's or victim's harness. The helicopter then lowers a "capture ball" with a float and two stainless-steel rings set at right angles to each other. The rescuer clips the hook on one of the rings, and then is lifted to safety.

Spider Rig. CMC Rescue and Star Flight of Austin, Texas teamed up for the Spider Rig, which allows up to four fully-equipped firefighters to be inserted and as many as six victims to be evacuated simultaneously. In a recent exercise Star Flight moved thirty victims in thirty minutes as opposed to seven during the same time using previous methods. Quick to rig and easy to store and operated, the rig requires only one trained operator on the LZ and one at the evacuation site. It is similar to the MERS system developed in Europe for ski lift and tram evacuation.

Helicopter Support

Helicopters can be invaluable for supporting rescue operations. If conditions permit, they are ideal for river searches. Often, the victim is moving downstream and is difficult to track with ground-based methods. A helicopter can keep track of the victim while other resources are being mobilized. Helos can also perform a quick sweep up and down the river to locate missing boaters. This is especially effective in areas of rough terrain. If adequately equipped with communications gear, a helicopter-based forward command post can be a great asset in a fast-moving situation.

If an LZ can be set up reasonably near the accident site, the helicopter can ferry in personnel and equipment, which can then be staged to the accident site. In rough terrain, this can save a great deal of time. When the victim is rescued, he can then be

Fig. 10.13 The Sky Hook™ system in use in a flood channel.

Fig. 10.14 The Spider Rig allows up to four personnel to be lifted simultaneously.

evacuated by helicopter, which can drastically cut evacuation time, with corresponding benefit to the patient. Since the helicopter lands and takes off normally, this is all relatively safe.

A helicopter can also sling load gear into tight LZs, even dropping it in if necessary. Since this is not a live load, the crew has considerable flexibility in jettisoning the load if necessary.

Line ferries. Another support activity is ferrying lines, as well as personnel and equipment, across the river as part of the rescue setup. Under favorable conditions this can be much faster than doing it with a boat, especially on wide rivers. It is relatively low-risk. Most of the line ferry techniques already covered for boats apply here also, such as using a lighter messenger line first. A major advantage of using a helicopter for line ferries is that it makes it much easier to keep the line out of the water.

Nevertheless, there can be problems. Helicopters are even more sensitive than boats to the drag of a line in the water. Therefore, the helo should be lightly loaded so as to have maximum power available. The rope must be attached to the helicopter in such a way that it will not interfere with any vital part of the aircraft, (e.g., the tail rotor), and so that it can be released instantly if the situation calls for it. As with a boat, there should be a rope team on the opposite shore to receive the rope and belay it.

Helicopter Deployment and On-Site Evacuation

Helicopters can be used to deploy rescuers directly into the water or onto the accident site. There are several methods of doing this, and while they vary in degree, all are more dangerous than a normal landing on a prepared LZ. It is difficult to list these methods in order of danger since so much depends on pilot and crew training and the type of helicopter and equipment used.

Rescue swimmers and watercraft. Rescue swimmers can be deployed directly from a helicopter by *helicasting*. The fully-equipped swimmers normally perch on the skids in pairs on both sides of the aircraft. The helicopter approaches low over the water and, at a signal, the swimmers drop into the water in spaced pairs from either side of the aircraft. The aircraft normally follows the *10 and 10* rule, meaning that the helicopter is 10 feet above the water, moving at no more than 10 knots airspeed (3 m x 18.5 km/hr). Most agencies use this technique only for lakes, ocean, and deep, slow-moving rivers.

Helicopters can also deploy personal watercraft and even rafts. The Indiana River Rescue School has successfully stuffed a personal watercraft inside a Bell 412 and deployed it in much the same way as a rescue swimmer. The watercraft is pushed out from a low hover, followed moments later by the crewman, who then climbs aboard and starts the engine. Sling loading is also an option.

Inflated rafts can be attached to the helicopter with an electrically-operated cargo hook and sling-loaded to the rescue location. Since this is not a live load, it is a somewhat safer option since it can be readily jettisoned if necessary. The raft is attached so that it flies vertically, bow up. Using a crew spotter, the helicopter lowers the raft into the water so that it enters the water right side up. The crew rides inside the aircraft and are deployed like rescue swimmers. They then enter the raft, unhook the boat if necessary, and begin operations. Sometimes a crewman rides attached to the cargo line, standing on the bow of the boat. While more dangerous, this does make for a quicker operation, since a crewman stays with the boat. Putting a raft in moving water like this is a difficult operation. If the rescuers start moving downstream, the helicopter has to match its speed to keep them near the raft.

Because of the danger of injury, these techniques have fallen out of favor in recent years as there are usually other, safer ways to insert rescuers. For example, a less risky option would be to set the raft down on shore (or in an eddy) and let the crew unload on solid ground, then put the raft in the water. Another option is to carry the raft uninflated. The helicopter then sets the rescuers down on the river bank, where they inflate it using an air bottle or similar device.

Single-skid landings. A skillful pilot can put one skid or wheel on an object and hold it there in a *single-skid landing*. Personnel can then enter and exit the aircraft. The pilot must be able to hold the hover despite the inevitable weight shifts this practice entails. Opinions vary as to the risk level of this maneuver. It is quick, simple, and requires no rigging or special equipment, nor does it require much skill on the part of the rescuers (other than the pilot). The helicopter is operating in a ground effect hover and so requires less power. On the other hand, there is very little margin for error, and any failure is likely to be catastrophic. This technique is also heavily dependent on factors like altitude and cross winds, not to mention pilot skill.

A variation of this technique is to have the helicopter put its skids in the water and have the victim climb aboard the aircraft, assisted by the crew. *This technique only works in still water or very slow-moving currents.* In swift current the water may catch the aircraft's skid or wheel, unbalancing it and causing a

Swiftwater Rescue

crash. Aircrews, like everyone else, have been caught off guard by the unexpected force of moving water.

Helicopter rappelling and fast roping. Rescuers can also get to the ground from a hovering helicopter by fast roping or rappelling. Both techniques are fast but dangerous. As such, they appeal primarily to the military, to whom casualties are more acceptable than to civilian organizations. However, they do offer a way to quickly insert (but not extract) personnel from a hover into areas where a helicopter cannot land.

Fast roping is extremely simple—the roper slides down a large-diameter rope from the helicopter with a pair of heavy gloves. Since he has no contact with the rope except his hands, a fall generally means injury or death.

Helicopter rappelling is somewhat safer, although still a high-risk option. The rappel itself is done much the same as normal practice. Using a friction device (e.g., a figure eight for simplicity, Sky Genie, etc.), the rescuer rappels to the ground from a rope attached to the helicopter. While this may sound simple, it is not. First, it is impractical to use a backup belay for helicopter rappels. Second, helicopters vary widely as to their suitability for rappelling. Some have wheels instead of skids, which complicates exiting the aircraft, and some of the smaller models (e.g. MD 500 series) require the rappeller to rappel from inside rather than outside the skid to avoid upsetting the aircraft's center of gravity. The locations of anchor points for the rappel rope also vary widely. Even with a stable aircraft with skids (e.g., 212/412 Bell 205 series), the rappeller must invert almost completely as he exits in order to avoid smashing his face against the skid. Third, the aircrew must be trained for the operation, and communications maintained between the rappeller and them. Last, there are the usual problems with any rappel, such as getting one's hair or clothing caught in the friction device, plus "flight stressors" like noise, vibration, and rotor wash.

Although rigging for rappelling varies from helicopter to helicopter, most aircraft have *hard points* inside the fuselage from which the rappel ropes can be secured. In a Bell 212/412, for instance, the hard points are rated at 2,500 lbs (1134 kg) each. Normally, two or three of these are used for a multipoint, self-adjusting anchor. Especially on smaller aircraft, they are often rigged in pairs on opposite sides of the fuselage to balance the load. Other helicopters have rappel brackets specifically designed for that purpose.

This is not to say that helicopter rappelling cannot be done efficiently and safely. However, it does require extensive training to reach the necessary level of proficiency, and sufficient practice to keep those skills honed. For swiftwater rescue, other methods

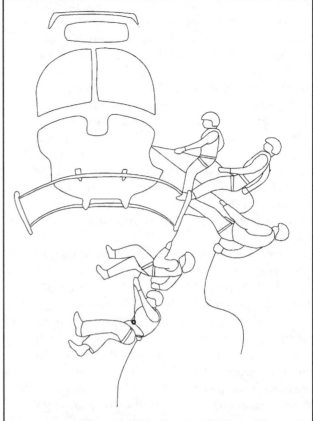

Fig. 10.15 *When exiting a skid-equipped helicopter, a rappeller must first invert almost completely to avoid smashing his face into the skid.*

usually work better, and the time required to train for rappelling can often be better utilized elsewhere.

Short-Hauls

A *short-haul (fixed-line flyaway, helicopter rescue slinging)* means flying a live load slung beneath the aircraft from one point to another. This may be a victim, a rescuer, or both. He is attached to a helicopter with a line varying in length from 50–200' (15–61 m), and there is no attempt to get him inside the aircraft. Instead, he is flown from a dangerous location to a safe one. This is considerably less complicated than trying to hoist the victim into a helicopter, and is generally considered safer than trying to hot-load a patient during a single-skid landing at the rescue site. Compared to hoist operations, short-hauls significantly reduce aircraft hover time. Unlike rappelling, it allows rescuers to both insert into and extract from a rescue site. As a result, it has been widely adopted by rescue organizations world-wide.

However, differences persist. American, Canadian, and European short-haul practices differ considerably, both in equipment and technique, and there are regional differences as well. As with anything else, awareness and training are the critical factors. Current U.S. FARs (federal aviation regulations) also present a problem. These require that any civilian helicopter used for external live load (i.e., short-haul) applications have two engines, and that it be capable of hovering on one of them in an emergency. Most readily available rescue helicopters do not meet these criteria. However, the FAA does allow deviation from this standard in an emergency. The result, as Jim Segerstrom put it, is that "you can do it, but you can't *train* to do it." Under current interpretations of the FARs, however, "public use aircraft" are considered exempt from these limitations, and national parks and fire departments routinely use short-hauls with single-engined aircraft.

Since most short-haul accidents can be traced to pilot error, realistic pilot training is imperative. For short-hauls, it is critical that the pilot know exactly what is happening below him. Helicopter designs vary, but in most cases it is difficult for the pilot to see straight down. American practice is generally to use a crew chief or spotter to advise the pilot—a definite advantage in tight quarters. The Canadians, on the other hand, prefer to let the pilot handle everything, overcoming vision restrictions by removing the aircraft doors or by using a bubble window.

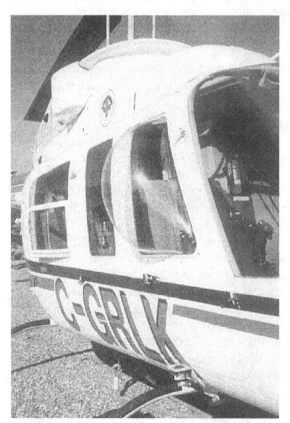

Fig. 10.16 This Canadian Jet Ranger sports a bubble window, allowing the pilot to observe short-haul operations directly under the aircraft.

There also needs to be communication with the person hanging from the line ("the dope on the rope") as well. He, better than anyone else, is able to judge his height from the ground and what is happening in his immediate vicinity. The recommended method is to give the rescuer a flight helmet hooked through an interface cable to a hand-held radio. Because of the rotor wash and vibration, other systems may not work as well and should be tested before use. If radio is not available (some agencies consider communications failure grounds for an automatic mission abort), hand signals can be used. In any case, hand signals should be used to supplement and back up radio communications.

Rigging: Short-haul lines may be purchased from various manufacturers or improvised. Many of the newer manufactured short-haul lines are made from 7/16" (11 mm) Samson Amsteel II, a synthetic Dyneema core rope protected by a polyester braid. It is extremely strong (14,600 lb/65 kN), floats, and has very low stretch under load. It knots poorly, however, and comes with spliced-in thimbles. Improvised lines are usually made from a doubled $^7/_{16}$–$^1/_2$" (11–13 mm) low-stretch rescue rope. While line lengths differ with agency and circumstance, the most common are between 70' (21 m) and 150' (46 m). However, lines of over 600' (180 m) have been used successfully. In general, with lines shorter than 70' (21 m) rotor wash becomes a significant factor, while with longer lines there are increased problems with oscillation and depth perception.

Attachment points vary with different helicopters, agencies, and missions. Most riggers prefer to attach the line to the aircraft by at least two points, both of which can be released or cut in an emergency. Canadians and Europeans prefer dual electrically-operated cargo hooks, but American practice is usually to attach one point to the cargo hook beneath the fuselage, and another to a yoke or belly band run through the fuselage. These bands are easily rigged on different helicopters, but do require that the doors be removed. All feature a quick release that requires two separate motions to activate. Yoke bands are now available with two sets of hooks and releases, making it possible to rig helicopters without cargo hooks for short-hauls. Attached to the line is a weight bag (20 lb/9 kg) to ensure that the line stays taut and out of the rotor.

Ken Phillips, SAR Coordinator at Grand Canyon National Park, also recommends that lines be clipped to the aircraft with spring-loaded auto-locking carabiners, since vibration may cause non-spring-loaded models to unscrew. Conventional screw-lock carabiners should be positioned with the gates down to prevent them from unscrewing.

The U.S. practice of live-load operations, including hoisting, recognizes that if the load destabilizes the aircraft and puts it in danger of crashing, the crew is entitled to cut it free. Canadians, however, discourage any such thoughts. In any case, it is to everyone's benefit to rig the short-haul properly so that the aircrew is never faced with that decision. If it must be done, the aircraft should have a quick-release system. The pilot can jettison the load with the electric cargo hook, but the crew chief or spotter must be ready to release the belly band backup also, either by cutting it with a knife (which he should have within easy reach at all times) or with a quick-release.

The rescuer normally wears either a full body harness or a sit and chest harness, and clips himself to the line with two short webbing tethers. Again, because of the vibration, either non-locking carabiners or auto-locking models are recommended. Victims, on the other hand, require more consideration. Uninjured, untrained victims should be lifted out in a secure device like the Bauman Bag, evacuation triangle or screamer suit. If these are not available,

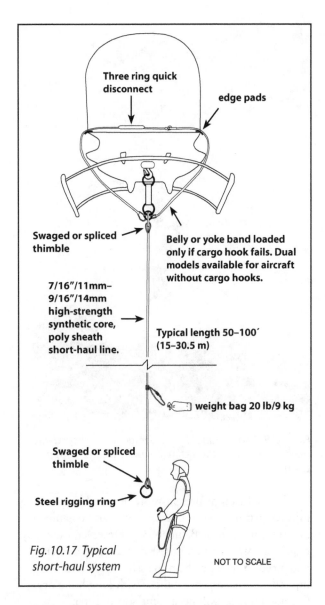

Fig. 10.17 Typical short-haul system

NOT TO SCALE

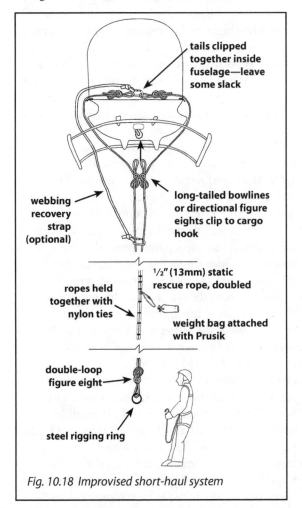

Fig. 10.18 Improvised short-haul system

they can be outfitted with a body harness or an improvised harness like the Hansen harness. For short-hauls, a cinching device like the Cinch Collar will work although it is extremely uncomfortable for anything but a short move.

As an example of how this process works, Grand Canyon National Park is frequently faced with the job of moving a dozen or so passengers from a large pinned motor raft to shore. At the start of the operation, they short-haul a ranger or two to the raft, who carries with him six evacuation devices, usually evacuation triangles. The ranger then supervises the passenger's donning of the suits and hooking to the line. When ready, the passengers are short-hauled to shore two at a time. The rescue proceeds with two passengers removing the suits (which are then clipped to the line and passed over on the next pass), two getting dressed, and two being short-hauled. In this way, a substantial number of people can be ferried over in a very short time. The spider rig, mentioned above, also holds promise for moving large numbers of people in a short time.

Fig. 10.19 A short haul. The rescuer (above) is clipped into the haul line, while the victim (below) is secured with a cinch collar.

Injured victims present more of a problem. Lightly injured victims, even if unconscious, can be hauled out in a screamer suit or Bauman Bag. However, they may require assistance to don these devices, and a rescuer may need to fly with them to assist. More seriously injured victims, especially those with long bone or C-spine injuries, will need some sort of litter.

"Flying" litters have been responsible for more than their share of problems, making litter rigging and patient packaging critical (these are covered in more detail in Chapter 14). One problem is an uncontrolled back-and-forth oscillation underneath the helicopter—something that can be overcome with pilot training. The other is litter spin, an uncontrolled counter-rotation caused by the aircraft's rotor wash. This effect is often associated with solid plastic litters. A Stokes-type litter, whose wire mesh construction catches less air, seems to work better. Most riggers prefer to give the litter a slight head-up attitude, and some rig a small drogue chute at the foot to stabilize it. Other agencies prefer to use the Bauman Bag with a rigid stretcher like the Ferno-Washington or Miller board inside.

There is also some disagreement as to whether an attendant should fly with a litter patient. Yosemite National Park, for instance, does not use an attendant.

Instead they package the patient sideways (lateral recumbent) in the litter to protect his airway. Grand Canyon National Park uses an attendant who carries a hand suction device to keep the patient's airway clear. Others rig the litter with a "barf line" attached to the litter to roll the patient sideways if needed.

NPS Ranger Ken Phillips describes a typical short-haul sequence of operations. "Once the helicopter is rigged for short-haul, the pilot and spotter perform a reconnaissance flight over the incident site. During this flight, crew members evaluate four important factors: outside air temperature, pressure altitude, wind speed, and rotor clearance. Following hookup, rescuers must be on the lookout for cross-loaded carabiners as the rope becomes taut. And, at all times, personnel working beneath the hovering helicopter must maintain constant visual contact with the aircraft. Once medical treatment and packaging are completed, the patient is extracted to the helispot."

Rescue Swimmers

The U.S. Coast Guard has used rescue swimmers in helicopter operations for some time after realizing that an untrained victim could not always be trusted to properly use the rescue equipment let down to him, or who might be injured and require assistance. A rescue swimmer might be deployed either as a free swimmer to assist the victim or on a tether from a helicopter. While Coast Guard rescue swimmers operate mainly on open water, the Los Angeles County Fire Department and other agencies have adapted the concept to swiftwater operations.

The "LA" technique is similar to Coast Guard practice in that the rescuer deploys on a short (30–50'/10–15 m) tether from the aircraft. This is attached with a self-equalizing anchor system to hard points inside the fuselage and controlled by a crewman using a standard friction lowering device, usually a brake rack. As the helicopter approaches the victim, it hovers just above the water, matching its speed to that of the victim. The rescue swimmer then deploys from the skid into the water. With his tether line slack, the rescuer swims to the victim, secures him, then clips him to the tether line. Once this is done, he signals the helicopter, which lifts them both out of the water and short-hauls them to shore, staying as low as possible. For this system to work, the helicopter must match the speed of the swimmer in order to keep the line to the rescuer slack. If possible, the helicopter attempts to put the rescue swimmer in the water downstream of the victim, which allows him to stall against the current and brings the victim to him quickly. This technique can also be used in calm water or the ocean, or to rescue stationary victims. A

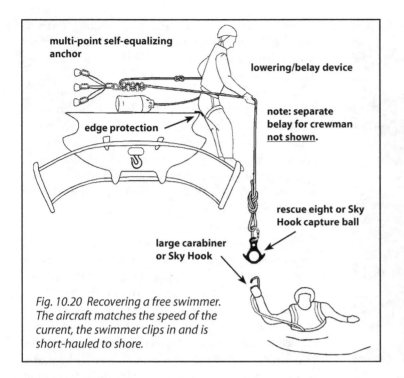

multi-point self-equalizing anchor

lowering/belay device

note: separate belay for crewman **not shown.**

edge protection

rescue eight or Sky Hook capture ball

large carabiner or Sky Hook

Fig. 10.20 Recovering a free swimmer. The aircraft matches the speed of the current, the swimmer clips in and is short-hauled to shore.

variation of this method is to use the aircraft's hoist to lower the rescue swimmer into the water and have that act as the tether.

A related technique uses a free rescue swimmer. Here, the rescue swimmer is not tethered, and deploys either from shore or by stepping off the aircraft's skid into deep water. He wears a harness with a short tether attached to a device like a large carabiner or Sky Hook™. After making contact with the victim, putting a cinch strap on him and securing him to his harness, the rescue swimmer signals the helicopter, which then matches the downstream speed of the rescuer/victim. The crew chief or spotter extends a belayed line about 30' (10 m) long to the rescuer. On the end of the line is an attachment ring (e.g., pear ring, figure eight, or a Sky Hook™ "capture ball") to which the rescuer clips his harness. Once all is secure, the helicopter short-hauls the rescuer and victim out of the water and to shore. They are lifted only as high as necessary to clear the water and any obstructions.

Both these methods, while risky, do allow a rescue swimmer to be deployed very quickly over long distances, and to actually chase down a victim in the current, even when it is very swift. They do, however, require extensive training for both rescue swimmers and aircrews.

There are also some concerns with rescue harnesses for these applications. As we have seen, a swiftwater rescuer needs a quick-release harness, while this is the last thing a rescuer wants while being short-hauled beneath a helicopter. Solutions vary—some rescuers hook directly into the PFD rescue harness,

which protects them while in the water but is risky to fly with. Others, including LA County Fire (who feel that getting dropped from a helicopter is the greater danger), use a climbing sit/body harness under the rescuer's PFD, which is safer to fly with but risky in the water. The risk can be reduced by: (1) keeping the rescuer low and the flight short, (2) giving both rescuer and crew chief/spotter a readily accessible knife, and (3) rigging the belay system in the aircraft so that the rope will run free if released. As of this writing there are PFDs in development that incorporate an integral harness.

A rescue swimmer must secure the victim quickly with some sort of hasty harness. One of the most widely used is the Cearley Strap, designed by retired LA County Fire Department pilot Rick Cearley, who also designed the Cinch Collar. While not suitable for long carries, it is simple, quick to rig, and easy to carry.

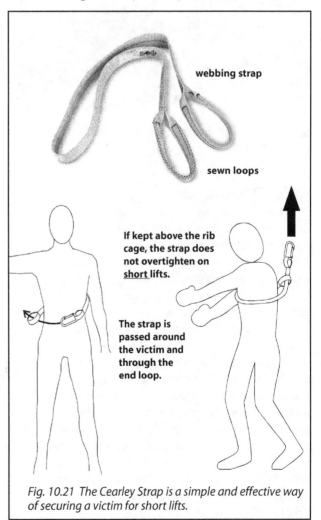

webbing strap

sewn loops

If kept above the rib cage, the strap does not overtighten on **short** lifts.

The strap is passed around the victim and through the end loop.

Fig. 10.21 The Cearley Strap is a simple and effective way of securing a victim for short lifts.

Hoists have disadvantages, too. Hoisting a person into the aircraft can be a rather slow process, which means that extracting multiple victims may require a lengthy hover time. Holding a hover for extended periods, especially over moving water, can be very hard even on experienced pilots. Military helicopters often have a co-pilot to share the load, but this is rare in civilian practice. Hoist failure, including cable breakage and "free-spooling," are not unknown. Some agencies routinely back up hoists with a rescue rope top belay. Most hoists are designed for a straight vertical pull and do not tolerate side loading well. However, because of the force of the current, side loading and overloading are always a possibility during swiftwater operations. Some newer hoists, however, are capable of handling substantial side loads.

As with short-hauls, secure victim packaging is a must. Ground personnel must also be aware that any device attached to an aircraft with a metal cable is likely to build up a static electricity charge capable of delivering a considerable jolt to anyone who grabs it. This can be avoided by letting it touch the ground first.

Some agencies have begun integrating hoists and short-hauls rather depending on either exclusively. Since more rescue helicopters are hoist-equipped today, this allows rescuers to select the best option for the situation. Agencies like LA County Fire go even further by using the hoist for brief short-hauls and for recovering rescue swimmers.

Fig. 10.22 A hoist mounted on a Bell 412, using a horse collar for personnel lifts.

Hoist Operations

According to the Airborne Law Enforcement Association, less than 10% of civilian aircraft are equipped with hoists or winches, and fewer still have crews trained to use them in swiftwater-specific situations. A typical military hoist (e.g. Goodrich 42315, 44301) has a maximum load capacity of 600 lbs (272 kg) for a straight vertical lift. Civilian models often have considerably less. Cable lengths vary from 100–250' (30–76 m).

The big advantage of a hoist is that a victim may be lifted inside the aircraft. If the aircraft is a considerable distance from shore—often the case with ocean rescues—this is preferable to short-hauling. If the patient is injured, it is much easier to care for him inside a helicopter (even though some helos are very short on space) than if he is dangling beneath it. The aircrew need worry less also about mishaps like dragging the victim through a tree, although litter spin is still a concern.

Fig. 10.23 A Bell UH-1H from the King County Sheriff's Department uses a hoist to rescue residents from the rooftops of Adna, Washington.

Fig. 10.24 Grand Canyon NPS MD-900 helicopter crew running a short-haul. The crew chief supervises operations but because of the small size of the aircraft the pilot is able to see as well.

Conclusion

Helicopters are wonderful, expensive tools that can solve many rescue problems, and can effectively be used in many swiftwater situations for both evacuation and rescue. However, like any other tool, effective use of helicopters requires training and an understanding by both the ground and air crews of their strengths and limitations. Although in general they remain a higher-risk tool, with crews properly trained and equipped helicopters can be used for a wide variety of swiftwater rescues.

Fig. 10.25 An MD 900 from the National Park Service short-hauls a patient over the Grand Canyon, AZ.

Special Situations
Chapter 11

Fig. 11.0 An important initial step in any swiftwater rescue is to get a PFD on the victim.

So far we have discussed rescues and equipment mostly in general terms. However, there are some swiftwater rescue situations that come up frequently or that require special techniques. These include:
- A victim stranded in midstream
- Foot and body entrapments
- Low-head dam rescues
- Bridge rescues
- Automobile rescues
- Flood channel rescues
- Swiftwater scuba operations

Each of these situations will be considered separately. However, each one is really just an application of the basic rescue principles already covered. These situations, taken together, represent a large part of the swiftwater rescues done each year. Naturally, they also account for a sizable number of failures as untrained personnel attempt to improvise rescues on the spot—sometimes successfully, sometimes not.

Since we are now discussing overall rescue technique, it's time to look at the rescue sequence. All swiftwater rescues generally go through four phases:

- Assessment (size-up, if you're a firefighter)
- Stabilization
- Extrication
- Evacuation

Rescuers arriving on the scene must immediately *assess* the situation, establish incident command, allocate resources, and begin the rescue. They must then attempt to *stabilize* the victim so that the situation does not worsen until they can *extricate* him from his predicament and *evacuate* him to safety and medical treatment. Obviously, in some cases the phases will overlap. Look for this sequence in each of the following rescues. However, we will save the overall organization of the rescue site and incident command for a later chapter.

There are two very important initial steps when beginning a swiftwater rescue. First, *get personal protective equipment to victims as soon as possible.* A PFD is the most important item, but the victim should also be given a helmet and thermal protection if needed. Second, *establish contact with the victim(s).* Reassure them, evaluate their mental and physical status, and tell them what the plan is.

Communicate the Plan

Communicating the rescue plan to victims is always important, but particularly so in swiftwater situations. News photographer Bill Perry witnessed a tragic example of a misunderstanding in Yosemite National Park.

"It was a desperate scene," he recalls. "Two women and three young children about 7, 5 and 3, perched on top of a car jammed against a giant tree trunk in the roiling, icy Merced River. They were terrified.

"I'm a non-swimmer and had no emergency gear, only a trunkful of Nikon cameras. I knew there was nothing I could do. 'Be calm,' I shouted. 'Stay where you are; you are safe. I'll go for help.'

"They seemed to understand. I jumped in the car and raced to the nearest campground office. I called for help and the park rangers said they were on the way. I turned around and sped back to the scene, for the first time thinking of my job as a photographer, about the image of those five terrified faces looking up at me from the river.

"The ranger was on the scene, I knew things would be under control. 'Now, do your job,' I told myself.

"I pulled my cameras out and reloaded with faster film and hurried to the river bank. Shooting with a long lens, I isolated on the elderly woman jammed between the driver's side door and car. The icy water swelled around her knees, a rope was tied around her waist, she cried, she shouted to shore, she put her hands together in prayer.

'The others must be inside the car,' I thought, and continued to shoot. The ranger was methodical. He called for backup and put on a wetsuit, he tied himself off to a tree and asked a passerby to hold the rope as he eased into the raging currents. I moved downstream to get a better angle of his attempted rescue. Through the long lens, I peered at the car.

'Where were the kids? Where was the middle-aged woman who was with them?' I asked a couple of people, and they shrugged.

"More help came, a helicopter swept through the trees lining both sides of the river. The older woman was rescued and I started asking questions.

"While I had been away calling for help, some young men had stopped. They had a rope; they tried to help. When they threw out the rope, the 7-year-old girl grabbed it and jumped in the river and they started to pull her to shore. The 5-year-old and 3-year-old thought they were supposed to jump in, too. They did. The deadly current swallowed them instantly, then their great-aunt jumped in to try to save them. The rangers found her body later in the afternoon. The boys have not been found and probably won't be until the end of summer when the water is lower.

"I know I couldn't have found help any faster than I did. I know the rangers got there as fast as they could. I know that three people would still be alive if they had waited for professional help.

"I don't think I'll ever forget those five terrified faces."

Fig. 11.3

Victim Stranded in Mid-Stream

Probably the most common swiftwater rescue situation involves a victim stuck on some sort of obstruction out in the current. This happens in a number of ways: a car runs off the road into the river or is swept off by floodwaters; someone flips or pins a raft or canoe and ends up on a midstream boulder, debris pile, or tree; a sunbather or fisherman is trapped by rising water from a dam; or a person is trapped by rising floodwaters. Whatever the cause, all these situations are similar—the stranded person is in no immediate danger, but must somehow be rescued from the middle of the river to the shore.

The initial size-up should include a quick survey of:

- **The victim.** Is he injured? What is his mental state? Does he appear to be able to assist? If he does not have a PFD, how can we get one to him?

- **The situation.** Most importantly, is the obstacle upon which the victim is stranded stable? A calm victim on a boulder with no other complicating factors (i.e., bad weather, darkness, rising water) can be dealt with in a fairly deliberate fashion. However, if he is on a car that seems likely to shift, or is hanging on to a tree branch, or if the water is rising, something must be done quickly to stabilize him until a rescue plan can be implemented. This might mean stabilizing the car with ropes or sending a rescuer out to assist him.

Pendulum Rescue. A simple variation of the basic throw bag rescue works well to rescue a victim stranded in midstream. If the victim is within throw bag range, the area is clear of obstacles, and the victim is uninjured and capable of helping himself, the rescuers can simply throw him a rope. If not, it may be possible to use one of the range-extending devices (e.g., line gun) to get a rope to him. When the victim gets the rope, he holds it over his shoulder, faces downstream, and enters the water below the obstacle. He should be instructed how to hold the rope so that his body assumes the correct ferry position. The rescuers assume a belay and the victim is *pendulumed* into the shore by the current. If the current is not strong and the rescuers many, they may choose to haul the victim in like a fish.

This system has the advantage of being very simple and safe for rescuers, who never enter the water. However, there is always the danger that the victim may be either too fearful to actually enter the water, or worse, that when in it he may let go of the rope. For this reason, backup throw bags and perhaps a rescue swimmer or live bait setup are a good idea, and this technique should be limited to victims who seem reasonably capable of carrying it out. In any case, the victim should be cautioned not to tie himself to the rope or to wrap the rope around any part of his body. If possible, he should also have a PFD sent to him if he does not already have one.

Fig. 11.2 The pendulum rescue. The rescuer throws a rope to the victim and allows him to swing into shore.

A variation of this rescue is to send a swimmer wearing a rescue harness to the victim. The rescue is done in exactly the same way except that the rescuer physically holds the victim and acts as a "human throw bag" for the pendulum. This is a more secure method since it depends less on the competence of the victim.

Continuous Loop. If the victim is near shore, a *continuous loop* rescue is quick and simple to set up. It was developed by the British Army for moving troops across jungle rivers. Once established, it is relatively easy to keep the loop moving, which makes it possible to move several victims to shore in a short time. It is, however, limited to fairly short distances and does not work well in very swift currents.

To set up the continuous loop, the rescuers first swim, ferry, or throw a loop of rope across the distance to the obstacle. The loop should be approximately twice as long as the distance to be crossed. One rescuer stands inside the loop on shore, while another wades or swims the other end of the loop across. Once the loop is across, another on-shore rescuer steps into the loop, forming a triangle with the apex on the obstacle. The victim is given a rope loop to hold (either an in-line knot like a butterfly or a Prusik loop) and is pulled rapidly across the river on the downstream leg of the triangle. The victim's security can be increased by putting him in a rescue jacket and clipping it to the moving loop.

Fig. 11.3 *Continuous Loop Rescue*

A rescuer wades out to the accident site.

After he establishes the loop, the second rescuer wades out.

Holding the loop, the rescuer and victim move back to shore.

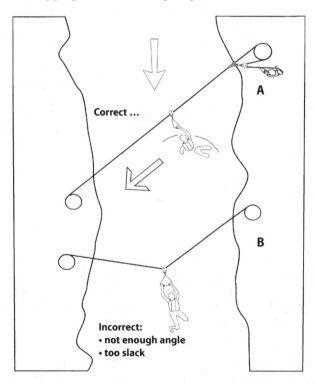

Fig. 11.4 *The tension diagonal or zip line (A). The line must be kept tight to avoid stranding the swimmer in mid-current (B). Using a pulley on the line and angling the swimmer's body in the ferry position makes the operation more efficient.*

Zip Line or Tension Diagonal. Another technique to consider is a *zip line* or *tension diagonal*. Here, a taut line is angled downstream from the obstacle to shore at an angle of 45–60° or greater. The victim or rescuer clips to the line with a carabiner, holding (*do not* tie or wrap the cord around the wrist) a short loop of rope or cord. A pulley makes the system even more efficient. He then enters the water and the current pushes him from the obstacle to shore. As with throw bag rescues, it helps if the person on the zip line angles his body in the ferry position. The tension diagonal can be used to move large numbers of people across the current quickly. Since the person on the diagonal moves at the speed of the current, there is much less water pressure than in some other types of rescue.

Fig. 11.5 To use a zip line, the swimmer holds a short tether attached to a pulley.

The line must be anchored at both ends and *tightly* tensioned with a haul system, or it will bow downstream and strand the victim in the strongest part of the current. Because of this danger, users *should not* tie themselves to the system. If the water is shallow, a tension diagonal can be used in conjunction with a wading rescue to get across.

Fig. 11.7 A zip line in use.

Earlier we saw that a common mistake on swiftwater rescues is to stretch a line at right angles across the river to catch swimmers. This invariably strands the swimmer in place as soon as they grab the rope. Unable to move, he eventually becomes exhausted and hypothermic and lets go. A tension diagonal, however, can be used this way, since a swimmer can work his way out of the current and toward shore hand over hand. This is the basis of the "shower curtain" and other rescue systems that we will consider in more detail in the section on flood channel rescues.

Rescue Ferry. Another common method of rescuing stranded victims is to use a boat to ferry behind the obstacle and pick the victim up. This is a fast and very effective method on wide rivers, *provided the boat crew has the boat-handling skills to do it safely.*

Fig. 11.6 A tension ferry.

A special mention needs to be made of strainers. Rescuers should approach the upstream side of a strainer with great caution. If at all possible, they should try to catch the eddy behind the strainer to avoid being swept into it. If there is no alternative but to approach it from upstream, rescuers should consider using a tethered boat system.

A related method is the *tension ferry.* The rescuers tension a rope across the river 3–4' (1 m) above the water at right angles to the current. The rescue boat is attached to the line with two pulleys and a short tether, one at each end. To make the ferry, the crew lets out the stern tether, and the current pushes the boat across the river. This is useful when the rescuers do not want the boat to go downstream. Although this method works well with crews whose skills might not enable them to make a ferry in a free boat, they must know enough to shift their weight downstream or the boat will capsize as it enters the current.

Rescuers can also use a boat on tether if time and resources permit. This gives good control and requires less boat-handling skill on the part of the boat crew. Normally, the boat would be lowered down past the obstacle and then pulled back up into the eddy behind it, which makes loading considerably easier. However, it can also be lowered to the upstream side of an obstacle such as a car or a strainer, although this is much more dangerous and should be done only if other options won't work. Rescuers

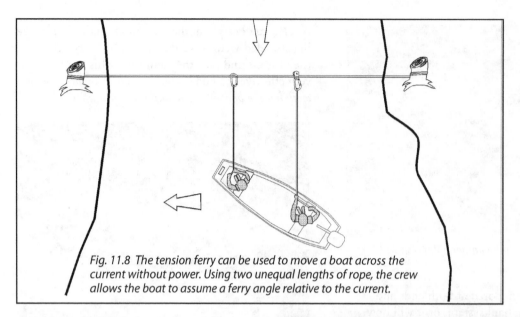

Fig. 11.8 The tension ferry can be used to move a boat across the current without power. Using two unequal lengths of rope, the crew allows the boat to assume a ferry angle relative to the current.

should also consider this system if there is a major hazard below the rescue site and they need to have positive control of the boat.

Aerial Ladder Pickoff. If the victim is close enough to shore, an aerial ladder mounted on a fire truck can be used for a reaching rescue as described in Chapter 7. However, the design of some ladders is such that they will support a two-person load with the ladder angled upward (e.g.,<30°) but not with it near horizontal. This precludes using it as a reaching device in many situations. A way around this limitation is to use the ladder as a crane, lowering a rescuer down to the accident site for a pick-off. The ladder is rigged with a simple haul system run through a directional pulley on the end of the ladder. Using the haul system, the victim is secured, then lifted up with the rescuer. The ladder is then traversed around to shore.

Helicopters. Another method of rescuing stranded victims is the helicopter, using the techniques described in the previous chapter. While this option is probably used more often than it should be, it can be very effective on wide rivers or in situations where a major hazard (e.g., a low-head dam) is just downstream of the obstacle.

Foot and Body Entrapments

A foot or body entrapment is a much more serious situation than a stranded victim. A foot entrapment occurs when a person's foot becomes jammed in a crevice in the riverbed, usually as a result of trying to stand up in the current. However, under some circumstances a swimmer's entire body can be entrapped by an undercut rock (see fig. 2.30).

Fig. 11.9 A foot entrapment victim screams with pain as rescuers attempt to free him.

The critical part of the assessment phase is to determine whether the victim's head is underwater or not. If it is, it is referred to as a *"head-down"* entrapment, while if he can breathe, it's a *"head-up"* situation. In most cases the time available to rescue a head-down victim is very short—two to four minutes—and the number of successful rescues from this situation has been small. Obviously, rescue efforts must begin immediately, and high-risk options may be justified. In other cases, however, a victim is able to get a breath periodically by dog-paddling to the surface, and there have been instances where victims have survived this way for over an hour. In any case, any foot-entrapment victim's time is limited for two reasons. First, because he is constantly pushing against the force of the water, any victim will quickly become exhausted. Second, with the victim stationary and the water flowing past him, the incidence of hypothermia

Fig. 11.10 A rescuer wades out on the upstream side of a fixed line. Since he is not attached, he can get free at any time by ducking under the line.

is dramatically increased, since moving water sucks heat from the body roughly 10 times as fast as still water of the same temperature.

If the victim cannot breathe, the rescuers must begin extrication efforts immediately. If he can breathe, he should be stabilized. If the situation permits, the best and fastest way is usually with a tag line, preferably one with a float on it. If the victim is underwater, however, the rescuers may want to use a bare or weighted line (a bare line is much easier to handle) to get the line under the surface and attempt to lift the victim's body up enough for him to get a breath. Even if the line can't be gotten under the victim, he may still be able to reach up and grab it.

Once the victim has been stabilized, extrication can begin. Almost always, the victim is going to have to be moved back upstream to be freed. In some cases the stabilization line may also be the extrication line. By pulling upstream, the rescuers may be able to pull the victim back enough for him to get free, or he may be able to push himself back against the stabilizing line. If this doesn't work, the rescuers may opt to use a second line, attempting to work it down the victim's leg towards his ankle, and trying to pull his foot free.

Another shore-based extrication technique discussed in Chapter 7 is the cinching rescue. There are several techniques of passing a line around a victim into a cinch, then pulling him upstream. This also has the advantage that if the victim is exhausted or semiconscious, the cinch will secure him so that the rescuers don't lose him.

A tethered raft can also be an effective tool to rescue a foot entrapment victim, assuming that there is enough time

to set it up. If so, it provides a secure platform from which to work.

If the water is shallow enough, wading rescues can be used. One wading technique is to wade out just in front of the victim to break the force of the water, either with a single person or a group. The victim may then be able to escape, or a rescuer may be able to assist by getting into the eddy behind the group. Some formations, like the wedge, lend themselves to this type of rescue, since a rescuer can operate from inside the wedge. The pivot will also work, since the downstream person can turn around to assist the victim while being held by his teammates.

If the rescuers have established a stabilization line to the victim, a rescuer can sometimes slide down the line to the victim, using the stabilization line as a tension diagonal. Depending on the speed of the current, he may be able to get in the eddy downstream of the victim long enough to lend assistance.

The rescuers may also be able to get a rescuer down to the victim by lowering a rescuer upstream of the victim, using either a bare line or, if the rescuer has a rescue PFD, a V-lower. As the rescuer gets closer to the victim, he creates an eddy, making the task of extrication easier.

Rescuing foot entrapment victims is one of the more difficult tasks rescuers face. The limited time, technical difficulties, and danger are all factors that work against them.

Low-Head Dams

Low-head dams, or weirs, also represent a serious rescue problem. These "drowning machines," which often look quite innocuous to the untrained eye, are very common in some areas of the country. Victims may be held in the backwash of the dam for extended periods of time, and rescuers may also end up there. Probably the most notorious low-head dam incident took place at Binghamton, New York, in September

Fig. 11.11 A kayaker caught in a low-head dam. Both the kayaker and a would-be citizen rescuer drowned.

1975, when three firefighters drowned (and three others narrowly escaped) in the backwash of a low-head dam.

The degree of danger in a low-head dam varies with river conditions. A dam that may be relatively benign at lower water levels may be a killer at higher flows. The design of the dam also affects the danger level, as do other factors. The degree of difficulty of rescue from a dam is based on a number of factors, including:

- The height and angle of the drop. The higher the dam, the greater the backwash. Sloping drops in the 45–60° range often cause worse hydraulics than vertical ones (however, this was not the case at the Martindale dam).
- The volume of water flowing over the dam. In simple terms, the more water, the greater the backwash.
- The width of the river. Wider rivers reduce the range of effective rescue options.
- The design of the dam. Some dams seem designed to catch and hold swimmers. Generally speaking, the best way to escape the backwash is to move sideways, but this is often inhibited by vertical concrete walls. Rescuers must determine whether there is access to the dam from both sides and/or from the river below, and if (and where) there are anchor points for rope systems.

Some dangers are not obvious. If the base of the dam is eroded, for instance, there may be debris or exposed rebar underwater.

Public safety agencies are well advised to inspect low-head dams in their jurisdictions *before* they need to perform a rescue there. Dams should be rated and rescue plans formulated. If possible, rescuers should conduct exercises at the site and coordinate with local officials about dam access. Often it is possible to improve the site by adding anchor points and removing obstacles. Preplanning a dam rescue can save precious time.

When the rescuers reach the site, the first thing to do, if possible, is to get the victim extra flotation. This depends, however, on the exact nature of the hydraulic. The backwash in some dams in relatively smooth, and victims can survive in it for some time if they have adequate flotation. The best float is something long and relatively slender, like a lifesaving "can." On other dams, the hydraulic is violent and aerated, and a victim will gain little benefit from a float. In this case, rescue efforts must begin immediately.

In a low-head dam hydraulic, pulling the victim upstream is virtually impossible. He may be lifted vertically or pulled downstream over the backwash, but this is also difficult. It is easier to pull a victim to one side of the hydraulic or the other.

Many shore-based rescue techniques (see Chapter 7) can be adapted to low-head dams. If the victim is near the side of the dam, he may be close enough for a reaching rescue, or for a simple throwing rescue using a throw bag or a tethered float. Rescues have even been improvised using automotive jumper cables. Some low-head dams have bridges or catwalks above them. If this is the case, the rescuers may be able to either lift the victim vertically using a haul system and a cinching device. An alternative is to extend a rope with a float to the victim and tow him to the side.

Another very effective reaching rescue device is the inflated fire hose. The hose is inflated and pushed out into the hydraulic. The rescuers roll the hose upstream as this is done, being careful to keep it in the backwash. If any part of the hose gets into the current, it will pull the rest of the hose out of the backwash. If the river is fairly narrow, a rope attached to the end of the hose from the other side of the river will help control the hose and keep it from washing downstream. Often it helps to tie the end of the hose in a "U" shape to aid the victim in holding it.

Fire departments can also use aerial ladders for reaching rescues on narrow rivers, or if the rescuers can get the victim close enough. Placing a rescuer on the end of the ladder may get him close enough to the victim for a throwing rescue with a throw bag or a tethered float. They may also lower a rescuer down from an aerial ladder for a vertical pickoff.

A floating tag line is often recommended for low-head dam rescues—and for good reason. It is simple, works well on narrow rivers, and keeps the rescuers safely on shore. The difficult part is getting the rope across the river. Typically, it is shot or ferried across with a float tied in the middle. The rope must be long enough to bring the float all the way to one bank or the other. In practice, the rescuers bring the float to the victim and pull him sideways to one bank or the other. If this is not possible, they must pull him downstream over the backwash—a much more difficult proposition.

If time and resources permit, the rescuers may also consider a high line with a boat on tether to put a rescue craft in the dam's backwash. In this case, the anchor line is strung over the dam, and the boat pulled *upstream* into the backwash. To keep it from going in, it is necessary to have at least one (and preferably two) tether lines to shore. While this system is effective, it is seldom used because of the time and resources needed to set it up.

Most of the techniques covered so far will not work on really wide rivers. The practical limit for most of them (including tag lines) is about 300' (91 m). For wide rivers, rescuers must use either boats or helicopters. Most boats don't work very well near low-head dams because conventional propellers don't "bite" well in water that is over 50% air. Jet boats, however, or boats with jet drives, can operate effectively in the backwash, although this puts them at the high-risk end of the rescue spectrum. Catarafts or "spider boats" can also operate near, and sometimes in, the backwash of a dam. Since the cataraft has two tubes and no floor, it has little drag in the backwash. On smaller dams a "cat" can enter and leave the backwash at will, and even park its tubes on the face of the dam! While this technique can be quite effective, it is also extremely high-risk.

Fig. 11.12 A two-boat tether in action. The lead boat stays just outside the boil line and throws a ring buoy to the victim.

For low-head dam rescues on wide rivers or where a lowering system was impractical, the Ohio Department of Natural Resources devised the *two-boat tether*. This system, a variation of the shore-based tether, uses two boats tied together with a 100–150'

(30–46 m) line. The upstream boat is usually an IRB, since this is less likely to swamp or capsize if it does get into the dam's backwash. The tether line is rigged to the IRBs stern with a bridle system and a small float to avoid snagging the tether line in the prop. The downstream boat is usually a rigid-hulled power boat.

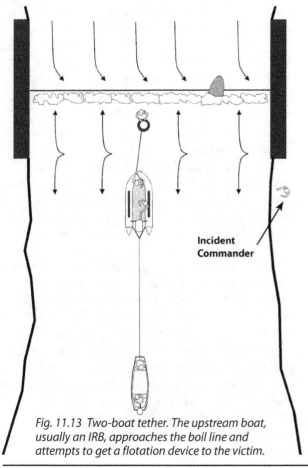

Fig. 11.13 Two-boat tether. The upstream boat, usually an IRB, approaches the boil line and attempts to get a flotation device to the victim.

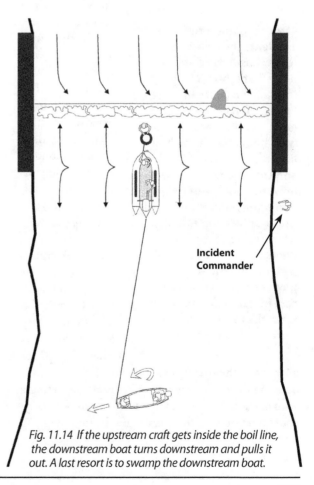

Fig. 11.14 If the upstream craft gets inside the boil line, the downstream boat turns downstream and pulls it out. A last resort is to swamp the downstream boat.

Fig. 11.15 The lead boat of a two-boat tether approaches the boil line. The tether line to the second boat is attached to the stern with a bridle to avoid fouling the prop.

Naturally, no boaters should be committed to this type of rescue unless they are properly trained and equipped. Communications are critical—particularly between the two boats. On narrow rivers, the incident commander usually stands on shore; on wider rivers he may have to command from a boat. It is essential that someone, preferably the commander or a control, be able to observe the position of the upstream boat in relation to the boil line, and to communicate this information to the boat crews.

Helicopters can also be used on wide rivers. The helo lowers a cinching float (e.g., Cinch Collar) to the victim and short-hauls him to shore. However, this technique is risky, since it essentially involves hooking up for a short-haul in the middle of a hydraulic—hardly the ideal environment. If this technique is used, the victim must be kept low and there must be backup boats in case he falls out.

We cannot totally rule out using a rescuer tethered beneath the helicopter like a tea bag to grab the subject. This may be the only way to get a semiconscious victim. This was the method used in the San Marcos incident. However, it must be considered a last resort.

Low-head dams are common and represent a challenge for rescuers. Rescue systems range from safe and simple to complex and dangerous. The choice of selecting the right one must rest with the incident commander based on the circumstances.

For a rescue, the two boats approach the dam's boil line from downstream. The upstream boat (the IRB) gets as close to the boil line as possible. A rescuer in the bow then attempts to throw a tethered flotation device (e.g., a ring buoy) to the victim and pull him to the boat. Under normal circumstances, the upstream rescue boat *does not* enter the dam's backwash. The second boat, downstream, performs two functions. First, it assists the upstream boat in maneuvering. The propeller on the upstream boat may not function well in the frothy water near the backwash, while that of the other boat will work well in the "dark" water downstream. Second, it acts as a safety. If the upstream boat does start to go into the backwash, the second boat quickly executes a power turn downstream, pulling it away from the boil line. Even in an extreme situation, where the upstream boat enters the backwash and swamps, the second boat can drag it free by swamping itself and acting as a sea anchor. At all times the downstream boat keeps enough tension on the line to keep it from snagging on an obstacle or a propeller.

After several training accidents the two-boat tether has to be classified as a high-risk activity. Overall two problems have emerged. The line length must be adequate to keep both boats out of the boil line in case of trouble, and the downstream boat must be capable of being swamped if necessary (and thus not be an inflatable). At least two of these accidents have occurred at high water when the size and power of the hydraulic was considerably greater than what the rescue agencies were used to training on.

Bridge Rescues

Bridges are somewhat of a paradox for rescuers. On the one hand, they represent a quick and safe way to cross the river and gain access to the middle of it. On the other hand, the bridges themselves cause a fair number of accidents and have serious drawbacks as rescue sites.

The major cause of bridge accidents is the nature of the bridge pilings. Unlike rocks, bridge pilings do not generate a "cushion" of water upstream to help push objects away from them. As a result, they are infamous for having boats pinned against them. This characteristic also makes them prone to collect debris against the pilings, causing strainers that can entrap boats or swimmers. High water may also reduce the clearance between the river and the bottom of the bridge to nil. For all these reasons rescuers should be very wary of operating upstream of bridges dur-

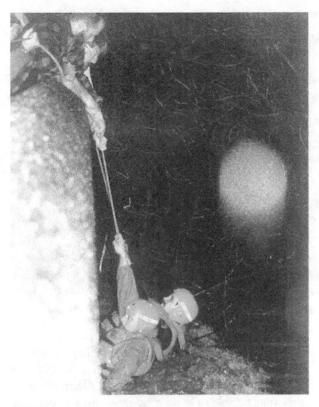

Fig. 11.16 Firefighters cling to a rope on the upstream side of a bridge after swamping their rescue boat. One was pulled to safety; the other washed under the bridge and was only rescued after a lengthy swim.

ing flood conditions, or of getting too close to the upstream side of bridge pilings at any time.

Rescuing a pinned boat or a paddler stuck or pinned on the upstream side of a bridge piling is relatively straightforward. If the victim is not entrapped or in the water, but merely stranded on a debris pile or the platform of a bridge piling, he can usually be rescued by simple methods such as having a fire ladder or rope ladder lowered to him. Another method is a raising system set up on top of the bridge. It is often possible to sling a pulley from the upper works of the bridge, on the end of an extended fire ladder, or from an A-frame for edge protection. An aerial ladder may also make an excellent rescue platform for a vertical pickoff or bridge lower.

Large rescue vehicles like fire trucks make excellent anchors and, in any case, the rescuers will want to set the anchor far enough back to give themselves some working room (e.g., on the downstream bridge railing). The victim should be given some form of security, such as a harness, a screamer suit, or a cinching device like the Cinch Collar. In most cases, the rescuers will want to send down one of their number to assist the victim. A potential complicating factor is the wire mesh that is often added to higher bridges to

discourage jumpers. This may have to be removed or cut before rescue operations can begin.

If the victim is pinned or entrapped, however, time becomes critical. In these cases, one or more rescuers must be sent down to the victim if this can be done without unduly endangering them. Sometimes there will be a platform on the piling from which to work; in other cases the rescuer may have to be lowered down and operate from a hanging position, or climb down on a rope ladder. This is extremely dangerous for the rescuer, and his teammates should ensure that he can get free of the lowering system if he goes into the water, and that there is a downstream backup. If the piling is close enough to the bank, rescuers may also employ shore-based rescues.

In some cases, the rescuers may be able to get to the bridge piling by boat. Some pilings, as mentioned before, have platforms on them, and rescuers can get to them by ferrying into the eddy behind the piling. If there is a large debris pile on the upstream edge of the bridge piling, that may also create a large enough eddy for a rescue boat to ferry into.

Another very common type of bridge rescue is one where rescuers on the bridge attempt to rescue swimmers being swept past in fast-moving water. If the river is wide, this may offer one of the few ways that the rescuers can get far enough out into the river to reach the swimmers. However, there are problems.

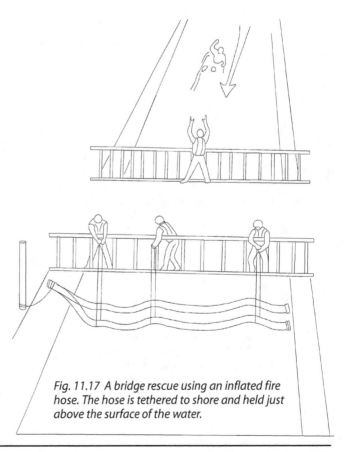

Fig. 11.17 A bridge rescue using an inflated fire hose. The hose is tethered to shore and held just above the surface of the water.

In the past, rescuers have lowered ropes to the swimmers from the upstream side of the bridge. What follows is quite unlike throwing a rope to a swimmer from shore, where the force of the water quickly pushes the swimmer to shore. A swimmer grasping a rope lowered from a bridge soon finds himself in a desperate situation. He is now under the bridge and stationary in the current, with its full force trying to break his grip on the rope. Charlotte Fire Department battalion chief Tim Rogers calls this "the bridge dangle-do."

As with a foot entrapment victim, exhaustion and hypothermia quickly set in. To be rescued, the victim must either climb the rope or be pulled up by the people on the bridge. Either option is extremely difficult. The longer he holds on, the colder and more tired he becomes, leaving him in a much worse situation when he finally does let go. This scenario, unfortunately, has been played out many times.

Fig. 11.18 An inflated fire hose set up for a bridge rescue.

If there are plenty of rescuers or if they are able to quickly set up a haul system, they may be able to pull the victim up out of the water. Alternatively, they may be able to lower a ladder—either a solid fire ladder or an improvised rope ladder—to the victim, and send a rescuer down to help.

The best solution, however, is to avoid fighting the force of the river. Instead of trying to stop the victim in the water and then rescue him, the rescuers should drop a tethered float to the victim *on the downstream side of the bridge* and pendulum him to shore. A spotter on the upstream side of the bridge keeps them in line with the victim.

On some bridges, the rescuers may run the rope *under* the bridge to the upstream side. This way, the rescuers on the upstream side of the bridge have clear view of the victim. As he passes under the bridge, the rescuers drop the float to him, and he is then hauled to shore or allowed to pendulum in.

An inflated fire hose also works well. The hose is tethered to one shore, and held just above the surface of the water by several rescuers standing on the downstream side of the bridge. A spotter stands on the upstream side to advise them of the victim's location. As the victim passes under the bridge, the rescuers drop the hose, the victim grabs it, and the shore-based rescue team pulls him in or allows him to swing into shore.

Bridges, in short, can be effective rescue sites *if* rescuers know their limitations and advantages.

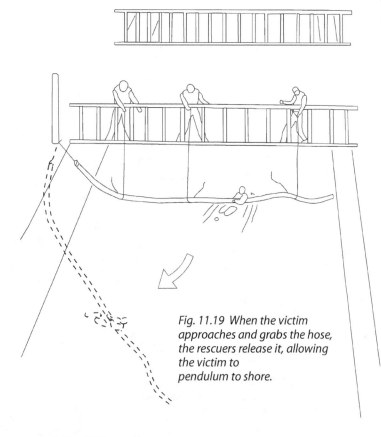

Fig. 11.19 When the victim approaches and grabs the hose, the rescuers release it, allowing the victim to pendulum to shore.

Vehicle in the Water

In developed countries more than half the fatalities in any given flood are drivers whose cars are either swept off the road or who run into the river, making vehicle rescue a necessary skill for any agency. Roughly 75% of these accidents happen at night or during periods of poor visibility, in part because judging the speed and depth of muddy water at night is extremely difficult, but also because a higher percentage of drivers have been drinking during the late evening hours. Many swiftwater rescues start this way: a driver, not wanting to take a lengthy detour, ignores a barricade and tries to cross a flooded street. The water is flowing swiftly across the pavement, but it doesn't look that deep. The car plows into the water, which quickly reaches the doors. Just when the car seems to be coming back up out of the water, the engine suddenly speeds up and the steering wheel goes slack. The car begins moving sideways with the current. As the driver watches helplessly, the car heads into the nearby creek, which is flooding out of its banks.

A rough rule of thumb is that each foot of water pushes against the broad side of a typical car with about 500 lbf (2.2 kN) of force and displaces about 1,500 lbs (680 kg). Thus, *two feet (.6 m) of water will float most cars*. However, a car can be washed away in less,

depending on variables such as the speed of the current, the design of the car, whether the car is sideways or end-on to the current, and the type of bottom. For example, where the current is swift, the bottom flat and slippery (e.g., pavement or smooth concrete), and the car's body low to the ground, as little as one foot (.3 m) of water with a current of 6 mph (10 km/hr) or 10 feet per second (3 m/sec) will move most cars. On the other hand, if the car is heavy and has a lot of ground clearance, the bottom is sand or gravel, and the current slow, it may take deeper water to move the car.

The type of river bottom has a lot to do with a partially-submerged car's behavior in the water. If the bottom is slick with no obstructions (e.g., pavement or concrete), the car is very likely to continue to slide or roll, especially if it is broadside to the current. This can be very dangerous to both the occupants and rescuers. Since a sudden weight shift to the downstream side can cause the car to roll, those in the car should practice the *reverse high side*, that is, keeping their weight on the *upstream* side of the car to keep it from rolling. If the bottom is soft, however (e.g., sand, gravel, or mud), the water will quickly excavate the soil under the tires so that the car's chassis rests on the riverbed. This results in a much more stable situation, one in which the car is much less likely to roll. If the bottom

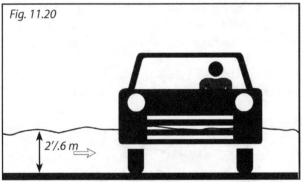

Fig. 11.20

2'/.6 m

Two feet (.6 m) of water flowing across a hard surface will float most cars.

Fig. 11.21

On a hard bottom, a vehicle is more likely to roll.

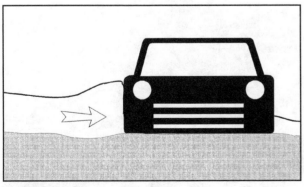

On a soft bottom, the vehicle settles down on the frame, resulting in a relatively stable situation.

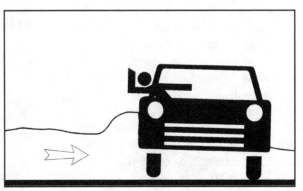

This can be prevented by a "reverse high side," where the occupant shifts his weight to the upstream side of the vehicle.

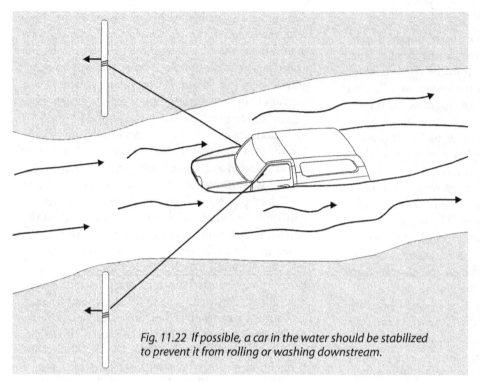

Fig. 11.22 If possible, a car in the water should be stabilized to prevent it from rolling or washing downstream.

car is bellied down on a soft bottom, the water may not be as deep as it looks. Rescuers should, however, exercise caution until the car is stabilized, especially on hard bottoms, since the car may slide or roll over downstream at any time. In most cases, however, the roll will be a slow one, giving an alert rescuer time to escape. However, since the rescuer's attention may be occupied with the task at hand, having a shore-based spotter watching the car adds an extra margin of safety. For the same reasons, rescuers should not go more than arm's length into a vehicle unless it is absolutely necessary.

is really soft (e.g., mud) the car often ends up buried engine-end first in the mud with the other end out of the water. Once the car stops moving, it then acts like any other river obstacle, except that it is much more likely to shift. Sometimes the car will wash up against something solid like a rock, a guard rail, or a tree, and this may help to stabilize it.

The rescuer's first thoughts should be to stabilize the vehicle so that it does not either roll or float downstream, and to get PFDs to the passengers. If the situation is marginal, the weight shift as the passengers are being rescued may cause the car to move. To prevent this, rescuers should attach stabilizing ropes to the car to keep it from moving. These need not be cables—rescue ropes will do. If possible, these ropes should go to both banks. Rescuers should also routinely dispatch a tow truck to the scene at the same time as the rescuers, rather than waiting until it is needed.

Once the car is stabilized, the rescue can begin. Sometimes the car can be reached by fire ladders or an aerial ladder, allowing the victims to climb to safety or be picked off. This does not, however, eliminate the need to stabilize the vehicle first if possible. If the rescuers need to get to the car, they should approach it from downstream, working from the eddy that the car creates. Even if the car is submerged, it may create enough of an eddy for the rescuers to get into. If the

Depending on the situation, the rescuers may access the car by wading, swimming, or by boat, ferrying into the eddy behind the car. Passengers can be evacuated the same way, although rescuers may have to break the car windows to get to them. Another method that has worked well is to evacuate the passengers by means of a tension diagonal set up from the car to the shore.

If the victims end up on top of the car, they can be treated like any other stranded victims, although

Fig. 11.23 Yosemite Park Ranger Dave Panebaker briefs a victim on the rescue plan.

Fig. 11.24 San Diego lifeguards use a rope to stabilize a flooded vehicle and as a hand line for evacuation.

Escaping From A Car In The Water

When a car enters deep water, usually after skidding off a roadway during a storm or at night, it will almost always float for a couple of minutes. This varies with the type of car, the integrity of the weatherstripping, and other factors. *This is the time to escape.* While opening the door against the pressure of the water is usually not possible, opening the window is. Testing has shown that electric windows will continue to operate for a period underwater, and a center punch or window-breaking tool stowed within reach will allow you break a window if the electrics fail. Overall your best bet is to open the window and crawl out immediately—even if you can't swim. If you can grab something that floats on the way out. Once the car goes underwater, your chances of escape or survival are minimal.

with the understanding that their car may not be a very stable refuge. Helicopters, if available and trained for the job, may be used to pluck victims from cars, and this is an especially useful technique for cars stranded in wide rivers.

A car that plunges off an embankment into deep water presents another type of problem. If the water is deep enough, most cars will float for 1–4 minutes—even with the windows open—giving a composed passenger enough time to roll down a window and get out. The car will float with the engine down, taking on water until it sinks, eventually ending up on the bottom. If the water is deeper than the car is long, it will generally invert and end up upside-down. In swiftwater terms, it will start as a top or floating load, then become part of the suspended load in the current, and eventually become part of the river's

Fig. 11.25 Yosemite Park rangers evacuating a victim from a car in the Merced River. The car has been stabilized by ropes to shore.

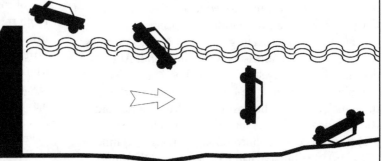

Fig. 11.26 A car entering deep water will float for a short time, then sink engine first. If the water is deeper than the car's length it will generally come to rest upside down.

bottom load. In strong currents, the car is likely to be totally wrecked by being rolled along the river bottom.

If the water is deep enough to completely submerge the car, and the victims are unable to get out on their own, the rescue options are limited. In fairly shallow water, a rescuer may be able to hook a tow cable to some part of the frame so that the car can be dragged or lifted out of the water, or he may be able to break a window with a diving knife or specialized glass breaking tool to allow the victims to escape. However, in strong current, the sudden decompression caused by breaking a downstream window may cause the upstream windows to blow out as well. If possible, crack the upstream windows first to relieve the pressure.

In many places, scuba divers are used for submerged vehicle rescues. While this works quite well in still water, it is *extremely hazardous* in moving water unless the divers have had specialized swiftwater training.

Incidentally, underwater tests have shown that a car's electric accessories, like power windows and door locks, continue to work even if the car is fully submerged. A new battery will power electrical systems underwater for 5–10 minutes.

Urban Swiftwater

According to the UN Population Fund roughly half the world's population lived in urban areas as of 2008, with over five billion people expected to live in them by 2030. Urbanization is a worldwide phenomenon, meaning that many if not most swiftwater rescues will be in urban or semi-urban areas, which adds a whole list of additional hazards and problems. Some of these hazards were mentioned in a general way in Chapter 2, here we will consider the specifics.

Flood Channels. The greater Los Angeles area has the largest flood channel system in the world. After decades of urban development, the Los Angeles River is almost entirely concrete-lined. The system includes three major rivers—the Los Angeles, San Gabriel and Rio Hondo—and is comprised of 100 miles (161 km) of main concrete-lined flood channels, 370 miles (596 km) of tributary channels, and over 2,000 miles (3,219 km) of ditches and storm drains. There are also 129 debris dams, 15 flood-control and water-conservation dams, and five flood-control dams. During

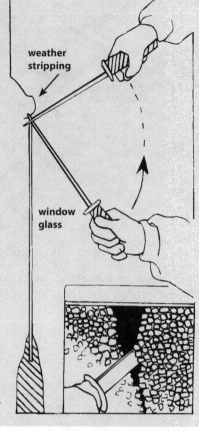

Breaking Windows

Breaking a car window, especially underwater, can be harder than it looks. Even when using a rock or heavy tool, the cushioning effect of water often keeps rescuers from striking the glass a really hard blow. One solution is to use commercial "center punch" tools made for the purpose. Another is to use a square-tipped, thick-bladed diving knife. This is wedged between the edge of the side window glass (this will not work on windshields) and the frame. The rescuer then lifts up and twists, shattering the window glass to gain access inside.

Since breaking windows is likely to be part of any vehicle rescue, count on having appropriate tools for the job with you.

weather stripping

window glass

Fig. 11.27

large storms the system carries more water than the Mississippi River. In an average year there are six flood channel drownings and over 40 major rescues. Many large cities in the American Southwest, notably Phoenix and Albuquerque, also have large flood channel systems and similar problems. However, it is a mistake to think that this phenomenon is limited to desert areas—creeks, irrigation ditches, arroyos, and canals anywhere are subject to the same problems, as are rivers that man has "improved" by smoothing and "armoring" the banks to reduce erosion.

From the engineer's point of view, the storm drain system is designed for efficiency—to get the water out of town as fast as possible to prevent flooding. From the rescuer's point of view, however, flood channels are a veritable shop of horrors. The water moves extremely fast (and so, consequently, does anyone in it), the channels often have slippery sloping sides with few anchor points and no eddies, and there are added hazards like low-head dams, underground sections, accumulated debris, strainers, and low bridges—not to mention chemical and fecal pollution.

The salient features of flood channels are their extremely high current velocity and the virtual lack of eddies. We saw earlier that in natural rivers, even in flood, the current velocity rarely exceeds 11 mph

Flood Channel Rescue

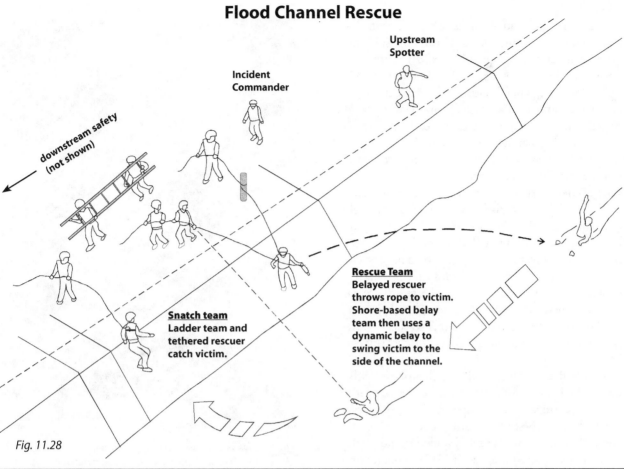

Upstream Spotter

Incident Commander

downstream safety (not shown)

Snatch team
Ladder team and tethered rescuer catch victim.

Rescue Team
Belayed rescuer throws rope to victim. Shore-based belay team then uses a dynamic belay to swing victim to the side of the channel.

Fig. 11.28

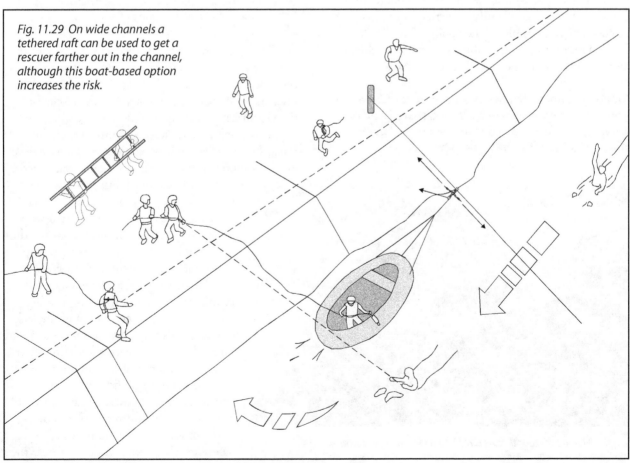

Fig. 11.29 On wide channels a tethered raft can be used to get a rescuer farther out in the channel, although this boat-based option increases the risk.

(17.5 km/hr). In flood channels, where all obstacles have been removed and the banks carefully smoothed, the current velocities frequently exceed 15 mph (24 km/hr) and have been measured at double that (30 mph/48 km/hr). There is little difference between helical and laminar flow as far as current velocity goes; the water pressure on the victim is nearly the same at the sides of the channel as in the middle. However, the helical current still tends to push the victim back into the middle of the channel, and in some cases this is deliberately designed into the channel to prevent debris buildup. Some ditches have slanted sides, while others have straight vertical sides. They vary in width from four to over 600 feet (1.2–183 m) and in depth from one to 34 feet (.3–10 m). Bottoms may be rock, debris, or concrete. Low-head dams (in some cases, inflatable models that can be moved around) are common, and there are frequently underground sections.

The assessment of flood channel rescues is heavily tied in with preplanning. We will discuss this further in Chapter 13, but for now it is enough to say that rescuers should familiarize themselves with the flood channels in their area before they are called to perform rescues in them. Specifically, they should know where the best places for rescues are, where the hazards like low-head dams and underground sections are, and what resources are available. At potential rescue sites they should identify, or perhaps even install, rescue-related items like anchor points. There will be little chance for assessment during an actual incident, and usually but one fleeting chance for rescue.

TIP: Where flood channels merge there is often a downstream eddy; where it makes a sharp bend or curve there is often an eddy on the inside; where the channel widens, the water slows down.

Fig. 11.30 A rescuer with a throw bag swings the victim in close enough for another rescuer to use a reaching rescue. Both are tethered to avoid being pulled in.

The rescuer's first major problem, because of the speed at which the victim is moving, will be to intercept the him. It is often the case that by the time the rescuers are dispatched and arrive at the rescue site, the victim has already gone past. Here, interagency cooperation and centralized communications are a must. Los Angeles County rescue agencies have developed computer software that lets them track a victim's position based on sightings and water speed. Helicopters can also be invaluable to vector rescue resources. It often works better to have rescuers already on-site during flood conditions than to try to deploy them once an incident is in progress.

The safety of rescuers is also a big problem. Because of the slick, sloping nature of the channel walls and the lack of eddies, it's very easy for rescuers to end up in the channel with the victim. Here, even more than in other situations, the rescuers should use shore-based rescues whenever possible. Rescue PFDs, which make it possible for rescuers to tether themselves, add an extra measure of safety for those who work in or near the

Fig. 11.31 One rescuer throws a rope to a victim while another waits downstream to catch him.

Fig. 11.32 A rescue team uses fire ladders to gain access in a vertical-walled flood channel. The victim has been pendulumed to shore by a rope positioned upstream.

ditches. Backups, such as rescuers with throw bags or a tension diagonal set up downstream of the rescue site, are also important.

Because of the speed of the current and lack of a good belay position, throw bag rescues must be done differently. If a rescuer simply stands on the channel walls and throws the victim a rope, he will almost certainly be pulled in. Because of the force, the victim is also unlikely to be able to hold on to the rope for long, and there are few eddies in which to swing him. The rescuer should tether or belay himself to avoid being pulled in, and his goal should be to swing the victim to the side of the channel as quickly as possible. If the tether is long enough, the rescuer should attempt to make as much of a dynamic belay as possible in order to reduce the amount of shock on the swimmer.

Just downstream of the thrower should be a "snatch team" of one or more rescuers whose job it is to physically grab the victim and get him out of the current, then help him to safety. These rescuers are also tethered and access the bottom of the flood channel by means of a ladder belayed to the top of the bank. A tethered ladder makes an excellent device for getting rescuers down to the water and giving them a place to stand and grip, as well as a way of getting the victim out of the channel.

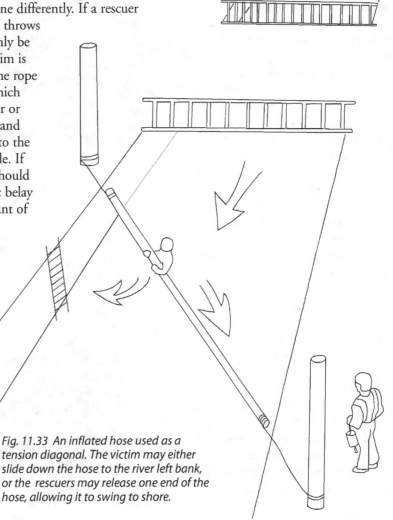

Fig. 11.33 An inflated hose used as a tension diagonal. The victim may either slide down the hose to the river left bank, or the rescuers may release one end of the hose, allowing it to swing to shore.

Throwing the swimmer a rescue can or ring buoy will also work, and gives him something more substantial to grasp. Victims in flood channels are seldom dressed for the occasion, and hypothermia often reduces their grip. Inflated fire hoses can also be used to good effect. The rescuers can suspend the fire hose just over the surface of the water, either from a bridge or using lines across the channel, then release one end of the hose when the victim grabs it, swinging him to shore.

While rescuers should *not* stretch lines across the channel at right angles to the current, they can use a tension diagonal placed at a downstream angle. Because of the forces involved, any tension diagonal in a flood channel requires a very taut line. The tighter the line, and the more downstream angle it has, the less chance there will be of "stranding" the victim in midstream.

Some victims may be able to slide down the line to safety, although they may get rope burns in the process. It requires strength and presence of mind on the victim's part. The County of Los Angeles and the Albuquerque Fire Department have developed a rescue net called the swiftwater rescue curtain, which is attached to a tension diagonal. This "shower curtain" is deployed across the channel just above the water and controlled by tag lines on either side. When the victim hits the net, the rescuers release the upstream tag line and pull on the downstream one. The combination of current and tag line rapidly gets the victim to shore. This solution does have its drawbacks, however. It takes time to set up, and in some cases has entangled victims.

There are other variations on the system. A simpler one is to combine the floating tag line and the tension diagonal. A ring buoy or similar float is suspended just above the surface of the water, clipped to a tension diagonal and controlled by tag lines on either side of the river. The tag line operators position the float in the victim's path. When he grabs it, they pull him to shore, assisted by the current. There have also been some designs with a series of rings tied together on the tension diagonal, each with a rope tied to it hanging down to the water.

Fig. 11.34 The rescue net in use. It is positioned just above the surface of the water at a 45-60° angle to the current.

Fig. 11.35 The net is held in position by a tag line on either side. When the swimmer makes contact with the net, the rescuers release the upstream tag line and haul on the downstream one . . .

Fig. 11.36 . . . pulling the victim quickly to shore. Because of the danger of entanglement, the victim must be brought to shore quickly.

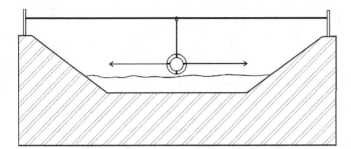

Fig. 11.37 *A very simple variation of the rescue net is a ring buoy mounted on a tension diagonal.*

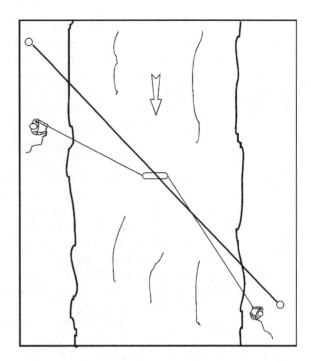

Bridge-based rescues often work well on wide channels. These use the same principles as discussed in the earlier sections. A fire hose or float is dropped to the victim, who is then pendulumed to shore. As with the throw bag rescues, a snatch team should be ready on shore to help the victim quickly out of the water.

The current in most flood channels is too swift for a 2- or 4-point tether with a raft, although a boat on tether can be set up to good effect. However, it is difficult to get a victim who is moving that fast into the raft. Often the tethered boat is better used to extend the reach of a shore-based throwing system. The rescuer in the raft throws the rope or float to the victim, who is then hauled in with a shore-based rope system.

Live bait systems can also be used to rescue exhausted or unconscious swimmers, but this poses even greater hazards than does the normal practice. An even more risky proposition is to use a free rescue swimmer in the flood channel. This has been done experimentally with rescuers using a large capture device like the Sky Hook™ and a tension diagonal. The rescuer secures the victim, then clips to the tension diagonal with the hook and slides to shore.

CLASSIFICATIONS OF VERTICAL WALL FLOOD CHANNELS

CLASS	SPEED	DEPTH	HAZARDS	
CLASS I	0–10 MPH	Less than 10'	No obstacles	No gradient
CLASS II	10–20 MPH	Less than 10'	Few obstacles	Has a gradient
CLASS III	20–30 MPH	10–20'	Numerous obstacles	
CLASS IV	30+ MPH	Greater than 20'	Stair-step channel, low-head or rubber dams present	

In classifying vertical-wall channels, three criteria have been established to qualify for a specific class. **ONLY ONE OF THE THREE AREAS (SPEED, DEPTH, OR HAZARDS) NEEDS TO BE PRESENT.** For example, a vertical-wall channel that is less than 10' deep, although it has a few obstacles, is classified as a Class II channel.

The classification system can also help you select the rescue option most likely to succeed:

CLASS I: Low-risk options will usually work.

CLASS II: Some low-risk options may work; however, "row" options work extremely well.

CLASS III: Higher-risk options are usually required. The use of personal watercraft, helicopters, Sky Hook Rescue System™, and tension diagonal boat systems work well.

CLASS IV: Extreme caution is needed in these instances. Rescue options depend on the channel characteristics and hazards present.

–devised by Gary Seidel, Los Angeles City Fire Department

There are certain places, however, where a free rescue boat or even a swimmer may be employed. If the channel has been scouted prior to the rescue and it is known for certain that no hazards exist downstream, the rescuers may launch a raft and chase the victim down, then ride the raft to the end of the channel, whether it ends in the ocean or a reservoir.

Helicopters, which are a quickly-deployed and flexible solution, are routinely been used for flood channel rescues in the Los Angeles area. They are also quite useful as an airborne command post or spotter. If helicopters are to be used as rescue vehicles in a flood channel, it is *absolutely essential* that the channel(s) be scouted prior to use so that helicopter "safe zones" can be established (i.e., zones that are free of obstructions like power lines, bridges, etc.) and so that rescue swimmers don't go in the water just above a major hazard in case the hookup fails.

The rescues are done as described in Chapter 10: the helicopter matches the speed of the victim in the flood channel, and either lowers a tethered rescuer on a long line, or inserts a free rescue swimmer. The rescuer then secures the victim, and the two are short-hauled to shore. A free rescuer will have to make a hook-up to the long line first.

Flood channel rescue is one of the most challenging types of swiftwater rescue. More than any other area, pre-planning and training are the keys to successful operations.

Storm Water Systems: In addition to flood channels, almost all urban areas have some sort of drainage system to allow storm water to drain away from populated areas. During heavy rains the storm water system may become *overcharged* by taking in more water than it can hold, in which case some must escape. A common indication is a mound or even a geyser of water spewing out of the system, which frequently pushes the storm water drain covers off. The danger begins when the water level drops and the process reverses. Water then flows back into the system through the now-open drains. *This inflow may take people and even vehicles with it!* These may be flushed down into culverts and underground sections of flood channels, where they are virtually impossible to rescue. Often this is hard to see and avoid. A good rule is to avoid walking or driving through any area where you can't see bottom, and to be alert for the circular swirl of water disappearing underground.

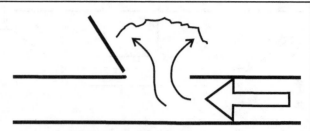

Signs of outflow: *a mound or hump of water boiling up, or water spraying into the air.*

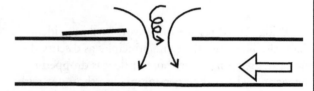

Signs of inflow: *a whirlpool or depression in the water. DANGER — use extreme caution!*

Fig. 11.38

Fig. 11.39 *A whirlpool like this means that water is being sucked underground into the storm drainage system.* **Use extreme caution!**

Other Urban Hazards

Low bridges: cars can be swept off them during floods, or they can be blocked with debris and become dams. These debris dams form dangerous strainers and can release large volumes of water downstream if they break. Bridge railings may also become hazards.

Under flood conditions, enough debris may accumulate under a bridge to obstruct the entire river, making passage underneath potentially deadly. In

Fig. 11.40 Low bridges such as this one often collect debris and may entirely obstruct a river, becoming a "debris dam." Operations upstream of them are extremely dangerous.

or otherwise contact with them. Boats may contact overhead power lines during high water.

Fires may break out during a flood, and in fact this is a fairly common occurrence. One common cause is propane tanks mounted outside of buildings that are washed away by flood waters. When the heavier-than-air vapors contact an ignition source a fire or explosion results. Other fuels include natural gas, gasoline, fuel oil and other hydrocarbon fuels. Access for fire-fighting equipment may be very difficult in a flood, as will the choice of proper PPE for rescuer/firefighters. In one case in the upper Midwest, fire equipment was brought in through flooded streets on high-clearance flat-bed trailers.

extreme cases, this may actually convert the bridge into a dam. These "debris dams" under bridges have also been known to collapse suddenly, sending a surge of water downstream. Rescuers must be extremely cautious about operating upstream of them in the water.

Strainers of all types are very common in urban areas as water flows through fences, etc. and debris collects across narrow areas.

> **TIP:** Beware of sharp objects in flood waters that may cut personnel and damage inflatable watercraft. All personnel should wear sturdy foot protection.

Fig. 11.41 Urban areas abound in strainers, such as this fence.

Electric shock is a significant hazard in urban floods. Rescuers should coordinate with utility companies to turn off power to flooded areas before entering. However, there may be some facilities like hospitals where power must remain on. Be aware that some places have backup generators that may power up areas that have been de-energized by the power company, and these may start automatically when the main power goes out. Use caution in flooded buildings—avoid contact with electrical appliances such as air conditioners and the like, which may be hidden under water or mud. Live power lines can also be covered with water or mud where you may step on them

Contamination and HAZMAT considerations. Any flood is by definition a HAZMAT incident, but urban areas often present a significantly greater contamination threat, including but by no means limited to:

- Human and animal fecal matter. Water treatment plants often stop working during floods, dumping raw sewage into the rivers.
- E. Coli bacteria and other waterborne pathogens
- Acids and caustics
- Chemical leaks and illegally dumped chemicals
- Dry chemicals leached out of storage areas and basements

- Hydrocarbon fuels (gas, oil, diesel, etc.)
- PCBs
- Propane, natural gas and other fuels
- Pesticides and agricultural chemicals
- Household chemicals (insecticides, fertilizer, paint, etc.)
- Dead people and animals

Industrial and agricultural areas should be carefully inspected for contamination and hazardous materials. Check:
- Placards and labels; container shapes and design
- Shipping documents, including Hazardous Materials Management Plans (HMMPs) and Material Safety Data Sheets (MSDSs)
- Quantity and physical characteristics of the material.

When responding to an urban flood incident, keep the following considerations in mind for HAZMAT protection.

HAZMAT Zones
- Hot—Incident Site and water
- Warm—Decontamination Area
- Cold—Safe Area for EMS Staging

HAZMAT Decontamination Principles
- Decontamination in warm zone.
- Decontamination prior to patient transport.
- Patient care by protected responders.
- No patient/rescuer allowed to leave without decon.

HAZMAT Decontamination Procedures
- Patient removed from hazard area by rescuers.
- Remove patient's clothing, jewelry and shoes.
- Gently brush away any dry materials.
- Blot away any visible liquids.
- Take care not to cause skin damage.
- Rinse patient with warm water.
- Keep spray gentle.
- Wash with mild soap (Tincture of Green Soap).
- Rinse patient.
- Contain runoff.

After decontamination: rest–hydrate–vital signs every 5 minutes. Hydration solution—1 tablespoon sugar in 1 gallon of water (protocols courtesy of University of Virginia Prehospital Program)

Before the incident:
- Establish a medical baseline for all rescue personnel.
- Insure that all inoculations (typhoid, tetanus, etc.) are up to date.

During the incident:
- Establish a medical monitor.
- Monitor water conditions at frequent intervals for contaminants. Periodically collect and analyze water samples; date and retain them for record use.
- Limit the number of personnel in the water, and the time they remain there.
- Use dry suits where possible to reduce skin exposure.
- Decontaminate all personnel *as soon after exposure as possible*. No exceptions!
- Include rescuers, victims and animals in your decontamination plan.
- Locate a decontamination point as close to the incident as possible. Establish a "decon gateway" and do not let rescuers, victims, animals, etc. leave the area without decontamination. Have a designated area to collect contaminated clothing, equipment, etc. as well as private areas for changing.

After the incident:
- Monitor the post-incident condition of all exposed personnel and follow it up at intervals.

Swiftwater Scuba Operations

At this writing, safe and effective scuba operations in swiftwater are the exception. It is much easier, unfortunately, to cite instances where divers have drowned while attempting rescues or recoveries in moving water. In most cases, this has been due to ignorance of the special dangers of swiftwater and attempts to use the same techniques for swiftwater operations as for normal scuba operations. Swiftwater scuba operations are a recent innovation whose equipment and doctrine are still evolving, and there are at present only a small number of people able to do it safely. Scuba is a specialized subject, and we can only mention some of the more important problems and trends here. Any diver contemplating swiftwater rescues *must* get expert instruction before attempting it. Unfortunately, in some agencies all water rescue calls go to the dive team, whether it has been trained in swiftwater rescue or not.

A scuba diver, encumbered as he is by tanks and a weight belt, is at a decided disadvantage in swiftwater. His equipment tends to catch the current much more so than surface rescuers, nor is most of his equipment designed for swiftwater. Problems include:

- Lack of quick-releases on equipment like tanks, weight belts, and especially tethers. Tethering divers in calm water is common practice, but may be deadly in swiftwater if the diver is unable to release himself. Most diving tethers are not designed with this in mind.

- Regulator problems. The pressure of the current on many common regulators will cause them to release oxygen at a much higher rate than normal. It is possible to empty a set of tanks in a few minutes under the right conditions.

- Communications problems. Most standard diver-to-topside signals will not work in swiftwater.

There have been several instances of divers in sit harnesses clipping themselves to fixed lines and either having their tanks emptied in a few minutes by water pressure, or having their masks and regulators torn off by the force of the current.

Nevertheless, it is possible to dive in swiftwater if the divers are properly equipped and know what they are doing. It is even possible to set up an underwater tethered lower, moving the diver back and forth in the river to search for drowning victims.

First, any diver contemplating this type of work should, in addition to his scuba and public safety diving credentials, get appropriate surface swiftwater rescue training. Second, he should investigate the proper equipment. In order to avoid the regulator problems mentioned earlier, swiftwater divers use a full-face mask with a built-in regulator. Divers like Rescue 3's Barry Edwards also recommend using the new diver-to-surface communications equipment so that the surface crew is instantly aware of any problems below. Any tethering device must be capable of quick-release in an emergency.

Because of the problems outlined above, swiftwater scuba operations are limited to fairly slow current. In addition, because of the relatively cumbersome nature of scuba equipment, rescuers should avoid setting any swiftwater dive site above a hazard. Instead, dive sites should be chosen with an open, hazard-free runout below. Setting safety is difficult, since an underwater diver is difficult to reach with conventional means like throw bags.

To set a search pattern, the dive team, after surveying the site, sets a conventional tethered lower across the river. The biggest difference is that the anchor line is set closer (1–3'/.3–1 m) above the surface of the water. In one version, the diver hangs on to a padded loop as he is worked across the current. A major advantage of this system is that it makes a grid search pattern very easy. In others, the diver uses a shoulder harness for a tether. As mentioned before, this harness must have a quick-release in case of an emergency.

Conclusion

Some swiftwater rescue situations—especially stranded victims, auto rescues, and bridge rescues—happen often, and are something that any swiftwater rescue team should be familiar with. Identifying these situations in your jurisdiction is part of the preplanning process discussed in later chapters.

Incident Command and Site Organization
Chapter 12

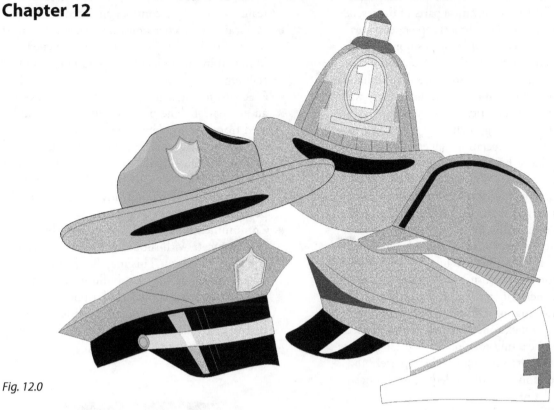

Fig. 12.0

So far we have discussed the duties of the individual rescuer and the mechanics of setting up various systems. Now it's time to see how to actually put this into practice at a rescue site. To perform a successful rescue, we need not only trained rescuers and equipment, but a way to control their use—that is, an incident command system. Incident command sets priorities, allocates resources, and organizes and provides leadership for the rescuers. The command system does not operate in a vacuum, however. Although the principles of incident command can be applied to any emergency situation, there are specific considerations for swiftwater rescue. There are also subjects that are an important part of any incident: critical incident stress, interagency coordination, and liability considerations.

This chapter is not, however, a rewrite of incident command publications. There is not enough space to go into the details of the systems in use today: fire incident command, the ICS system more familiar to search and rescue organizations and the NIMS (National Incident Management System) used at the federal level. Other countries have their own systems. Instead, we will look at the duties and functions of team members and the organization of a typical rescue site. With this information, the responsible agency can integrate these functions into its own incident command system.

Team Organization and Functions

Leadership/Command. There must be a single person in charge of any swiftwater rescue incident. There is no time for democratic discussions about techniques or arguments about jurisdiction. The structure is a hierarchy—the overall person in charge of the incident is the *incident commander* (IC). In addition, there may be subordinate *team leaders* who report to the incident commander. It is the IC's overall responsibility to ensure the safety of the rescuers, choose the correct rescue systems, organize the rescue site, and supervise the rescue and subsequent evacuation.

In short, the IC is responsible for the big picture. That means he must rely on delegation of authority to get the actual rescue done. While he can delegate some of his *authority* to subordinates, the *responsibility* of command remains his. The most common mistake ICs make is to become so involved in the details of the rescue that they ignore the larger aspects of site management. This, as we will see, can be quite complex, with a large number of things happening simultaneously.

The IC should position himself where he can best see the entire incident and exercise overall control. He must provide himself with the means to communicate his orders to subordinates and have them relay information back to him. The IC himself need not be the

---

Swiftwater Rescue

most skilled or knowledgeable person at the rescue site. However, he should ensure that qualified people are in charge of the more technical parts of the rescue. A technique often used in technical rope rescue is to designate a *control;* that is, someone positioned to do the actual supervision of critical tasks like raising and lowering litters, leaving the IC some freedom to handle other problems. For the same reason, the IC may want to delegate hands-on authority to *operations officers* who have specialty training, such as a swiftwater operations officer, a medical operations supervisor, or an air operations officer. Each one of these officers might in turn have team leaders who report to them. The maximum span of control that any team leader can be expected to effectively supervise is four to six people.

The IC is often assisted by a *staff* who advise him on specialty matters or take care of business that would otherwise distract him. While the composition of this staff varies, some positions of great value on any swiftwater incident are:

- **Public Information Officer (PIO).** Handling media personnel, concerned relatives, and spectators can be a full-time job. A PIO can insulate the IC from pressures like pushy reporters and distressed family members, letting him concentrate on the rescue.
- **Safety Officer.** In the confusion and hurry of the typical swiftwater rescue, a safety officer can be invaluable. He should be given no other duties but to roam the rescue site looking for unsafe practices (no PFD, rope abrasion, knots improperly tied, rescuers becoming tired or hypothermic, etc.). He should either correct these problems on the spot or advise the IC of the situation. It is the rescuers' job to concentrate on rescuing the victim; it is the safety officer's job to concentrate on the safety of the rescuers.
- **CISD/Family Advocate.** Crisis counseling, whether by clergymen (e.g., volunteer chaplains) or secular counselors, has been used by law-enforcement agencies for some time. These serve both rescuers and victim family members. Unfortunately, they are still rare for EMS agencies.

In addition to the IC, the following functions must be addressed. They may be done by different teams or several functions may be performed by the same person.

- **Extrication.** Some team members, preferably the most knowledgeable and experienced in the type of rescue being performed, should be assigned for the actual rescue. They may be supervised by a team leader or by a control appointed by the IC.
- **Medical.** The rescuers must be prepared for treatment of their patient once they have rescued him, within the limits of their equipment and training.
- **Evacuation.** There are generally two stages to an evacuation. The patient must first be evacuated from the accident site to a place of relative safety and stabilized, and then evacuated to a medical facility like a hospital. During the first stage, care is usually limited to stabilization, while in the second, advanced life support (ALS) is usually available.
- **Communications.** The IC must be able to communicate with subordinate rescue units— and they with him. This may be with such diverse means as whistles, bullhorns, hand and arm signals, radios, cellular phones, pyrotechnics, or signal mirrors. In addition, the IC must be able to communicate with outside agencies and resources.

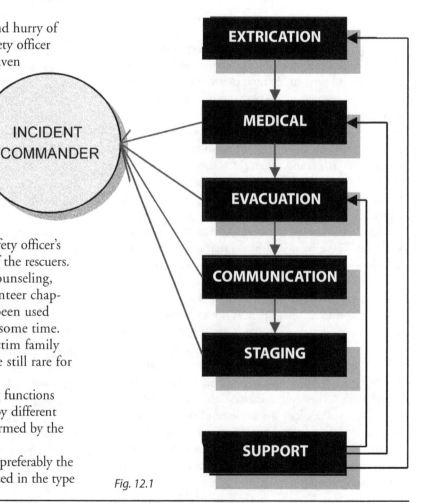

Fig. 12.1

- **Support.** Logistics are vital to any rescue. Personnel must be monitored and accounted for, and warm-up facilities provided for rescuers, as well as food, drink, and shelter during extended operations. Gear, sometimes a lot of it, must be transported to the rescue site. Sometimes this can be done by trucks or aircraft; other times it must be done by "mules"—i.e., on the backs of rescuers. After the rescue is over, the gear must be broken down, accounted for, transported back to storage facilities, cleaned, and stored.

To get an idea of how this works in practice, we will look at the Los Angeles City Fire Department Swiftwater Rescue Response Teams. These teams, similar to the "strike teams" employed in wildland firefighting, are deployed for the actual rescues. The overall incident command and logistics support come from the host agency. The teams are often organized on an interagency basis, typically with four firefighters and two Los Angeles County lifeguards. The team leader is appointed from the agency having jurisdiction.

The actual size and organization of the team varies according to the mission. The minimum size is two persons.

- **Two-person:** leader and rescuer
- **Four-person:** leader, rescuer, technical specialist, downstream safety.
- **Six-person:** leader, rescuer, technical specialist (2), downstream safety, upstream safety.

In addition to the rescue teams, there are also *search teams*, whose only job is to find the victim (e.g., helicopter units), and *first responder teams*, who are limited, by reason of their training to shore-based rescues, technical support, and search functions.

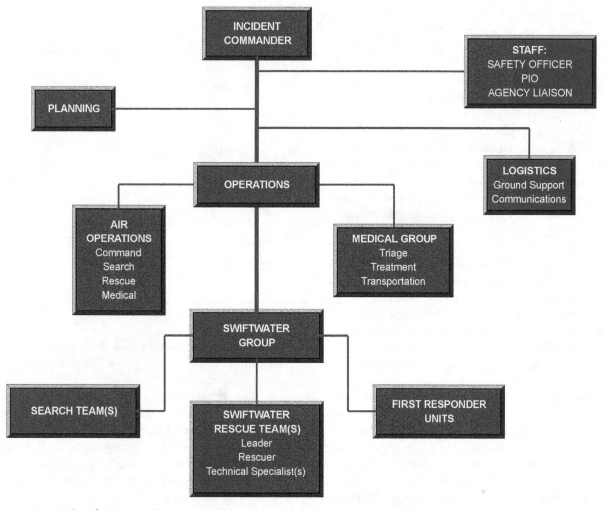

Fig. 12.2 A typical swiftwater incident command structure.

Swiftwater Site Management

It is up to the IC to effectively manage the rescue site so that victims are rescued safely and efficiently and rescuers kept safe. He is assisted in this by the safety officer, who constantly reviews safety options. Some of the most important of these site management functions are:

- **Upstream Spotter.** His job is twofold. First, to stop anyone from coming down the river and interfering with the rescue. Rescuers often have lines stretched across the river that could "clothesline" river traffic. In some cases the civil authorities may close the river to commercial traffic like rafts. Second, for those things he is unable to stop, like logs or swamped boats, the spotter can at least warn the rescuers that they are coming.

- **Downstream Safety.** Both rescuers and victims may get into the current and end up being washed downstream. Rescue systems, whether throw bags, rescue boats or a combination, should be set up downstream of the accident site so that *no one gets downstream of the last rescue system.*

Naturally, the IC should exclude anyone from the rescue site who has no business there. This includes media personnel, bystanders, and even rescuers if there are too many of them. If personnel are few, the IC may want to accept citizen volunteers. There are some jobs, such as pulling on a rope, that can be done by volunteers without unduly endangering themselves. There may also be people available on-site with needed specialties, like doctors or paramedics. If possible, these volunteers should be given a safety briefing and a PFD. Two groups of potential volunteers who are frequently overlooked are professional river guides and recreational whitewater paddlers, who may have excellent river skills.

Another important part of site management is the concurrent setup of rescue backup systems. Here, we mean not just downstream safety, but alternative rescue systems. For example, one team might be trying a quick system like a rescue ferry with a boat, while another is busy setting up a boat on tether, and the IC is vectoring in a helicopter for a short-haul attempt. It is the IC's job to stage manage all this, and generally to balance the conflicting demands of time, rescuer safety, available resources, and the needs of the

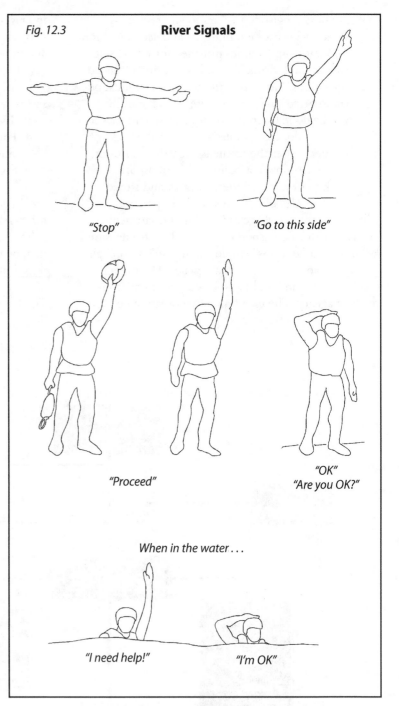

Fig. 12.3   **River Signals**

"Stop"

"Go to this side"

"Proceed"

"OK"
"Are you OK?"

When in the water . . .

"I need help!"

"I'm OK"

victim. He must constantly ask three questions: How *long* will it take? How *dangerous* is it to the rescuers? How much *equipment* will it take?

The personal safety of the rescuers should also be emphasized. Anyone within 10 feet (3 meters) of the water should wear a PFD (the 10-foot rule). When working around ropes in the water, personnel should take care to stay out of the bight of a rope so that they do not become entangled, and should also not stand downstream of a rope with an in-current load like a throw bag.

Upstream spotter

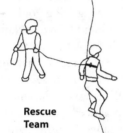

Incident commander

Support/ backup

Rescue Team

Downstream safety

Both the IC and the safety officer should be constantly checking the rescuers for signs of fatigue and hypothermia, and rotating them for a rest or a warm-up as needed. They should not forget to check the spotters, who may be out of sight up and downstream. If the circumstances warrant it, the logistics coordinator should set up a place for warming up and provide hot drinks and food. It is not unusual for rescuers to be so focused on the victim that they neglect themselves and each other.

Especially in an urban setting, the public information officer (PIO) is also an essential part of the incident command team. The PIO's job is to reduce the pressure on the IC so that he is able to make calm, rational decisions. While it is the PIO's job to convey information, he should limit himself to facts and not speculate, especially when dealing with reporters. Speculations reported as facts have a way of coming back to haunt the speakers. At the same time, PIOs and ICs should recognize that the public does have a right to know what is going on, and that the media has a right to report it. They do not, however, have a right to be in the way of rescue operations. PIOs should be especially careful not to reveal the identity of victims while the rescue is ongoing or if a fatality has occurred. The victim's name should be released only after notification of the next of kin.

The victim's family, some of whom may be under extreme emotional stress, may also be a source of pressure. It is important that the rescuers realize that this is a difficult time for any family and act accordingly. Someone must buffer the two—giving the family an honest account of *what* is being done and *why* while acting in a sensitive and sympathetic manner, and at the same time protecting rescue team members from appeals to "do something." In some organizations this will be done by the PIO, but a better solution may be to use a family advocate such as a clergyman or crisis intervention counselor.

Nancy Rigg, a Los Angeles screenwriter, remembers her feelings after her fiancé, Earl Higgins, was swept away while attempting to rescue a child in the flood-swollen Los Angeles River in 1980. Even though public safety agencies soon responded, Higgins went 30 miles downstream past rescuers who had neither the equipment nor training to rescue him.

Left to stand alone in the rain, with no one to talk to and no offer of a blanket, comfort, or shelter, Rigg was stunned when officials simply called off the rescue and indicated that there would be no follow-up search for Higgins's body. "In terms of disaster management," Rigg recalls, "I now know that there is a big difference between 'rescue' and 'recovery.' But when Earl was swept away, no one was willing to communicate with me, let alone trust that I could understand things, so I concluded that life held little value in Los Angeles."

Fig. 12.4 Organization of a typical swiftwater rescue site.

Agency Coordination

Very few rivers flow entirely in one jurisdiction. Most run through several and often they form the boundary between two jurisdictions. For example, the Chattooga River forms a state border (Georgia and South Carolina), and the Rio Grande an international one (the United States and Mexico). Often, little thought is given to what will happen if an emergency happens on the boundary. A good example of this type of jurisdictional confusion followed the airliner crash into the Potomac River in Washington, DC, in 1982.

Another recent example was the drowning of 15-year-old Adam Bischoff in Los Angeles in 1992. Bischoff fell into a flood channel and was swept downstream at high speed. Rescuers deployed before the TV cameras to save him, but by the time the word had been passed from one jurisdiction to another, the boy had passed through. On a more positive note, this high-profile incident resulted in a thorough review of rescue plans, and eventually a model interagency action plan. This is more of an accomplishment than it might at first seem. Los Angeles is not a single city, but rather an aggregation of over 80 smaller cities. In all, there are over 115 agencies in the greater Los Angeles area (police, fire, EMS, lifeguard, etc.) that have some responsibility for water rescue. If it can be done there, it can be done anywhere.

Pride, unfortunately, often takes a hand, as do the normal perversities of human nature. Some agencies are loathe to admit that they need help; other times there is friction or bad blood between agency heads, turf wars, and funding battles that limit or prevent cooperation. Preplanning is an essential, and sometimes lengthy, part of this process. However painful, it needs to be done *before* an emergency.

Los Angeles County, for example, now uses interagency swiftwater rescue teams comprised of both firefighters, law-enforcement rescue-paramedics, and lifeguards. When conditions dictate, the teams are deployed under an unified interagency incident command system.

While we are on the subject of getting along, it is worthwhile to consider "the other professionals," the raft guides and boatmen. Commercial whitewater rafting, and river sports in general, have grown greatly in popularity since the late 1970s. In many areas of the country almost every stream, it seems, has an outfitter. In many cases these people have had extensive experience on whitewater rivers and rescue training as well. Many outfitters, in fact, regularly run river rescue clinics and sometimes even "rescue rodeos." They also have equipment that is suited for swiftwater operations. Yet working together is often a problem. Why?

Some of the problems come from attitudes on both sides. Rescue squads sometimes tend, often quite unconsciously, to regard the rafters more as part of the problem than of the solution. There is also the culture-clash aspect: it is often difficult for a crewcut firefighter to take the raft guide, who has a pony tail and gold earrings, seriously.

The mutual benefit can be illustrated by an incident on the Tuckaseigee River in western North Carolina. Following the drowning of a kayaker, the local rescue squad and a river outfitter worked together to search for the body. By mutual agreement, all on-river operations were handled by the outfitter, while the rescue squad and the U.S. Forest Service provided incident command, communications, and logistical support.

Critical Incident Stress

"We were getting goofy," recalls Jim Traub, a ranger at Grand Canyon National Park. "It was the third body recovery in a four-day period, and we couldn't readily recall what the number of major SAR incidents was for the season. While the operation was going OK, it wasn't the smooth and polished event we had practiced. A look and a few words with Ken (another ranger) confirmed that he felt the same way. We added an extra level of caution to our actions of rigging the raising and lowering systems as we realized that the season was catching up with us and we were stressed."

The cumulative effects of incident stress, both physical and mental, are something that any rescue professional must deal with. The effects are both immediate, as in the above incident, and long-term. It affects rescuers, victims and families, and the unit cohesion of agencies.

In addition to rescuers, the victim's loved ones, particularly those who may have witnessed or somehow participated directly in the incident, can become secondary, "invisible" victims. "When they suddenly called off the search, there was no one on scene to help me understand *why*," Nancy Rigg recalled. "Earl's 'case' was then dumped in the laps of homicide detectives who knew nothing about water rescue, let alone how to mount a useful search for remains, or the very compelling need to do so. I was in limbo for nine months until Earl's body was finally recovered, *by chance*, during a routine dredging operation. Even more than watching Earl get swept away, it was these numerous 'secondary injuries' which led to my developing a serious case of PTSD."

These emotional traumas are called *critical incident stress* and can result in both physical and psychological disability. Since the effects often surface well after the incident, it is also called *post traumatic stress disorder.* While this is not a book on psychology, rescuers should realize that injuries like this can be just as disabling as physical ones like hypothermia, and act accordingly. Sleep disorders, flashbacks, and feelings of guilt are common, as are physical symptoms like hypertension. It can also cause a reduction in job efficiency, even to the point of endangering both rescuers and victims.

Like hypothermia, critical incident stress should be dealt with on an ongoing basis. Defusing and dealing with incident stress is best done with *critical incident stress management* (CISM) practices, integrated into the command structure. Trained peer support personnel, operating either under the safety officer or IC, assist at every stage of the rescue process.

Sherrie Collins, coordinator of one of the National Park Service's critical incident debriefing teams, defines their role during a search or rescue incident. "They act as the eyes and ears for the incident commander on field conditions, behavioral problems or changes noted in crews, and signs of fatigue, strain, and stress in workers. Their focus is prevention of stress-related problems." According to Collins, they interact in the following situations:

- **On-scene Support.** Available during emergency operations, particularly lengthy technical rescues, body recoveries, disasters or mass casualty situations, large-scale searches, or situations involving the death or injury of emergency workers.
- **Demobilization/Defusing.** These are short information/rest sessions and small group discussions, respectively, that take place immediately after an incident.
- **Debriefings.** Formal debriefings utilize a critical incident stress debriefing (CISD) team after the incident as needed.
- **Peer Support.** This is an ongoing process where individuals are trained in peer counseling and provide support as a "listening post" for fellow workers.

SOPs and SOGs

Standard operating procedures (SOPs) or guidelines (SOGs) establish guidelines for emergency operations. They may cover everything from incident command to which knots should be used in certain situations. A good SOP/SOG,, properly integrated with a training program and the preplanning process, can be invaluable, saving the IC many small decisions. However, they can also cause problems, such as:

- **The paper plan SOP.** These are thick, dusty SOPs moldering in various offices that haven't been updated since the Cuban Missile Crisis. No one ever reads them.
- **The nitpicker SOP.** The document is huge, and tries to cover in exact detail everything that could possibly happen. Since it cannot do this, the only effect is to tie the IC's hands, preventing him from making any real-world adjustments to the situation as it unfolds.
- **The new age SOP.** This is so vague that it provides no guidance at all. Instead it offers only self-congratulatory platitudes.

A realistic SOP/SOG should cover ways of handling specific situations that are likely to arise. At the same time, the IC must have enough flexibility to improvise where necessary, since each swiftwater rescue situation is unique. Some areas that most SOPs should cover are:

- Personal protective equipment
- Team equipment
- Incident command and site organization
- Training standards
- Incident risk management
- Standard methods of implementing rescues (i.e., boat on tether, two-boat tether)
- Rigging

Liability Considerations

What follows is intended as a general discussion of liability. While there are common concepts of liability that apply pretty much universally, it is also true that liability law varies a great deal from one area of the country (and the world) to another. And while it is a good idea to get advice on specific situations from a lawyer, you should remember that lawyers like to stay out of trouble and tend to be rather conservative in what they tell clients.

The overall concept of *negligence* is simple: if you have a *duty* to rescue someone (which most agencies do) and you *fail* in that duty, and someone suffers *damages* (i.e., is injured, killed, or suffers financial loss) as a result (the proximate cause), then you can be held *liable*. All these elements must be present for a judgment of liability. For example, you can totally fail in a rescue, but if it were determined that the victim died of unrelated natural causes, no liability would result.

It is impossible to second-guess the legal system, yet rescue agencies sometimes spend a great deal of time and money trying. Often this leads to elaborate "CYA" policies written by lawyers rather than rescuers. The best defense in a negligence suit is usually common sense and knowledge—"we did it that way because, considering our training and the circumstances, we felt that it was the best and safest way to do it." Be prepared to explain the rationale for what you did or didn't do. When it comes down to a question of rescuer safety, consider the axiom "it is better to be judged by 12 than to be carried by six."

The level of competence rescuers are expected to show depends on two things: what might loosely be called "the overall standards of professional competency"—sometimes called the *standard of care*—and their individual level of training. The first applies to the agency as a whole and the second to the rescuers themselves.

Standard of care is a somewhat murky area, and stems from the general proposition that rescue organizations are supposed to know what they are doing.

For example, if a rescue squad undertakes a swiftwater rescue, they will be assumed to know generally how one is supposed to be carried out. Their performance will be compared to other squads in similar situations nationwide.

As for the rescuers themselves, there are several levels of training and proficiency. As these increase, so does liability exposure.

- **Awareness/familiarization.** An ordinary prudent person has a low level of liability exposure, and is responsible only up to his level of training. The standard is what a *reasonable and prudent* person would have done under the same circumstances.
- **Professional/technician.** A person who has attended a specialized training course (e.g., Swiftwater Rescue Technician I) and has met the minimum level of competency (i.e., a standard course of the instruction's objectives), but is not necessarily *proficient*.
- **Specialist/expert.** Has had extensive training *and* experience to attain a specified level of proficiency (e.g., doctor, paramedic, firefighter, registered nurse). This is a professional *certification*, and proficiency is expected. Consequently, there is more liability exposure.

In a typical rescue organization, most team members will be technicians, with some members having specialist qualifications (paramedic, scuba, etc.).

Conclusion

Incident command forms the framework for rescue operations, allowing rescuers to integrate the various rescue functions—extrication, evacuation, and medical treatment—into a coherent whole. Incident command must, however, be combined into the larger operational framework, including cooperation with other agencies, resolution of jurisdictional disputes, training, and preplanning.

Preparation, Planning & Response
Chapter 13

Fig. 13.0 Fire and flood at Johnstown, Pennsylvania, May 31, 1889.

Many flood responses fall into the "too little, too late," category. The scenario is distressingly familiar, especially for flash floods: as the flood hits, residents and local emergency workers struggle to survive and rescue others. Emergency managers struggle under a staggering call load, trying to separate the desperate from the trivial. They call outside resources, but by the time they arrive on scene the flood is mostly over. Or, in floods that are prolonged such as Hurricane Floyd in North Carolina and Katrina on the Gulf Coast, the rescuers are unable to make a timely arrival because many of the roads are blocked or underwater. In the aftermath the response is generally justified with sentiments like "we did all we could" and "there's just no way you can plan for something like this." Yet many of the problems were due to using an antiquated and inadequate response model. In this chapter we will explore a more functional response model as well as the mitigation, planning and preparation measures that go with it. But first, let's look at some of the characteristics of floods.

Floods generally come in two varieties: *flash floods* and *river floods*. Many have characteristics of both, beginning as flash floods and ending as river floods.

In general, a river flood is just what the name implies: too much water in a river's watershed, usually caused by a sudden storm or a physical event like a dam break. It often affects a wide area, quite often an entire region, and generally has a slower onset and retreat than a flash flood. This leads to a different time problem than a flash flood, which comes and goes in a hurry. A major problem in river floods is exhaustion of the rescuers over an extended time period.

A flash flood, on the other hand, tends to be an intense but short-term event. The entire incident may last only four to six hours from start to finish. One of the salient characteristics of a flash flood is that there are a very large number of life-threatening incidents in a very short time (more about that later). A flash flood may affect an entire region—or it may be localized to only a few city blocks. Time is always the rescuer's enemy, and the major management problem is that there are never enough rescue resources to go around.

All floods, generally speaking, have three phases, which often overlap:

Swiftwater Phase: this initial phase is where the most fatalities (and the most rescues) occur. Large numbers of people, including those in rescue agencies, are caught unaware, often in darkness. Many are swept away. In developed countries the two largest groups are people killed are 1) those driving through moving water and 2) children playing near flooded creeks and flood channels. There are large numbers of incidents, which tends to overwhelm available resources.

Flood Phase: the situation stabilizes somewhat, but there is standing water everywhere and rivers may continue to rise, further restricting access. Those who are able have rescued themselves, at least gotten to positions of less immediate danger like the roofs of houses or the tops of cars. Because of the flooded roads rescue units cannot access many locations. Large numbers of people and animals may have to be evacuated from flooded areas and are displaced to shelters. In a river flood, this phase may be prolonged, sometimes for months.

Recovery Phase: the water begins to recede and people return to their homes. The dangers of this stage are more indirect—downed power lines, debris,

Fig. 13.1 The Mississippi surges through the levee near Columbia, IL, during the flood of 1993.

disease, contaminated water and food, etc. There are major infrastructure problems such as washed out roads, loss of power, and lack of clean water. Rebuilding begins.

Mitigation

The first lines of community defense are mitigation and prevention of swiftwater incidents. Obviously, the safest and easiest rescues are those that we don't have to do. Although it is hard to draw an exact line between mitigation and preparedness, mitigation is concerned primarily with measures taken well before disasters and incidents to lessen their effects. As such they call for the active participation of potentially affected communities in public policy questions involving land use planning, including building on flood plains and the role and design of flood control structures (dams, levees, flood channels); hazard analysis and abatement; building codes; and the role of insurance. An effective flood mitigation program, for example, requires some historical research, since all regions have dry and wet weather cycles.

People have always built on the river's flood plain. "Bottomland" is flat, fertile, valuable—and subject to flood. Some have compared the flood plain to a time-share condominium. Man occupies it most of the time, but at others the river demands its share, disregarding control structures like dams and levees. The hydrology is simple: constrain the river and it will rise higher and flow faster, increasing its destructive power. Creeping urbanization covers more and more land with concrete and asphalt, preventing water from

The Great Flood of '93 in the Midwest

- **Flooded more than 70,000 homes and 5,000 businesses.**
- **Killed 38 people.**
- **Caused damages estimated at $12–$16 billion.**
- **Cost the federal government more than $6 billion in disaster aid.**
- **Breached 800 of the 1,100 levees built by farmers and towns.**
- **Overtopped or breached 34 of the 275 federal levees and floodwalls.**

Source: American Rivers, Associated Press

soaking into the earth. Efficient and cleverly designed storm drain systems suddenly dump vast amounts of extra water into the rivers, making them rise still higher and more quickly. "Hundred-year" floods now come every few years. As the devastating floods of 1990s have proved (with two "500-year" floods in 18 months on the Mississippi), this unstable equilibrium cannot be sustained. When a part of the flood control system such as a dam or a levee fails, the river reclaims its own. For the river, this is natural and predictable behavior. For humans, however, it is a natural disaster. This unfortunate cycle has been repeated worldwide. Control of rivers, it seems, isn't as easy as we once thought.

The easiest method of flood mitigation is simply not to build on a flood plain, but this is often politically impractical. Following the Mississippi floods of the 1990s, however, some towns were relocated to higher ground. The river must be allowed somewhere to go when in flood. Planners may choose to protect high value areas like towns, while allowing wetlands, forests, and fields to flood. Some towns have bought flood-prone areas and turned them into parks, allowing them to flood when necessary. Over 100,000 acres of flooded Midwest farmlands were converted to wetlands following the flood of 1993. In all, intelligent zoning and land use planning can often take the place of extensive (and expensive) flood control

structures. Restrictive building codes for structures on flood plains have been adopted in many locations, and are in some cases mandated by law. Communities should remember also that flood plains change with development, that flood plain maps must be periodically redrawn, and that "fixing" the problem in one area often just moves it somewhere else.

Another part of the mitigation program should be a survey of potential flood hazards such as dams, low bridges, and the like. Communities should also consider an active program of hazard abatement, that is, the removal of such hazards, where possible, before they become a problem. Examples of this might be the removal of old dams and river debris, or raising low bridges.

Prevention (P-SAR)

Prevention programs (Preventive Search and Rescue) tie in with the general mitigation effort. They are of two kinds: *positive reinforcement* through educational programs, and *negative reinforcement* through law enforcement. The two complement each other. It is not enough for the public to be told not to do something, they must be told *why* they should not do it.

Public education about swiftwater takes many forms. It might be a news spot advising drivers not to go though flooded areas; brochures and flyers from

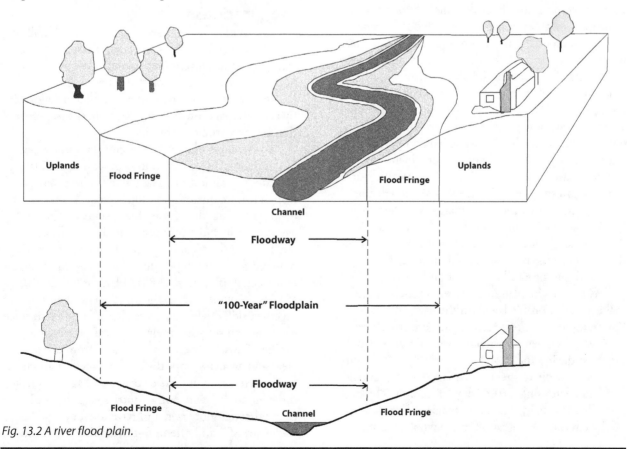

Fig. 13.2 A river flood plain.

---

organizations like the Red Cross and American Canoe Association explaining the dangers; an educational film like *No Way Out* detailing the dangers of flood channels; warning signs on hazardous areas; lectures by public safety officials in schools and churches; and "courtesy patrols" along the river during times of high water. One example is the Ohio Department of Natural Resources program of marking low-head dams to keep recreational boaters out. This includes signs and lines of brightly-colored buoys above the dam, as well as well-marked portage trails. Another is the National Weather Service's "Turn Around Don't Drown" (TADD) public education program aimed at educating people not to drive through flooded streets and across creeks.

On the enforcement side of things, laws limiting public access to dangerous areas like flood channels should be passed *and enforced*. Cars on flooded streets are another very common problem, and some thought should be given to a positive way of closing low streets off and fining those who ignore barriers. Putting up barriers and leaving them is usually ineffective because people will simply drive around them—they should be monitored by personnel. Enforcement should be proactive: it is much easier to stop someone from entering the river than to rescue him once in it. Arizona passed the so-called Stupid Motorist Law (Arizona traffic code Title 28-910) which provides for prosecution, fines, and the cost of a rescue to be assessed to motorists crossing flooded washes and roads, especially those with barricades. The local media has cooperated as well, frequently interviewing drivers after a rescue and asking them why they did such a dumb thing. After several high-profile incidents, Los Angeles public safety agencies adopted an aggressive approach to keeping potential victims out of flood channels. During actual and potential flood conditions, fire and police units (including helicopters) patrol flood channels and keep people out of them, arresting them if necessary. Although this method requires considerable manpower, they have found that in the long run it is easier and cheaper than rescues. It also puts rescuers out closer to where they are most likely to be needed, instead of having them waiting for a call at the station house.

As part of the planning process, agencies should make a victim profile based on the historical data. For example, the Great Falls of the Potomac River area averages seven drownings a year. Fifty-six people drowned during the period from June 1, 1975 to July 7th, 1990. Most of them were male, 15-25 years old, and had been using either drugs or alcohol. The biggest killer is usually ignorance, making the case for a public education program. On the whole, the most common contributing factors in accidental drownings are alcohol, drugs, and hypothermia, and the group most at risk are males 17-25.

Another contributing factor to accidents is the Delusion of Personal Invulnerability, or DPI, which may be expressed as "accidents happen to the other guy, not me." This is most obvious in young males, but also plays a part in floods and other disasters, when many residents who ought to evacuate insist on remaining in their homes to "ride it out."

Before leaving this subject we should consider adventure sports and the "right" of self-endangerment. While it is tempting for managers to close rivers at high water to prevent accidents, there may be expert kayakers and other river runners who are quite capable of running the river at high levels and will sneak on anyway. This has led to unintended side effects bordering on the comedic: the paddlers are precluded from taking necessary safety precautions because of the need for secrecy, and the authorities have in some cases spent more time trying to arrest people they didn't have to rescue. These chases put everyone in greater danger than if nothing had been done. Restrictions on rivers (not flood channels or low-head dams) should be implemented with common sense. Even better, agencies should consider recruiting expert river runners for volunteer assistance.

Preparedness

Preparation is the basis of a community's ability to react to a crisis such as a natural disaster. Unlike mitigation and prevention programs, preparedness includes active rescue efforts. It starts with the training and awareness of the individual rescuer followed by team preparation and then awareness and preparation by the entire community.

For rescuers, the process begins with individual credentialing, an administrative process for validating someone's qualifications as a rescuer. These may be from a private organization like Rescue 3 International or to meet a standard set by local, state, or federal agencies. It includes training, experience, physical and medical fitness and any certifications or licensing required for each job position. Credentialing is important to establish a common standard of training. As long as everyone is from the same agency and trains together, skill qualifications are less of a problem since everyone can see what their mates can or can't do. Problems arise, however, when teams from different areas and agencies work the same incident, particularly when large numbers of teams and agencies from different areas, states, and countries who do not know each other are mobilized and have to work together under stressful and often life-threatening conditions.

Because of problems that arose in several large-scale mobilizations (e.g. The New York World Trade Center on 9/11, Hurricane Katrina) the US Federal Emergency Management (FEMA) implemented the NIMS (National Incident Management System) and began *typing* rescue resources—the idea being to know what they were getting in a mobilization. Otherwise managers often had little idea of what the capabilities of various teams were when they arrived on scene. Did a "water rescue team" from another area of the country have proper PPE, boats, and the like? What level of *swift*water training and experience did they have, if any? Of what use was a heavy structural collapse team in a flood? It was not the sort of thing to be sorted out in the midst of an emergency.

A typed resource was to be defined with three areas of information:

- the category of the resource e.g. search and rescue, fire/hazmat, EMS, etc.
- the kind of resource (e.g. swiftwater rescue team)
- type of resource (Type I, Type II, Type III or Type IV)

Rescue teams were to be typed I though IV—Type I being a fully qualified and equipped team able to perform all necessary rescue functions of its specialty, through Type IV—the least qualified and equipped. Thus a fully-qualified and equipped swiftwater rescue team would be categorized under search and rescue, then identified as Swiftwater Rescue Team Type I.

Prewater map – Los Angeles River from Owensmouth to Sepulveda Flood Control Basin

Channel widths – from 70' - 120' / Channel widths – from 18' - 27'
Distance from Owensmouth to Sepulveda Flood Control Basin–5.5 miles
Average peak flow from Owensmouth to Reseda–16 fps. = 11 mph
    Distance – 3.75 miles, Travel time – 12 min, 30 seconds
Average peak flow Reseda to Sepulveda Flood Control Basin –26 fps. = 18 mph
    Distance – 1.75 miles, travel time - 9 min, 27 seconds
Total travel time at peak flow for entire section is–21 min, 57 seconds

The process starts with common individual credentialing (somewhat like the wildland firefighting Red Card) and ends with a listing of qualifications and equipment for different types of teams, which would then go into a state and nationally accessible database. In this way an IC would know exactly what he was getting, no matter how big or far-flung the mobilization. While at the time of this writing this process is far from complete, it has been implemented in states such as Texas, North Carolina, and California. Mobilization will be covered in more detail below. Team typing needs to be periodically reviewed since personnel and equipment changes may move a team's typing up or down. All teams and agencies need an ongoing swiftwater training program, both to train new personnel and allow everyone to maintain proficiency. There should be periodic individual recertification as well (for Rescue 3 International, for example, recertification is required every three years).

Managers and political leaders also need to remember that there are other people whose duties take them into the flood zone in an emergency. These include not only designated water rescue teams but law enforcement (police, sheriff, etc.), EMS personnel, storm water workers, electrical repair crews, and others. Although these people, to whom rescue is at best a secondary function, do not need to be trained in all aspects of swiftwater rescue, they should at the very least have awareness training of the dangers of moving water so that they do not unnecessarily endanger themselves and should have access to basic safety equipment like a PFD. Some, like law enforcement personnel, may also want to carry some simple rescue tools like a throw bag.

Floods are not only interagency but interdisciplinary. A strong working relationship should be established beforehand with the media and with agencies such as National Weather Service, US Geological Survey, Army Corps of Engineers, etc., as well as with private agencies like power companies. Everyone needs a consolidated mapping system that is simple to understand, works without power and when wet, and a communications system that will ultimately allow all agencies to exchange information (including radios that work when wet). Severe

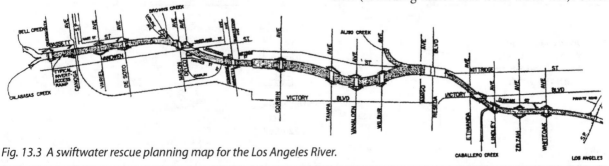

Fig. 13.3 A swiftwater rescue planning map for the Los Angeles River.

Example Flood Channel Site Pre-plan

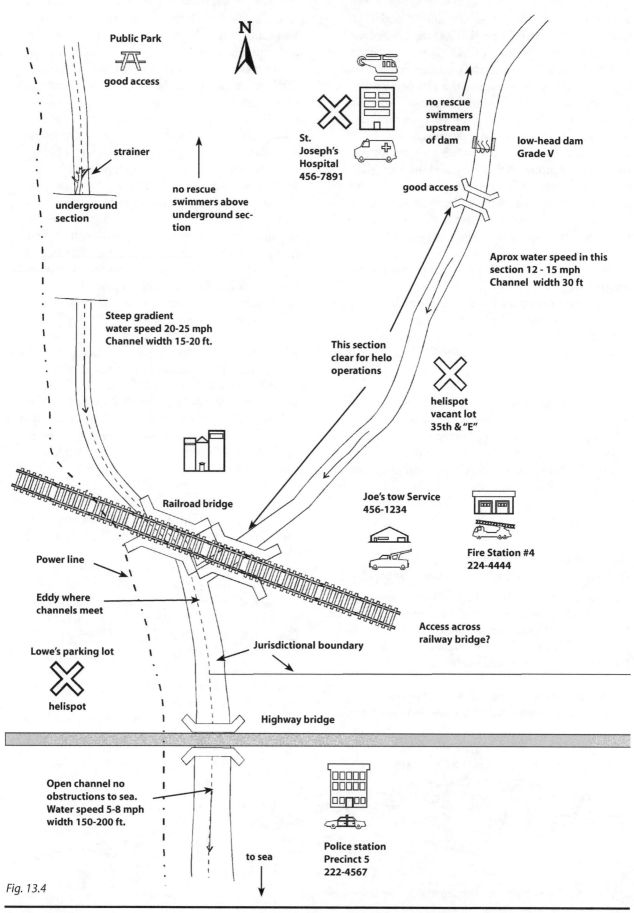

Public Park
good access

N

strainer

underground section

no rescue swimmers above underground section

St. Joseph's Hospital
456-7891

no rescue swimmers upstream of dam

low-head dam Grade V

good access

Aprox water speed in this section 12 - 15 mph
Channel width 30 ft

Steep gradient water speed 20-25 mph
Channel width 15-20 ft.

This section clear for helo operations

helispot vacant lot 35th & "E"

Railroad bridge

Joe's tow Service
456-1234

Fire Station #4
224-4444

Power line

Eddy where channels meet

Access across railway bridge?

Lowe's parking lot

helispot

Jurisdictional boundary

Highway bridge

Open channel no obstructions to sea.
Water speed 5-8 mph width 150-200 ft.

Police station
Precinct 5
222-4567

to sea

Fig. 13.4

flooding may render cities and countrysides unrecognizable, making the best maps less useful and confusing even locals, not to mention out of town units deployed there. In that case teams may have to rely on GPS to find their way around.

One of the most important questions flood managers and planners must answer is: where does the water in your jurisdiction come from and where does it go, especially when there's a lot of it? As we've seen before, watersheds quite often do not correspond with political boundaries. The answers are there but the information may be scattered over several public and private agencies who under normal circumstances seldom talk to each other. These, however, are usually the people you want to get to know beforehand anyway. If there is new development and construction, how does that affect the flow of the water? How current are your flood plain maps? (the answer in most cases is—not very).

Schedule periodic interagency exercises so that everyone gets an idea of how the system works and meets each other in a non-emergency situation. As Charlotte Fire Battalion Chief Tim Rogers notes, "many emergency managers at the county and state levels spend most of their time planning and little as IC's or Ops Managers. Many times these folks will be hurled into the middle of the event into one of these roles because some document says so. Many have little 'real time/real ops' experience." Planning should also include:

- An overall community survey of flood-prone areas, including actual and potential swiftwater hazards: dams, creeks, flood channels, and bridges, with changes made by new development
- A historical review of swiftwater incidents in local jurisdictions. This will identify trouble spots, as well as alerting agencies to the most likely types of rescues
- Site surveys, incident preplans, and interagency operations plans
- Equipment caches where needed
- Evacuation plans

For example, Los Angeles County conducted a detailed survey of its flood channel system, measuring widths, gradients, and current speeds, as well as identifying potential rescue sites. Using this knowledge, they developed a computer model to track a victim's progress in the channel based on his last known position. The county also developed an interagency pre-deployment and response plan so that integrated rescue teams were dispatched *ahead* of the victim.

Finally, based on the survey, they designated rescue zones for different types of rescues. Helicopter rescue zones required clear space free of bridges and power lines, while in-water rescue zones were located in areas that did not have downstream hazards. Potential rescue zones, like eddies and areas where the current slows, were carefully noted, as were access areas.

Once problem areas have been identified, agencies may also make on-site equipment caches and some physical modifications to the site. For example, anchor points might be added to a dam so that a tethered boat rescue system could quickly be set up. Flood channels can also benefit from rescue-specific modifications. Eyebolts might be added at several locations at designated rescue sites to act as belay points, changes of direction, or anchors for systems like swiftwater rescue curtains or inflated fire hoses. There might be several sets for different water levels. Another method is to install anchor holes for precut pipes, which are then capped. The first responding unit then pulls the cap and installs the pipe to use as an anchor. All this saves time and ultimately lives in an actual rescue.

The public at large must also be given appropriate information for safety, shelters, sheltering in place, rescue, and if necessary, evacuation. This includes preparation for the emergency to include food, water, and other supplies to last for at least 72 hours (and preferably a week), since utilities may be cut off and food and water in short supply. Citizens also need information about how to cope with personal emergencies, such as what to do if their car ends up in the water (see Chapter 11). Finally, those with animals in their care need to be reminded to make provisions for them, including extra food and provisions for evacuation if necessary (i.e. pet carriers, leashes, trailers, etc.). In a city like Los Angles this may have to be done in several different languages with due consideration for cultural differences.

CERT (Community Emergency Response Teams) teams are currently a much-underutilized resource, especially in floods. These community-based teams can be valuable adjuncts to professional rescuers, providing local information, checking on elderly and special needs populations, distributing flood and water, and assisting in evacuations. However like everyone else they need at least awareness-level training to keep themselves out of trouble. At present there are no flood or swiftwater training modules, although these have been implemented at the local level.

Mobilization and Management

Floods are by their very nature multi-agency and multi-jurisdictional events, making an integrated command structure vital. When local resources are overwhelmed outside help must be brought in. For localized events this may be handled with mutual aid agreements, but larger and more geographically extended events require more. As mentioned above, the US government has adopted NIMS to provide a common command structure, although this has yet to be completely implemented. At a higher level, states may avail themselves of EMAC (Emergency Management Assistance Compact), a federally authorized structure to facilitate interstate mutual aid, and for truly large interstate events federal agencies like FEMA coordinate mobilization. All these have been relatively recent developments spurred by a series of large-scale disasters in the last years of the twentieth century and the first years of the twenty-first.

Mobilization

Typing resources, as mentioned above, is only the first step in mobilization. Resources identified must then be made accessible to managers in a common database as being available. These resources are then prioritized and fitted into a mobilization framework called a Time Phased Force Deployment Data List (TPFDDL or "Tip Fiddle"). This is a list of resources to be mobilized, arranged by the type of resource, location, time needed to mobilize, order of mobilization, and expected place of mobilization and deployment. Using it a managing agency can conduct a reasonably smooth sequential mobilization, notifying appropriate units who then show up at critical points as needed. While in the real world things seldom go as smoothly as planned, the use of a TPFDDL is a vast improvement over a scattershot mobilization where managers simply go down through a list of agencies, resulting in a collection of teams of unknown strength and capability at random points.

The manner of mobilization is important also. There are two common systems in use—"push" and "pull." The pull system is the most common in the rescue field, and is driven by demand. Once a disaster occurs, it is sized up and resources are identified to handle the problem. These are then requested through proper channels, alerted and dispatched. This system has the advantage of being very efficient in terms of resources and lets rescue units remain in a central location until needed. The cycle of size-up, request, approval, selection, alert, and dispatch is called a decision loop, and it takes time. Once alerted and dispatched, it takes more time to actually move. Volunteer units, for example, must first come to station, don their gear, draw their equipment, then proceed to the incident location. In a large-scale disaster, outside resources are often not requested until local units realize that they can't cope with the situation, causing additional delays. In a fast-moving flood or swiftwater rescue situation there is simply no time for the identification–dispatch–deployment loop—rescuers must be as nearly in place as possible when they are needed. Experience has proved two things—things happen in a hurry and large areas are quickly cut off by rising water. This has meant in practice that the people who are on the spot make the rescues, ready or not.

The push system, on the other hand, is driven by anticipated demand. Floods are one of the few natural disasters having some degree of predictability. Hurricanes, for example, are now reliably tracked by satellite and both their intensity and landfall can be estimated with some exactness. Once the hurricane strikes, however, access to affected areas may be impossible because of wind damage (e.g. trees down across roads) and flooded roads. Even flash floods, once so unpredictable, may now be forecast with some degree of accuracy. As we've seen in previous chapters, events in swiftwater happen quickly and time is of the essence. The pull system stresses pre-deployment, i.e. deployment based on *anticipated* need. It is a proactive rather than a reactive system. Its advantage is speed, its big drawback is inefficient use of resources, since there will obviously be times when managers guess wrong—more resources are deployed than are needed or they are not quite the right ones, or the expected emergency does not materialize. Nevertheless there is no substitute for having rescuers in place at the right time, and only the push system can provide this.

Managers at state and federal level should practice strategic weather intelligence—that is having meteorologists available to make forecasts as the data arrives, allowing mobilization managers at all levels to anticipate trouble spots and stage resources accordingly. In a hurricane, for instance, rescue units might be staged just outside of the path of the storm, ready to enter the area after it passes. In other cases they might be left in place in the storm area so as to be there when needed. This is more commonly done for anticipated flash floods than for larger storms like hurricanes.

Another useful technique pioneered in Texas is the *rolling mobilization*, in which units are sent toward the incident "on the fly" as they become available. Local agencies such as fire departments tasked for mobilization needed to maintain coverage of their home areas and could only respond with part of their personnel, and it was often difficult to find enough extra equipment for these "expeditionary" teams. Local agencies also wanted to keep their people together and work in fire company-sized unit of 4-5 personnel when possible. The state of Texas agreed to provide each deployed strike team with an equipment trailer with enough equipment to make a 24-person strike team self-sufficient for 72 hours. This included boats, PPE (including life jackets), sleeping gear, emergency rations, water, tents, and rescue gear. The basic tactical unit was to be the 24-person strike team, comprised of four squads of five trained individuals each plus a headquarters. The team headquarters consisted of a team leader and an assistant team leader; a logistics specialist, and technical support specialist.

Each squad consists of:

- Strike Team leader
- assistant Strike Team leader
- paramedic in each rescue squad
- logistics specialist
- technical support specialist—someone to act as the communications specialist, small tools repairman, and jack of all trades.

This incident analysis was developed by Battalion Chief Tim Rogers of the Charlotte, NC, Fire Dept.

SEA DEPTH = **S**ITUATION
EGRESS
ACCESS
DEVELOPMENT
EXISTING RAINFALL
PREDICTED RAINFALL
TOPOGRAPHY

SITUATION Determine the Mode of Operation:

- ALERT—notify and prepare personnel and equipment. … *OR*

- INVESTIGATE/INTELLIGENCE-GATHERING— Predeployment of resources. Road closures/ evacuate flood prone areas. … *OR*

- EMERGENCY RESPONSE—Evacuation? Rescue? Search and Rescue? Recovery? Search and Recovery?

If Emergency Response Mode then determine Strategic Mode: Locate / Access / Stabilize / or Transport.

With Strategic Mode determined then do a Size-Up:

Time and Temperature
Energy and Equipment
Movement and Measurement
Pre-Plan and Personnel
Operate

Once Size-Up is completed then determine Tactics:

TALK
REACH
THROW
ROW
GO & TOW
HELO

EGRESS If my teams go in, can they come out? Is the public losing egress because of rising water?

ACCESS Do I still have access to the problem or am I losing it? Lose access, you lose egress. Pre-deployment and pre-evacuation are essential elements to a successful operation.

DEVELOPMENT How developed is the area that I am responding to? What is the population and how has the development affected water run off? More development equals more people, more roads, more business, more HAZMAT, and more run off which equals faster rising water that is in turn moving faster.

EXISTING RAINFALL What is already on the ground and historically, how much rain and how long does it take for a flood to develop? (One inch of rainfall per one square mile produces over 17,000,000 gallons of water.) Is the ground already saturated from previous rainfall?

PREDICTED RAINFALL More means everything is impacted. Beware of flood phase changes.

TOPOGRAPHY This can determine water speed, how fast the event will develop, and in some cases, how long the event will last. Do not confuse topography issues with development issues, although they significantly affect each other.

HAZARDS Have all potential and existing hazards been clearly identified and communicated? Storm water management systems change all of the rules as well as flooding in light and heavy industrial areas. Remember, floods by their very nature are hazmat/public health events. Can you decon evacuees prior to sheltering?

Management

Once units have arrived on scene they must be put to work. We have discussed the tactics of team level rescues in the previous chapter; here we will concentrate on the bigger picture.

A flood manager's first step is to get as accurate picture as possible of what is going on in the flood zone and even more importantly, what will be happening in the short and long-run future. It is essential to stay one step ahead of the flood, because it's nearly impossible to catch up once you get behind. Therefore planning must continue as the event unfolds. Since most floods are caused by rainfall, up to date forecasting is essential. Meteorologist John Weaver coined the term "tactical meteorology" to describe the process. Like an intelligence officer in a military operation, a tactical meteorologist sits at the elbow of the incident commander, keeping him up to date on current and forecasted rainfall totals. Other specialists monitor the flow of water, especially as it affects access to various areas in the flood zone. Water flow controls both ingress (can my teams get into the flood zone) and egress (once in, can they get out, and can I evacuate civilians in there). This must be monitored continuously during the flood event, and managers alert for flood phase changes. Obviously there must be a means of monitoring rainfall and water depths at critical points.

Dispatch is also a critical function and one often overlooked. In normal times dispatch handles only a few incidents at a time. In a flood, however, a dispatch call center may be overwhelmed by hundreds of calls for help—some trivial and some extremely urgent—in a few minutes, overloading the system and in some cases effectively bringing it to a halt. Meanwhile flood managers parcel out resources call by call until all are used up. The first order of business, then, is to triage incidents. A family trapped on top of a car in a flooded wash needs immediate help; someone with a flooded basement does not. Obviously dispatchers have to have some basic knowledge of swiftwater rescue and how the rescues happen. To assist them the City of Fort Collins Office of Emergency Management, the National Oceanic and Atmospheric Administration (NOAA) and the Cooperative Institute for Research in the Atmosphere (CIRA) developed a series of National Disaster Information Cards (www.fcgov.com/oem/pdf/ndic. pdf). These color-coded cards act as a flow chart and enable a 911 dispatcher to quickly get the relevant information from a caller to allow the appropriate action to be taken in various types of disasters, including floods.

Dispatch also needs to be scalable and have a ramp-up capability for large incidents that come up quickly like flash floods. The increase in call volume can be seen graphically in the flash flood that engulfed Fort Collins, CO, on 27-28 July, 1997. The 911 call volume roughly tracked the flooding and at one point overwhelmed the system.

With water covering large parts of the affected area, access restricted, and large numbers of incidents in progress, the traditional process of dispatching resources to an identified incident may become impractical. Units on the way to one incident may in the process come across several others that must be dealt with. Here managers and ICs may need to switch to a sector mode. Control becomes more like that of a search or a wildland fire, with teams being assigned sectors to search and rescue anyone in it.

Air space management is a vital issue too often overlooked that can cause big problems if left unattended. Since helicopters are mobile and easily dispatched to a disaster, they often show up in large numbers. For instance, at one point in Hurricane Floyd in North Carolina in 1999 over 60 helicopters of all descriptions were in use from agencies including active-duty military, National Guard, US Coast Guard, NC Highway Patrol, and others. There were problems with communications, common procedures, and the like. It is beyond the scope of this book to deal with this in any detail, but managers should plan for it and get it under control early to avoid problems and accidents.

And finally, while flash floods are usually over in a matter of hours, a river flood or one caused by a hurricane can linger on for weeks or even months. In such cases exhaustion of both rescuers and managers becomes a factor and must be allowed for.

Conclusion

Preplanning is a vital part of any rescue operation, but especially swiftwater. Mitigation and prevention are by far the most cost-effective way of dealing with swiftwater emergencies or floods. The time saved can also be critical. Writing a clear, comprehensive swiftwater interagency response plan, including typing of resources, should be the goal of all communities and states. Once this is done, planning has to be implemented with realistic training exercises and management scenarios to prepare for the day when the inevitable happens.

Rainfall Accumulation

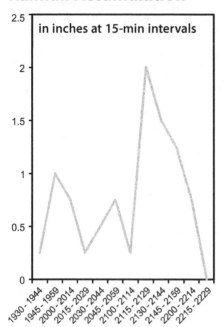

in inches at 15-min intervals

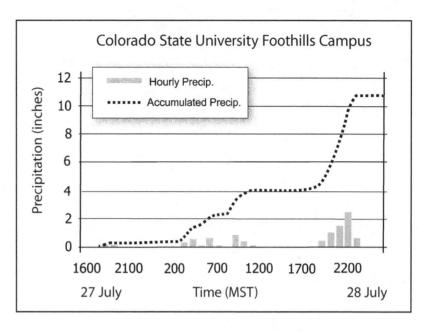

Colorado State University Foothills Campus

Hourly Precip.
Accumulated Precip.

Fig. 13.5 Rainfall at Fort Collins, CO, the night of July 27-28 1997.

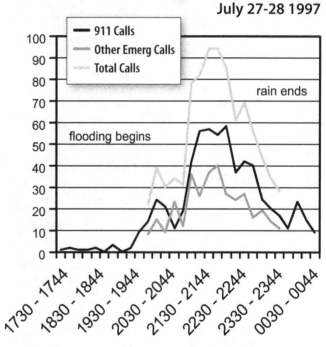

Fig 13.6 Rainfall patterns and totals at Fort Collins.

Ft. Collins Flood Emergency Call Volume
July 27-28 1997

911 Calls
Other Emerg Calls
Total Calls

rain ends

flooding begins

Fig. 13.7 This graph of the call volume shows how a dispatch system can be overwhelmed.

Medical Considerations
Chapter 14

In a medical sense, swiftwater rescue is a relatively straightforward field. Drowning is the main danger for both rescuer and victim with hypothermia a close second. Traumatic injuries, including C-spine injuries, are much less common than in other rescue fields. In many places, however, the effects of polluted or contaminated water must also be taken into account. Since most of what agency rescuers do is governed by their local medical protocols, this chapter presents only an overview of field medical considerations with detailed discussions of only those topics that directly affect swiftwater rescue or where they differ from normal practice. The second part of the chapter deals with patient packaging, litter rigging, and evacuation considerations.

Drowning

Drowning is an inherent risk when working in a liquid medium and occurs when a human is submerged for an extended period of time. The definition adopted by the 2002 World Congress on Drowning (www.drowning.nl) was: "Drowning is the process of experiencing respiratory impairment from submersion/immersion in liquid. Drowning outcomes are classified as death, morbidity and no morbidity." In order to provide a common definition, the Congress strongly discouraged such terms as wet, dry, active, passive, silent, and secondary, near drowning, parking-lot drowning, etc. as being too vague to be of much use. Instead, it recommended the following:

> The drowning process is a continuum beginning when the patient's airway is below the surface of the liquid, usually water. This induces a cascade of reflexes and pathophisiological changes, which, if uninterrupted, may lead to death, primarily due to tissue hypoxia. A patient can be rescued at any time during the process and given appropriate resuscitative measures in which cases, the process is interrupted.

In about 15 percent of drownings there is a laryngospasm that keeps water out of the lungs, even during fairly lengthy submersions, making the prognosis for recovery much better. Aspiration of water may cause complications such as pulmonary edema that can cause death hours or even days later.

Also of concern to rescuers is the inhalation reflex caused when the body suddenly enters cold water (70° F/21° C or less). Especially in big waves, this reflexive gasping can cause a swimmer to aspirate enough water to cause drowning. Sometimes called "flush drowning," it can happen even to experienced swimmers.

Although an immersion of 3-6 minutes is usually fatal, there are exceptions, and individuals have been resuscitated after lengthy periods. One six-year-old boy in Austria was revived after 65 minutes under 36.5° F (2.50° C) water. Most authorities seem to agree now that hypothermia can, under the right circumstances, reduce the body's need for oxygen and prolong underwater survival time.

For the rescuer, however, the medical details of drowning physiology are less important than the immediate response to it. Medical protocols require that CPR (cardiopulmonary resuscitation) be started immediately and continued until the victim can be transported to an advanced life support (ALS) facility. This can make for a difficult decision, since it can be quite difficult to move a victim, especially from a mid-river accident site, while delivering effective CPR. Rescuers must also consider potential complications. Although a victim may seem to be fine after being rescued and insist that there is nothing wrong with him, most authorities now recommend that all victims who have aspirated any water should be evacuated to an ALS facility for evaluation no matter how well they might feel at the time. This is doubly true for anyone who has had to have CPR.

The first 10 minutes or so of treatment after recovery of a drowning victim are the most critical. Rather than trying to breathe the victim in the water, rescuers should instead concentrate on moving him to a hard surface where effective CPR can be performed. A typical protocol requires:

- Clear the airway. Especially in storm drains and flood channels, aspirated debris can be a significant problem.

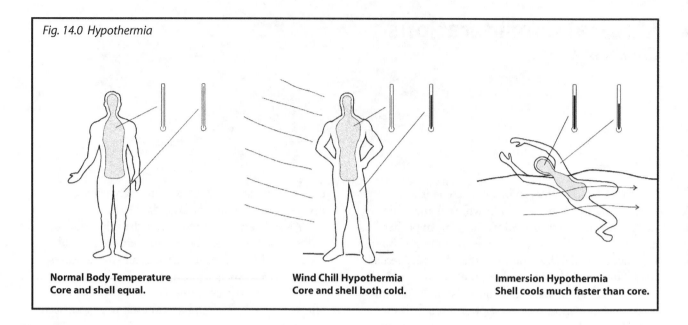

Fig. 14.0 Hypothermia

Normal Body Temperature
Core and shell equal.

Wind Chill Hypothermia
Core and shell both cold.

Immersion Hypothermia
Shell cools much faster than core.

- Perform effective CPR per medical protocol.
- Administer oxygen by mask (not nasal cannula).
- Initiate passive warming measures.
- Establish IV access if indicated.

Although the effectiveness of CPR decreases rapidly after the first 10 minutes, there have been exceptions. One kayaker was revived after 45 minutes of riverbank CPR.

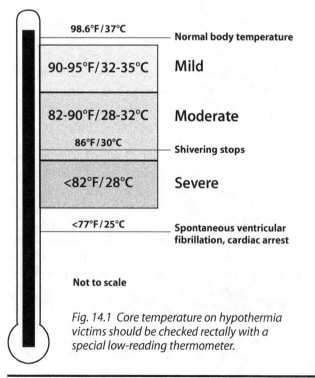

98.6°F / 37°C — Normal body temperature

90-95°F / 32-35°C   **Mild**

82-90°F / 28-32°C   **Moderate**

86°F / 30°C — Shivering stops

<82°F / 28°C   **Severe**

<77°F / 25°C — Spontaneous ventricular fibrillation, cardiac arrest

Not to scale

Fig. 14.1 Core temperature on hypothermia victims should be checked rectally with a special low-reading thermometer.

Hypothermia

Generally speaking, hypothermia has three stages: mild, moderate, and severe.

- *Mild hypothermia* occurs when the body's temperature drops much below 95° F (35° C). There is vigorous shivering and may be some mild changes in mental status, particularly in areas of fine coordination and judgment. Field treatment is to remove the patient from the cold environment and give him dry clothes, warm drinks, and sugary, quick-calorie foods.
- *Moderate hypothermia* begins at about 90° F (32° C) and produces noticeable mental status changes including incoordination, withdrawal, disorientation, and loss of judgment. At this point, field treatment is still possible and the effects of hypothermia may be reversed under favorable conditions, although the patient has limited ability to rewarm himself without outside assistance. This is especially true in immersion cases where the core temperature remains higher than the extremities. Shivering continues but stops at about 86° F (30° C). Toward the lower end of the scale the patient may become unresponsive. Evacuation to a warm environment should begin as soon as possible, but gentle handling is mandatory to avoid ventricular fibrillation.
- *Severe hypothermia* (< 82° F/ 28° C) puts the body in a "metabolic icebox." Shivering stops and the patient is unresponsive. At this point, field rewarming is impractical, and transportation to an ALS facility mandatory. Patients may, however, survive quite a long time like this even with no discernible vital signs. The general rule,

therefore, is that no one is declared dead unless they are *warm* and dead; that is, until after they have been clinically rewarmed. Severely hypothermic patients should be packaged to prevent further heat loss and transported gently because of the danger of ventricular fibrillation.

Accurately assessing a hypothermic patient's core temperature can be difficult. A quick field assessment is to feel the patient's skin under their armpit with a bare hand. If it feels warm, hypothermia is doubtful. Oral thermometer readings can be misleading, especially in immersion cases, and conventional thermometers often do not have a scale that reads low enough to be useful. A better solution is to take the patient's temperature rectally with a special low-reading thermometer adapted for hypothermia applications, but accurate measurement of low body temperatures may require an esophageal probe, which will not often be available in the field.

Hazardous Environments

In addition to protecting themselves from the river, rescuers must also protect themselves from what is *in* the river. This includes both water-borne pathogens and chemical pollutants. Intestinal problems and skin rashes are common, and more serious infections like hepatitis a possibility. In addition, there is always the possibility that a patient may have a communicable disease, ranging from the common cold to AIDS.

Up-to-date immunizations are in order for any water rescuer, and some agencies routinely use immune system boosters like gamma globulin. Every rescuer should carry and know how to use a CPR mask and exam gloves to reduce the possibility of infection from patients. In addition, rescuers should take routine precautions against coming into contact with the patient's blood or body fluids.

In a contaminated environment, a swiftwater rescue also becomes a HAZMAT problem. Rescuers can mitigate the problem somewhat by staying out of the water as much as possible and wearing protective clothing to minimize their exposure. A drysuit, since it seals out water, protects much better than a wetsuit. Rescuers should develop the habit of washing thoroughly immediately after exposure and cleaning any open wounds which usually requires setting up a decontamination site at or near the rescue site.

In-Water Spinal Management

The treatment of most traumatic injuries follows standard medical protocols. In-water spinal management, however, requires some special techniques.

Although the number of C-spine injuries in swiftwater is relatively small, they do happen. While normal C-spine "packaging" techniques apply to a patient who is in a readily-accessible location, a rescuer may be faced with a situation in which he will have to get a victim with a suspected C-spine injury out of the current to shore. Here, obviously, the goal is to effect the rescue without further injuring the patient. Normally, this requires an in-water contact rescue of the victim. The rescuer, using one of the techniques described below, brings the victim to a shallow eddy near the river bank, then transfers him to a spine board or litter. These techniques are intended *only for the emergency situation of getting the patient out of the current and over to shore*; they are not intended to replace normal procedures for spinal management.

Underarm/Lifeguard Technique. This has been used for many years by lifeguards in surf conditions. The rescuer moves behind the patient and places his arms under the patient's armpits. stabilizing his head with a hand on either side of it. By rolling on his back, the rescuer also keeps the patient's head out of the water. While this is an effective stabilization technique, it requires the full use of both of the rescuer's hands, making it difficult for him to swim or grab a throw rope.

Head Splint or Canadian Technique. This technique uses the patient's own arms to provide stabilization to the head. The rescuer pulls the patient's arms up so that they lie alongside his head, then clamps them together in the crook of his own arm. This allows the rescuer the use of his other arm.

Fig. 14.2 Underarm/Lifeguard head stabilization technique.

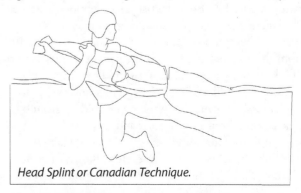

Head Splint or Canadian Technique.

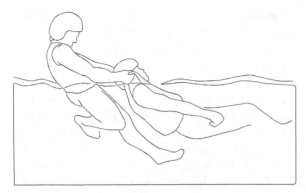

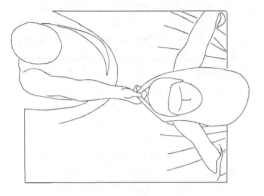

Fig. 14.3 Life jacket head stabilization technique.

Life Jacket Technique. In swift current, neither of the two methods previously described work very well. One method that does work in swiftwater is to use the patient's PFD for a rescue handle and to stabilize his head. The rescuer grabs the shoulder straps of the patient's PFD and pulls them up behind his head. With most PFDs this will form a pocket that stabilizes the head. Since the rescuer is using only one hand to stabilize the patient, this allows him considerably more latitude for swimming and other rescue activities.

Evacuation Considerations

The choice of evacuation method rests with the incident commander, who normally makes the decision after consulting his medical staff. The IC must balance the sometimes conflicting factors of the estimated time required, the patient's comfort, and the severity of his injuries. Helicopter evacuation is the fastest method, but may be risky. As mentioned before, if possible, the helicopter should operate out of a prepared helispot or LZ, with the patient transported from the accident site to it. If the patient's injury is not life-threatening and evacuation would require a high-risk procedure like being hoisted into a hovering helicopter or short haul, it is often better to do a ground evacuation even if it is difficult and lengthy. There have, unfortunately, been several incidents where lightly injured patients have been severely injured or even killed during high-risk helicopter evacuations. As with

rescue options, ICs should select the lowest-risk method consistent with the patient's need for treatment.

Patients who do not require immediate evacuation can also be transported by water. If there are no severe rapids, this can be a surprisingly quick and efficient method of evacuation, especially with a power boat.

Lightly injured patients can be walked out, although some may require assistance. If the patient is able to do this without risk of further injury, this is much faster and easier than a litter evacuation. Sometimes, a lightly injured patient can ascend the river bank with no more than a scurry line, but it is usually better to have a rescuer assist him and have a simple belay for both. If the bank is steep, rescuers can set up a simple haul system to pull them up the bank.

More severely injured patients require a litter and sometimes a spine board. Most common litters require six people to move them—normally, five to six bearers and a litter captain, who coordinates the litter's movement. For lengthy evacuations, the rescuers should establish a rotation system to get fresh bearers in and tired ones out for a rest. Where conditions permit, a wheel attached to the bottom of the litter makes evacuation much easier.

River banks are frequently very obstructed, and moving the litter along them can be difficult. It is usually done by means of a *caterpillar pass*: a double line of rescuers continuously passing the litter along hand-to-hand over terrain too rough for conventional evacuations. If the bank is not too steep, the litter can simply be walked or passed up. However, a belay rope on the litter is a good idea.

Rescuers may sometimes need to transport litter patients across the river. There are several methods of doing this. Probably the easiest is to find a calm spot, put the litter in a boat or raft, and ferry it across. If there is a danger of the boat going downstream, the rescuers may opt for a tethered system like a 2- or 4-point system, a tension diagonal, or boat on tether. If no boat is available, the litter itself can sometimes be used. Some have an optional flotation package for in-water use. This can also be improvised. One team successfully used 5' (1.5 m) capped sections of plastic sewer pipe lashed to the sides of the litter. This method obviously increases the risk for the patient. He should be given a PFD and not be strapped into the litter. There should be in-water litter tenders as well. Rescuers may also opt to pass the litter across on a high line.

The evacuations described so far are *non-technical* evacuations. Technical evacuations, and the differences between these and non-technical evacuations, are discussed below.

Fig. 14.4 Rescuers use an improvised kayak "trimaran" to transport an injured kayaker across the river.

Patient Packaging

"Packaging" a patient means preparing him for a safe litter evacuation. The patient should be secure in the litter for the type of evacuation planned (low-angle, high-angle, etc.), and protected, as far as possible, from further injury. The major considerations are:

- **Physical Protection** of the patient from hazards like falling rocks and debris during the evacuation. The patient should have padding like sleeping bags or foam pads above him for protection against falling rocks and debris, and below for protection against the litter itself. For technical evacuations he should also have a helmet and eye protection, and perhaps a plastic litter shield for his head and shoulders if available.

- **Thermal Protection.** Especially in river environments, the patient (who is stationary and not producing much heat) must be protected from the effects of cold just like the rescuers. The patient's face should also be protected against sun and rain.

- **Airway Protection.** There is always the danger of a litter patient vomiting and compromising his airway, aspirating vomitus, or both. Since he is firmly strapped into a litter, his ability to deal with this problem, even if he is alert and conscious, is limited. There are several courses of action when the patient vomits: the rescuers may physically turn the litter over; they may use a suction device; or they may package the patient in a sideways (lateral recumbent) position.

Fig. 14.5 Basic patient packaging

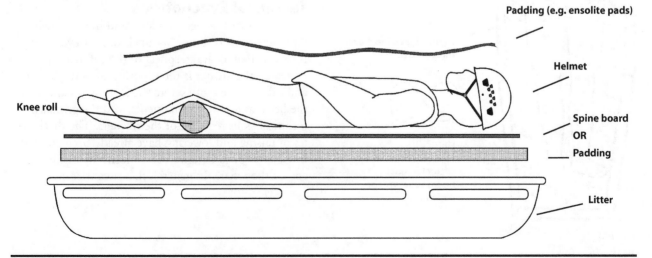

Padding (e.g. ensolite pads)

Helmet

Spine board

OR

Padding

Litter

Knee roll

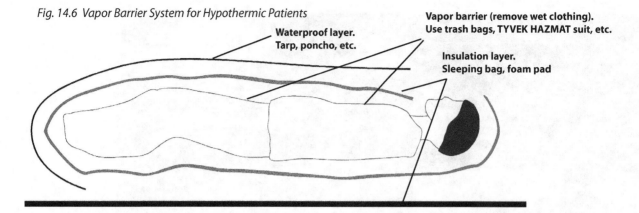

Fig. 14.6 Vapor Barrier System for Hypothermic Patients

Waterproof layer.
Tarp, poncho, etc.

Vapor barrier (remove wet clothing).
Use trash bags, TYVEK HAZMAT suit, etc.

Insulation layer.
Sleeping bag, foam pad

For most types of evacuations, the patient should be securely tied into the litter. The main exception to this rule is if the patient is being transported across water, where the rescuers may choose *not* to secure him. For aerial or technical evacuations, rescuers often put a sit or body harness on the patient as a backup. If the patient is conscious and alert, it's usually best to leave his hands free to combat feelings of helplessness and allow him to do simple things like scratch his

Yosemite Packaging Technique

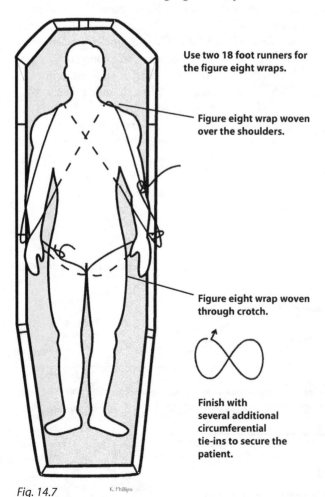

Use two 18 foot runners for the figure eight wraps.

Figure eight wrap woven over the shoulders.

Figure eight wrap woven through crotch.

Finish with several additional circumferential tie-ins to secure the patient.

Fig. 14.7

K. Phillips

nose. Patients, especially those with C-spine injuries, need to be securely tied so that they do not shift sideways if the litter has to be turned suddenly. The most common mistake when tying the patient in is to run the straps over him from one side of the litter to the other. A much better solution is to tie him in with an "X" pattern across his body.

The patient should also be secured against shifting downward toward the foot of the litter, especially on steep-angle evacuations. Some litters have foot boards, and in other cases it is necessary to use foot loops made from webbing to keep the patient from sliding down. Rescuers should also take care not to put any tie-downs under the patient's chin, since this could cause strangulation if he does slip downward.

Since it is specifically designed for the purpose, a spine board works much better for C-spine stabilization than does a litter. However, the two also work very well together. The patient is first secured to the spine board and the board then placed inside the litter. This adds rigidity to the litter and makes it much easier to transfer the patient in and out of it. If the board is X-ray transparent, the patient can be quickly moved from the litter to the X-ray table at the hospital without risk of further injury.

Technical Evacuations

The exact line between low- and high-angle litter evacuations is somewhat unclear. However, what is clear is that as the steepness of the evacuation increases, the greater is the exposure of both patient and rescuer; the more severe the loading on anchors, the litter, and other components; and the greater the technical skills required of the rescuers. The mechanics of raising and lowering have already been covered in Chapter 5, so here we will concentrate on the litter end of the rescue. In general, the categories are:

Nontechnical Evacuations

- **Nontechnical Carry out.** The terrain is flat or gently sloping (>15°). The patient is transported in-line (axially) by six litter bearers. There is no exposure of the patient or rescuers, and no belay rope. Almost any litter is acceptable.
- **Low-Angle.** The terrain is steeper, (15–40°) but still there is little exposure. A single belay rope may be used, but all weight is carried by the bearers. The patient is carried in-line (axially) by six litter bearers who are not tied in and who face the direction of travel. Litter selection is not critical. Helmets are recommended but not required.

Technical Evacuations

- **Steep-Angle.** The evacuation is steep but not vertical (35–60°). Litter selection is critical. Since the exposure is great, the litter is secured with both a working rope and a separate belay, or with a dual haul system. The patient is transported in-line. There are normally three to four litter bearers, who are tied directly to the litter and face the anchor. Helmets are mandatory.
- **High-Angle.** This is a vertical or near-vertical raising or lowering of the litter at an angle of 60° or more. The full weight of the litter hangs from the anchors. There are no bearers, although there may be one or two attendants to assist the patient. The exposure is grave, and the litter is secured with both a working rope and belay, or with a dual-haul system. The litter is normally rigged to allow it to ride horizontally, with the patient in a supine position. Helmets are mandatory.

Litter Rigging

Not all litters and stretchers are suitable for technical rescue. Some, particularly older plastic ones, are intended only for nontechnical carries. Since the loads go up with the angles, any litter used in steep- or high-angle rescues must be extremely rugged and strong.

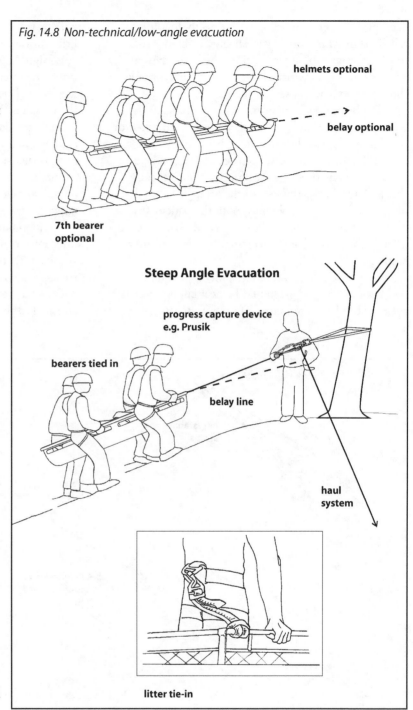

Fig. 14.8 Non-technical/low-angle evacuation

helmets optional

belay optional

7th bearer optional

Steep Angle Evacuation

progress capture device e.g. Prusik

bearers tied in

belay line

haul system

litter tie-in

This usually means a metal Stokes-type litter. For example, in a typical steep-angle evacuation, the litter has to hold not only the weight of the patient but that of three to four bearers as well.

Steep-Angle. For steep-angle evacuations, the litter is normally rigged with a *yoke* (harness, bridle) on the head, to which the rescuers attach the working and belay lines. While commercial harnesses are available, they may also be fabricated out of rope or webbing. The haul and belay lines may also be attached directly to the litter, either by tying the rope directly on to the head of the litter or by clipping it directly to a litter rail. This often requires a special large carabiner to fit over the rails. In any case, since the entire system depends on the integrity of the litter, the line or yoke should be attached at several points.

The bearers, who wear harnesses, can attach themselves to the litter rails with web straps girth hitched to the rail (an adjustable strap works best), daisy chains, or Prusik loops. Normally, there are either two bearers on each side (for a total of four), or one on each side and another at the foot of the litter (for a total of three). For additional safety, both the patient (by means of harness) and the bearers can also be attached directly to the belay line, reducing the effect of litter failure.

In any technical evacuation a medically qualified bearer or attendant should be near enough to the patient's head to take immediate action in case the patient vomits. The usual actions are to either tip the litter or use a suction device.

High-Angle. For vertical or near-vertical evacuations, the litter (which *must* be a sturdy, high-strength model) is rigged with a *spider* or "pre-rig" attached to the rails. The legs of the spider can be adjusted individually to allow leveling of the litter. The spider is attached to a yoke—usually a steel ring—which also accepts the working line and the belay. An attendant (sometimes two) is also attached to the yoke. The attendant helps maneuver the litter and monitors the patient's airway. Some attendants use a short line with a foot loop ("barf line" or "B line") attached to the litter rails to allow them to tip the litter if the patient vomits. Others use a suction device, and still others package the patient sideways. For added safety, the patient can be secured with a harness, which is also attached to the yoke.

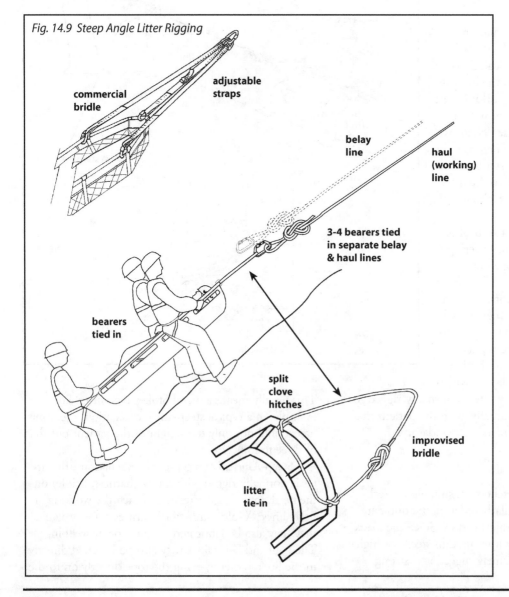

Fig. 14.9 Steep Angle Litter Rigging

commercial bridle

adjustable straps

belay line

haul (working) line

3-4 bearers tied in separate belay & haul lines

bearers tied in

split clove hitches

improvised bridle

litter tie-in

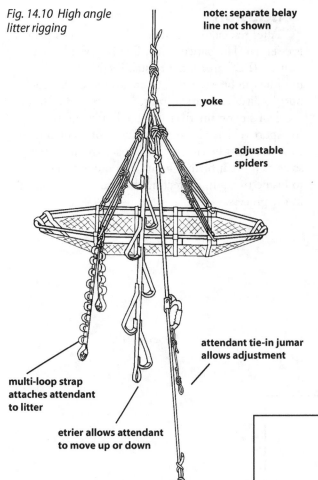

Fig. 14.10 High angle litter rigging

note: separate belay line not shown

yoke

adjustable spiders

attendant tie-in jumar allows adjustment

multi-loop strap attaches attendant to litter

etrier allows attendant to move up or down

tie-in to pt. harness

Litter rigging for helicopter short-hauls and hoist operations is very similar to that used for high-angle evacuations. However, there are some extra considerations:

- For helicopter hoist operations, the total height of the yoke and spider assembly must often be reduced, otherwise the winch may not be able to get the litter all the way up to the door. For example, the normal length of the spider legs are 3–6' (1–2 m), but for a Bell 212/412 series helicopter using an external hoist this must be reduced to about 1.5' (46 cm). This dimension varies, depending on the helicopter and hoist.

- Some litters fly better than others. Solid plastic litters have been blamed for several accidents. A solid litter seems to catch the rotor wash more readily than a wire-mesh basket, sometimes causing it to spin uncontrollably. Consequently, most authorities recommend using mesh Stokes-type litters rather than solid plastic ones. Other recommended fixes are to rig the litter with a slight head-up attitude and use a small streamer or drogue chute at the foot.

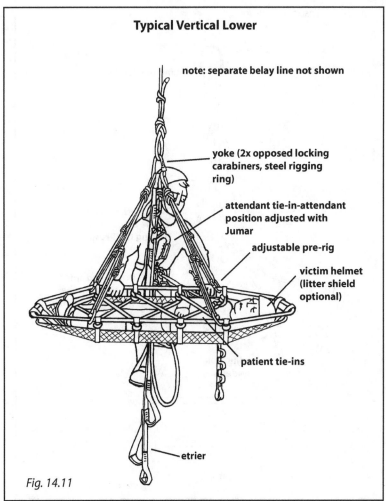

Typical Vertical Lower

note: separate belay line not shown

yoke (2x opposed locking carabiners, steel rigging ring)

attendant tie-in-attendant position adjusted with Jumar

adjustable pre-rig

victim helmet (litter shield optional)

patient tie-ins

etrier

Fig. 14.11

Rescuers must also decide whether they will use a single-rope raising/lowering system with a belay to back it up, or use a dual-haul system. The litter can also be rigged with a separate yoke and spider at either end of the litter. This allows the rescuers to adjust the litter's attitude in mid-operation, raising or lowering each end as needed. This can be very useful in tight spots.

This type of rigging can also be used if the litter needs to be put across on a high line. A cross-river high line, while it takes time to set up, can be used to get a patient across a gorge where using a boat would be impractical. The litter is rigged with a spider at each end, with each one attached to a high-line carriage. The carriages are then tied together with a short length of rope or webbing, and tag lines attached to the carriages. If an attendant goes across with the patient, he should be at the head of the litter.

Conclusion

After rescue, proper medical treatment of patients is essential. The patient must first be given basic life support (BLS) treatment to stabilize his condition until he can be evacuated to an advanced life support facility. The patient must be evacuated from the accident site on the river to a medically-equipped transport vehicle like an ambulance, often on a litter. Rescuers must be competent to package the patient so as to prevent further injury and move him quickly to his destination, protecting themselves from injury in the process.

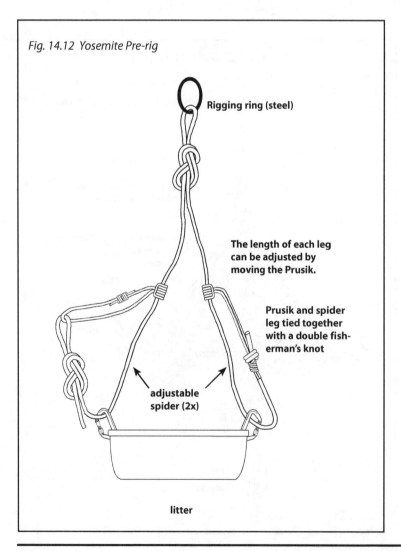

Fig. 14.12 Yosemite Pre-rig

Rigging ring (steel)

The length of each leg can be adjusted by moving the Prusik.

Prusik and spider leg tied together with a double fisherman's knot

adjustable spider (2x)

litter

Searches and Recoveries
Chapter 15

So far, we have been concerned mainly with rescue and have assumed that the victim's location is known with some degree of certainty. Obviously, however, this will not always be the case. This is where the search function comes in. There are four phases of search and rescue as expressed in the acronym **LAST**: rescuers must **L**ocate the victim, gain **A**ccess to him, **S**tabilize him, and **T**ransport him to safety or medical care. A number of excellent books have been written about the search function, so we will concern ourselves primarily with those techniques that are specific to river searches.

While we have already covered the mechanics of site management and personnel preparation, it might be a good idea to review a few things here. First, only trained rescuers should be put on the water, and even those who stay on the banks should have basic self-rescue training and adequate equipment. Boat and in-water rescues demand an especially high level of skill and training. Second, remember to ensure the safety of the rescuers first, especially if it is obvious that the situation is one involving recovery and not rescue.

River Search Organization

All searches are divided into two phases, which overlap to some extent: the *passive phase* and the *active phase*.

Passive Phase. Generally, passive measures are those that do not require active searching by the rescuers. All searches begin with the gathering of pertinent information on the missing subject. This usually involves interviewing a *reporting party*, or RP—the person who saw the subject last. Of primary and immediate importance is locating the *point last seen* (PLS) of the missing subject. Sometimes, the RP will have no direct knowledge of what happened. For example, the RP may be a family member reporting a group of canoeists overdue at a takeout. Other times, the RP will have witnessed a river accident like a raft overturning or a fisherman being swept away, and will be able to give a description of the victims and a fairly exact PLS. RPs should be interrogated for all information they might have about the victim, to include physical description, clothing, destination, experience, time the incident occurred, and any other details that might help the search (e.g., the type of shoes to aid trackers).

Fig. 15.0 A searcher from the Tuolomne County Rescue Squad, hanging from a cross-river high line, probes the upstream side of a boulder on California's Cherry Creek for a missing rafter.

The RP should not immediately be released, since the need for other details might arise later.

If the RP is released, searchers must ensure they can contact him if necessary. Searchers should also not rule out the possibility of foul play, or that the RP might be the guilty party. It is also a good idea to keep checking with the subject's next of kin and friends to avoid a "bastard search," i.e., one where the subject is—unbeknownst to the rescuers—not lost but safe in another location.

Once basic information about the victim has been determined, a hasty search should be started immediately, since time is usually of the essence in river searches. In many cases, the victim's location will be obvious and he will be easily found with a hasty search. Hasty teams sweep quickly down the river and the banks, attempting to locate the victim(s) for rescue teams. This might be done with a helicopter, vehicles driving down a road near the river, or by sending a watercraft downriver. Often, a combination is best.

Sometime during all this there should be a more formal initiation of the search process. An incident commander must be appointed, a temporary command post (CP) set up, and initial agency coordination made. Swiftwater rescue "strike teams" should be made ready in case the subject needs to be rescued after being located. The IC, his leaders and staff, should make a quick map recon to set the boundaries of the search perimeter.

The top of a river search perimeter is normally the PLS (unless there is some reason to think that the subject might have gone upstream) and the bottom is determined by multiplying the speed of the current by the time since the last sighting (e.g., 3 mph/4.8 km/hr current x 30 min. since last sighting means a zone 1.5 miles/2.4 km deep). If it has been only a short time since the subject's disappearance, rescuers may want to drop a marker in the river to approximate the victim's drift (a chemlight in a milk jug works well). The sides of the zone depend on the situation. If a subject is able to get ashore and walk, the sides of the zone must be expanded correspondingly. As the hasty search teams are being deployed on the river, so too should teams be deployed to patrol the perimeter. These might include foot and mounted searchers, barriers, and track traps in suitable locations, as well as obvious signals like smoke and flares.

As part of their map study, searchers will also want to make a *probability of detection* (POD) analysis. This is a planning tool that allows them to allocate search resources based on the POD. It includes:

- **Victim psychology.** Different types of people react in different ways when lost or in danger. Various authorities (Matson, Syrotuck, et al.) have made detailed studies about the behavior of lost persons. Lost hunters, for example, act differently than small children. By classifying victims according to categories and consulting the probability charts, searchers can divide the search area into zones with different PODs. This makes for a much more efficient use of resources than a random search.

- **River-specific high POD areas.** In addition to the normal search PODs, rivers have their own. Most of these correspond to river hazards already discussed. The upstream sides of boulders in the current, strainers, likely foot entrapment spots, low-head dams, and the like, are all river high POD areas, as are islands and other areas where a victim might have been able to get out of the water.

- **Consensus of the search leaders.** This is based on experience and common sense.

Searchers should also consider the behavior of a person in the water. A live person will usually try to get to shore or on a midstream obstacle as soon as possible, stopping his downstream progress. A victim without a life jacket who drowns, however, will assume a slight negative buoyancy as his lungs fill, and sink to the bottom. In most cases, he will stay there, since there are eddies on the bottom of the river as well. Therefore, for most drowning victims there will be a high POD area immediately downstream of the PLS.

After making this assessment, search leaders will normally divide the search area into areas ranked from high to low POD. Normally, high POD areas are searched first, with searchers working down toward lesser POD areas as additional time and resources permit.

Active Phase. As the search becomes more formalized, it enters the active phase. The searchers establish a permanent command post, a base area, and a formal incident command structure, and make detailed coordination with other agencies. As we have seen before, since rivers often form jurisdictional boundaries, cross-boundary coordination is essential. As search and rescue teams arrive, they are assigned tasks by the IC based on their training and capabilities. The IC and his staff conduct a more complete POD analysis based on the results of the hasty search and issue a formal *search order*—a detailed plan for continuing the search.

The searches at this point are detailed, formal searches—not hasty ones. It is better to have small, trained groups of searchers thoroughly search an area repeatedly than to do it with large groups of untrained searchers, since these frequently trample more evidence than they find. As the search progresses, the IC should frequently debrief team leaders and revise the search plan as necessary.

The search order uses the five-paragraph military format, or **SMEAC**—Situation, Mission, Execution, Administration, and Command and communications. Leaders should also be careful not to overuse planning time, since each leader in the chain requires time to issue his own order. For example, let's take a situation in which there are three levels of command: the IC, an operation or group leader, and a team leader. There are three hours before the operation commences. If the IC takes 2 ½ hours to issue his order, there will be no time for the operation and team leaders to do their own planning and preparation and issue their orders. To avoid this, all levels of command should observe the *1/3-2/3 rule*. Divide the available time at your command level until the start of the operation into thirds. You have one third of this time to issue your order; two-thirds of it must be given to subordinate units for them to prepare or to issue their orders. In this example, the IC has one hour to issue his order to the operations leaders. With two hours remaining, the ops leaders then have 40 minutes to issue their orders to team leaders, who must issue their orders in the next 27 minutes. This leaves some time (about 50 minutes) for individual preparation.

The search is continued until the subject is found or until it is called off. At the end of the operation, regardless of the results, all search personnel and equipment must be accounted for. The search personnel must also be debriefed, both for information and experiences that can be used to advantage on future SARs, and for critical incident stress where needed.

Body Searches and Recoveries

Body searches and recoveries on river SARs can be both frustrating and dangerous. Frustrating, since they can be very hard to find; dangerous, since they often lodge into inconvenient places in the midst of river hazards. Searchers should keep safety priorities in mind, and not risk the safety of live rescuers in a recovery. Unfortunately, this has not always been the case, and there have been a number of incidents where rescuers have died during recovery attempts.

Because of this, one of the things that needs to be determined at the outset of an operation is whether this is a rescue of a live victim or a recovery of a dead one. Because of medical advances this is not as simple as it may at first seem, since some victims have survived after lengthy submersions. However, the concept of the "golden hour" remains valid—any person who is known to have been underwater for over an hour has a very small chance of survival.

Searchers should first search the river high POD areas already mentioned: strainers, undercuts, and similar areas. This can be done both visually or by probing with pike poles or hooks. Reducing the river's flow, if it is a dam-controlled river, will make

Fig. 15.1 *A body recovery on the Clark's Fork of the Stanislaus River, California. Jim Segerstrom and Larry Gibson used a cross-river high line to tether a diver close enough to attach a rope to the body which was wedged in an undercut.*

both searches and recoveries much easier. On natural rivers, search and recovery efforts may have to wait until the water level drops. In some cases it may be possible to build a temporary dam to reduce or divert water flow. For example, in a recovery on California's Cherry Creek, SAR personnel lowered a plywood barricade to block the narrow channel in which the victim's body had become wedged. This reduced the flow enough for them to attach lines to the body.

One popular technique that should be relegated to the history books, however, is the practice of throwing grappling hooks into the river in a random attempt to snag a body. Instead of snagging on a body, what usually happens is that the hook snags on something else, cannot be recovered, and ends up becoming a hazard in its own right, especially if it has a rope trailing from it.

Fig. 15.2 *Do not use grappling hooks in moving water!*

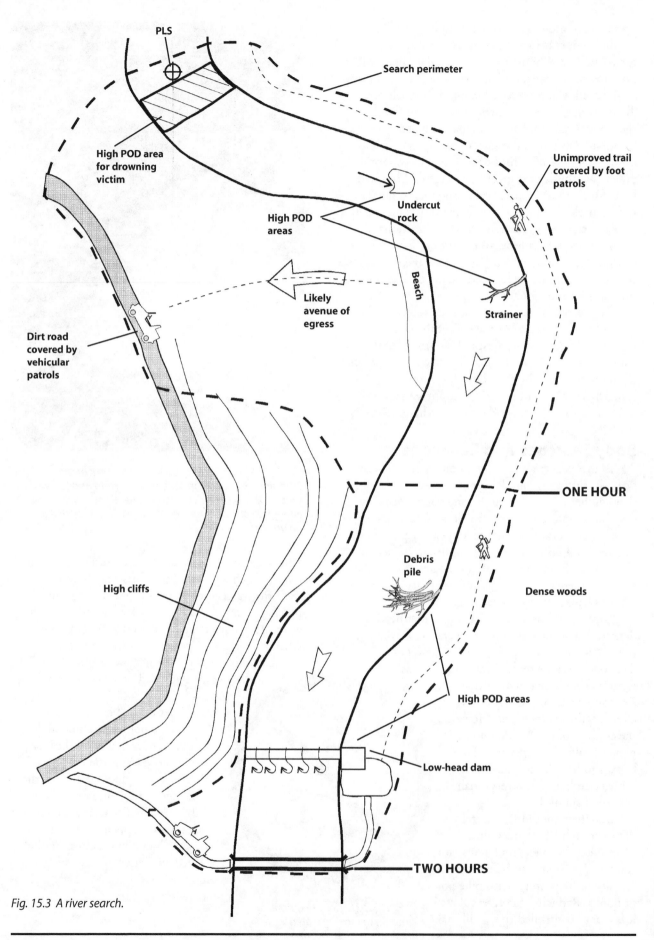

PLS

Search perimeter

High POD area
for drowning
victim

Unimproved trail
covered by foot
patrols

Undercut
rock

High POD
areas

Beach

Likely
avenue of
egress

Strainer

Dirt road
covered by
vehicular
patrols

ONE HOUR

Debris
pile

High cliffs

Dense woods

High POD areas

Low-head dam

TWO HOURS

Fig. 15.3 A river search.

Search dogs can also be used for body searches in moving water. This may sound surprising, but there have been good results with putting a dog trained in river searches in the front of a boat or raft and drifting slowly down the current. Apparently there is enough scent escaping from the body underwater to alert the dog on the surface.

In cases where there are few hazards, the water is relatively clear, and the current slow, divers wearing scuba or snorkel gear can conduct downstream searches *if* they have been trained in swiftwater operations. Divers can be held in slow current for searching with a 2-point tether or even a Tyrolean. However, these techniques are very dangerous, particularly for divers without swiftwater experience. Rescuers should also be very wary of putting in divers upstream of river hazards.

As noted before, drowning victims often do not travel far, even in fairly swift current; therefore, searches should initially be concentrated just downstream of the victim's PLS. If a high POD area can be identified, one technique that works very well for searching is the boat on tether using a raft or boat. It provides a stable, relatively safe platform for the searchers to work from, and it is easy to set up for a grid search since it can be worked across the current, stopping every few feet for probing. When the other side of the river is reached, the raft can be lowered down a bit and then worked back across. A boat on tether can also be used to inspect river hazards like strainers: the safest way is to lower it down past the obstacle and pull it back up into the eddy behind it, but it can, if absolutely necessary, be lowered down to the upstream side.

High lines can also be used for searching difficult areas like the upstream side of boulders. A searcher can be lowered down to the front of the boulder to probe with a pike pole. Since it requires a lengthy setup (not to mention the danger to the rescuer), this is best limited to high POD areas.

Flood searches present another problem. Searching during a flood is often difficult or impossible. Post-flood searches usually have to cover a much wider area than normal, and because of the amount of debris carried by a flood, the remains of victims may be buried under sand and debris piles. If they were in a car that was swept away, the car may have become an unrecognizable mass of wreckage. Search dogs are often a great help here, and much of the

Fig.15.5 Search dogs can be an effective tool in river search operations.

searching will take place on the banks. Searchers must also remember to look in trees, where victims are often carried by floodwaters.

The victim may also recover himself. After drowning, the action of the bacteria in a person's abdominal cavity will eventually produce enough gas to bloat the body and float it to the surface. In warm water this may take only a couple of days, while it may never happen in very cold water. For this reason, it is generally a good idea to continue hasty river searches once or twice a day if the body has not been recovered sooner.

If the victim has been wedged into a strainer or an undercut, however, his body will be held by the force of the current and will not float. In these cases the rescuers may have to attach lines to the body to pull it free. This can be a ticklish and difficult task. The force of the water makes attaching a recovery haul line physically difficult, and there is always the danger that the rescuer will be sucked in as well.

Generally, the best bet is to tie a line around the extremities and then use normal extrication techniques. One technique that works well is for the rescuer to tie a clove hitch over his own arm and grasp the victim's hand, then slide the knot down over the victim's arm and pull it tight (see fig. 15.4).

Fig. 15.4 A clove hitch may be slipped over an extremity.

If the dead person has been a victim of foot entrapment, the cinching methods described in Chapter 6 can be used to good effect.

Once the body is recovered, it should be treated with respect. It is normally put in a body bag or covered with a tarp or poncho. It is usually a good idea to do this as soon as the body is recovered, even if it means doing it in the middle of the river. Rescuers should also be alert to preserve physical evidence on the scene. In many jurisdictions any body recovery is considered a crime scene until ruled otherwise, and foul play is a factor in a significant number of cases. A death investigation checklist is included in Appendix 3.

Night Searches

Night searches on the river merit some special considerations. They are, unfortunately, a fact of life, since this is a prime time for tired and sometimes drunk drivers to run their cars off into the water. The resulting limitations on visibility affect both the rescuer's communications with each other and their search for the victim. Darkness also compromises the rescue leaders' ability to keep track of their people, whether they are on the river bank or in the water.

The searchers can take passive measures to increase their visibility, such as wearing reflective patches on their PFDs and helmets, and reflective armbands and vests for shore personnel. They can also use active lighting measures, like strobes or chemlights. Strobes are distracting for regular search work, and are best used to show an emergency in progress. Chemlights, on the other hand, come in different colors and so work well for marking various rescue functions. Foot searchers might be blue, for instance, with a red stick for those with a radio. The leaders might be green, with in-water searchers yellow. All this is a great help to the IC when trying to run a search on a dark river bank. Some searchers carry both a strobe and a chemlight: the chemlight provides normal marking, and the strobe indicates an emergency or that the victim has been found. Searchers also have the option of using portable high-intensity light sets or vehicle headlights for searching high POD areas.

The minimum size of a search unit at night should be two people and, ideally, each separate unit should have a radio. Individual searchers should not be sent out alone on night searches.

The Search Order

Anyone who has ever had any kind of military leadership position has issued a five-paragraph field order. This standardized format, developed over the years, allows leaders to communicate information to subordinate units quickly and efficiently. This has been adapted with few changes to the search-and-rescue field, where it is often called the SMEAC after the first letters of each paragraph. Each commander receives an order from his superior, and then issues his own order to subordinate units, using the 1/3-2/3 rule for allocating the time needed to issue his order.

The search order consists of five paragraphs:

Situation. What are the events leading up to the accident? What is the river like? Banks, water level, any special hazards a rescuer might face? Are there any other agencies assisting, and what are the jurisdictional boundaries?

- Mission. *What* is the search team's mission? Search, rescue, or both? *How many* people are we looking for? *Where* is the search area? Are there any limitations? *When* will it begin?

- Execution. This is a detailed description of exactly *how* the mission will be carried out. How are the teams organized? What sectors will they search? Where are the high and low POD areas? How will assets like helicopters be used?

- Administration. This should include the location of staging areas, warm-up tents, medical facilities, and the like. It should also cover food, water, and medical supplies for extended operations, public information coverage, and other administrative matters.

- Command and communications. Where will the command post be? How will units communicate with the IC and where will he be? What are the radio frequencies and other signal arrangements (chemlights, smoke, etc.)? How will the operation be terminated? What are the emergency signals?

Conclusion

If the victim's location is unknown, a search must be initiated. Searchers should allocate resources first to search high probability areas, working downward to lower probability areas. If the subject is found alive, he may have to be rescued. If not, searchers may have to mount a body recovery operation, which can be a lengthy and hazardous process.

Animal Rescue
Chapter 16

Why animal rescue? To some people this will seem obvious, but if you are firefighter or other emergency services worker, it may not. "We got better things to do than rescue grandma's cat," is the often heard refrain. However, the public knows that firefighters do two things—put out fires and rescue cats from trees. So animal rescue is part of the job, at least in the public's mind, whether we like it or not.

In the U.S. the matter came to a head in the wake of Hurricane Katrina, which struck the Gulf Coast in 2005. In the hard-hit city of New Orleans an estimated 50,000 pets and companion animals were left behind to cope with the flooded streets alone. Most of the human population was eventually evacuated, but evacuees were forbidden to bring their pets with them. Since the city remained flooded with broken levees, owners were not allowed to return. Although volunteers eventually rescued some 10,000 of them, many animals perished miserably without food or water and only an estimated 2000 were eventually reunited with their owners. The plight of the abandoned animals led Congress to pass the PETS Act (Pets Evacuation and Transportation Standards) of 2006, which requires animal evacuation plans in order to receive federal emergency funding.

Animal rescue and evacuation was an important area that had simply been neglected. Yet well over half of all U.S. households have a pet of some kind, and many have several. Nor are pets the only animals at risk in a flood. Livestock losses totaled "several thousand dairy cows in the 1991 Snohomish Valley, Washington floods; 1200 dairy cows in Tilamook, Oregon, in 1996; and approximately 90,000 beef cows in the Dakotas and Minnesota in 1997." (Heath, Sebastian, *Animal Management in Disasters*, St. Louis: Mosby 1999). During Hurricane Floyd in 1999 over 30,000 hogs, two million chickens, and

nearly a million turkeys perished in the Carolinas—a major economic loss in anyone's book. The problem is even more acute in the developing world, where loss of livestock may mean the loss of a community's livelihood.

The most compelling reason, however, is that the rescue of people and animals is tied together and can't be separated. People often refuse to evacuate without their pets, even if the evacuation is mandatory. One study of the 1997 floods in Marysville, CA, showed that while the overall evacuation rate was about 80%, just under 80% of those who did *not* evacuate owned pets. (Heath 1999) Seniors, especially, are reluctant to evacuate without their pets, since these may be "family."

Reluctance to evacuate is higher for pet owners without children. The likelihood of non-evacuation approximately doubles for these owners with each additional pet. (Heath 1999) Heath also notes that "approximately 20% to 40% of people leave their pets behind when they evacuate, and about half of them try to rescue their pets later." About one-third of these attempted "pet rescues" come during the hours of darkness, greatly increasing not only the risk to the would-be rescuers but to those who might have to rescue them. "Animal ownership," he concludes, "may be the single most important cause of human evacuation failure." This also applies to ranchers who want to feed, check on, or rescue their stock. If the pets and livestock are not evacuated or rescued, emergency crews may well have to go back in and rescue their owners (or, as the case may be, to recover their remains).

Animal rescues also generate an extraordinary amount of publicity, both good and bad. The success or failure of high-profile animal rescue may have a long-term effect on an agency's image (and funding). It also falls under the general heading of "customer service" and maintaining good relations with the citizens who pay the bills. Finally, it's the right and humane thing to do, consistent with the considerations we'll discuss shortly.

Generally speaking, there are two kinds of people who may be called upon to do flood and swiftwater rescues of animals. One is animal welfare and control personnel, who know a lot about animal behavior but usually not much about moving water, and rescue personnel. These may have knowledge of swiftwater but may not know much about handling animals, although in both cases this obviously varies with the individual. A century ago most Americans lived on farms, but today most are urban dwellers who have little idea of how to handle animals, especially large ones. There is also the "Bambi" syndrome—movies and television shows that give the impression that animals think and act like humans. They do not!

Most of the animals you will be expected to rescue are domesticated animals; that is, they are used to being around humans. However, you will sometimes encounter "exotics" who are not.

General Considerations

Mitigation. It makes sense to try to mitigate the effect of disasters on both human and animal populations, before, during, and after the disaster. This reduces the danger to both and reduces the work load of emergency services personnel. Some effective mitigation measures are:

Evacuate pets along with people. Except in emergency situation where evacuations must be done in a hurry, in general it is essential to evacuate pets and owners together. A good rule of thumb is that if the situation is dangerous enough to require evacuation of people, the animals should come out too. The most common reason given by owners not evacuating their pets with them was the owner's belief that they would not be gone long. Other reasons included poor owner attachment and ownership practices; not having a place to take the animals, not being able to catch them, and not having a carrier. (Heath 1999)

Public education and owner preparedness. As with many other areas, a good public education program will reduce problems later. Some suggestions are:

Pet Owners
- Pet owners should be educated that they, not EMS, are ultimately responsible for their pets.
- Owners should be encouraged to make up a pet emergency kit (see Appendix 5) for each pet and be prepared to shelter in place for a minimum

of 72 hours. Include medications and veterinary records.
- If possible make prior arrangements to board animals outside the danger area, such as in pet-friendly motels and with relatives. Although the rules have changed since Katrina, most disaster shelters still cannot accept animals except service animals for the disabled.
- Small animal owners (cats, small dogs, etc.) owners should purchase a carrier. Many households have several small pets but only one carrier, because under normal circumstances only one at a time needs to be transported (e.g. to the vet). For a disaster, *each* animal needs a carrier. Have enough leashes for all dogs.
- Use identification collars, tags, microchips, tattoos or similar devices for identification. A photo of each animal is also useful for identification. Consider using a two part tag system for pets and owners if they must be separated. Matching tag numbers is easier than locating pets and owners by description. Identified pets are about 12 times more likely to be reunited with their owners.

Farm and livestock owners
- Livestock and horse owners should prepare for the possibility that they will have to evacuate

Fig 16.0 Evacuate animals with people whenever possible. Here a San Antonio firefighter helps a resident and his dog escape flood waters.

their stock from flooded areas. They should scout evacuation routes to high ground and insure that adequate transportation is available. Moving livestock any distance will require adequate trailers, hitches, trucks, etc.

- Make prior arrangements as to where evacuated stock will go.
- Mark animals where possible: branding, ear tags, microchips, etc. Horses should be photographed, marked with leg tags, or with temporary markings such as the owner's name and telephone written on the hoof with indelible marker.

Emergency Services

- EMS should expect and encourage people to evacuate *with* their pets.
- Set up a public education program for pet owners as described above, both before and during the disaster.
- Coordinate with concerned citizens, humane agencies and groups (HSUS, etc.) and veterinarians to establish DART (Disaster Animal Response Team) Teams to work with EMS agencies. Integrate trained animal handlers into rescue strike teams, and conduct animal rescue classes for EMS personnel.
- Set up a pet hotline for lost and separated pets.
- Make reasonable accommodation for pets and livestock at or near shelters. Fairground exhibition buildings, for instance, have been used to house both pets and livestock during emergencies.
- Provide for marking both pets and owners for reunification. One successful system is to use a plastic wristband coded by color and number—one for the owner's wrist, the other for the animal's neck.

Philosophy of Animal Rescue

The general philosophy of swiftwater animal rescue remains the same—to rescue people and animals while exposing the rescuers to as little danger as possible. However there are two other considerations in animal rescue situations:

- Human rescues come before animal rescues.
- No unreasonable risk to humans when rescuing animals.

When dealing with animals RETHROG has to be modified.

Talk/Lead: Most animals cannot understand verbal commands like "swim this way." However, they may

be led to safety; or driven or induced to swim in certain directions.

Reach and throw: These rescue options are often limited, since most animals cannot grasp an extended pole or grasp a throw rope. However, there are some specialized tools like nets and snare poles that may work.

Row: Boat-based rescues are a high risk option with animals. Nevertheless this is a common solution and works fairly well in calm water. Rescuers must decide whether they need to get animal in boat, lead it or get it to swim alongside, or attempt to herd it with a boat. Larger animals obviously increase the risk.

Go/Tow: This option is usually too risky for humans, especially with larger animals. However, some in-water options like live bait may be okay for smaller animals if the risk is determined to be low. This leaves the hands free and provides a positive means of recovery.

Helo: High risk option. The noise and downwash are frightening to almost any animal. Small animals must be restrained, caged and/or sedated before loading inside a helicopter. Large animals can sometimes be sling loaded underneath with commercial or improvised harnesses. However, *this is a job for experienced handlers only.*

Options for Animal Rescue

Swimming animals

- Getting a swimming animal out of the water can be difficult. Most animals, especially larger ones like cows and horses, require a short ramp or incline to get out of the water. Others, like cats, are good climbers.
- Most animals can swim, but may not swim in a safe direction. Some can be induced to swim towards safety or follow a boat. Others can be herded by boats or other means. It is usually safer to minimize contact by leading, driving, herding, or guiding large animals to safety. Animals can often be convinced to swim where you want them by using barriers, such as an inflated fire hose or pool floats in the water.
- Horses can sometimes be led with a halter behind a boat. However, the rescuers should be sure they can release if need be.
- Small animals such as cats and dogs can be scooped up with a pole net from shore or a boat. Scoop from the rear so as to keep the animal's head above water. Beware of breaking the net as the animal is lifted out of the water. You may have to leave a heavy animal in the water and pull or lead him over to shore.

- Use great caution when trying to get an animal in the boat, or when transporting animals in a boat. Adding a large animal to a boat in swift water invites trouble. Large animals may raise the center of gravity of the boat or upset the weight distribution. They will not know how to high side.

Stranded animals
- If the animal is no immediate danger, approach it as described below. Remember, a stranded animal may think its retreat is cut off and decide it must defend itself. It may also decide to retreat by swimming. Try to minimize your threat profile.
- Especially in a flash flood, where the water drops rapidly, consider placing food and water and leaving the animal in place. This should only be considered for short periods of time when doing so will not place the animal in danger from rising water.
- Try to channel animals where you want them to go. Use barriers (fences, traffic barriers, inflated fire hoses, etc.) to exclude them from some areas and direct their movements into others. Use positive reinforcement (e.g. food) to get them into areas where you do want them.

Communication: Animal Behavior

The ability of humans to communicate with animals varies widely. Animals can't understand verbal directions in human languages, except for perhaps a few simple words like "outside." Even human gestures that we interpret as friendly may be interpreted

Fig 16.1 Nine-year-old Crystal Reynolds plays with her hamster, Rowdy, the family's sole pet survivor of the 1998 Kansas City flood. "Pocket pets" like hamsters, mice, gerbils, and the like, are becoming increasingly popular.

by an animal as threatening, especially during periods of stress.

This leads to a basic rule of animal behavior: *if an animal is threatened with harm, or what it perceives as harmful, it will defend itself as best it can.* The weapons vary widely but the reaction can be swift, violent, and occasionally deadly. In short—you can get hurt if you're not careful.

It follows, then, that the best way to deal with an animal is to convince it that you mean it no harm. This may be difficult if the animal is already in a life-threatening situation, such as in a flood. Since logical discourse with an animal is not an option (unless perhaps you have Dr. Dolittle along), you can deal with them by either physically restraining them or by convincing them to do what you want them to.

For large, heavy, powerful animals like horses the preferred method is to get them to cooperate. For small, skittish animals like cats, restraint is the best option.

As a rescuer, you need to understand enough about the animal's instincts, physical characteristics, and psychological makeup to make an intelligent decision about what your options are.

Animal behavior is driven far more by instinct that is that of humans. In disasters animals follow their instincts rather than what we would consider reason. Like humans, they may panic and exhibit non-purposeful behavior. For the rescuer, it is important to know how different animals react.

Body language is much more important to animals than words. Much of animal communication is done by species-specific postures, gestures and actions. These may communicate such varied themes as sexual interest, dominance, or threat display. When approaching an animal remember:
- *Move slowly.* Quick, decisive movements are usually interpreted as an attack. The profile of the person advancing also is important. Many animals use a "threat display:" that is, they make themselves look larger and more threatening than they actually are to ward off attackers or cow rivals. Thus a cat arches his back and makes his hair stand up, a dog's hair stands up around his neck and back. This is a warning: push me and I'll attack. So it's usually a good idea to lower your profile when approaching a small animal to make yourself look less threatening. You can squat or approach sideways to make yourself look smaller and less threatening.
- *Avoid eye contact* with any animal, wild or domestic, since this is usually interpreted as an invitation to a confrontation. Especially true of

dogs, since this is a way of establishing dominance. Look away from the animal or toward the ground.

- *Speak softly*: shouting at an animal generally gets a negative response, since harsh sounds are generally associated with an attack. In general, keep your voice low and pleasant, using a soothing, non-threatening tone. How you say it is much more important than what you say. *However*, some domestic animals, particularly dogs and horses, can often be controlled with familiar commands like "sit."

SMALL ANIMALS

Fig 16.2 Firefighter Stan Kinnard smiles after rescuing three puppies from the 1997 floods in Cincinnati, OH.

Dogs

Attack profile: bites.

- Dogs are pack animals. They like to be with other dogs and will chase and attack something small that runs away or screams i.e. children, other dogs.
- Many dogs will defend their territory even when it puts them in danger. There have been numerous cases of guard dogs "protecting" burning houses from firefighters. Dogs bred for guarding or fighting (Chows, Mastiffs, Pit Bulls) are the most dangerous, but do not assume any dog is "safe" until he is under positive control.

- Most dogs signal an attack by barking or by a bluff charge, but some do not. Beware the dog who attacks without barking or showing any of the body language covered below.
- The "kill zone" is about halfway between the front door and your car. It is a good idea to announce your arrival.
- Keep something like an EMS bag or pack between you and a dog you aren't sure of. If he attacks give it to him to bite. Leave yourself an escape route.

Approach: lower your profile; avoid eye contact; talk in a low, soothing voice. Approach slowly.

Dog language

- Head up, teeth bared, hair up on back and neck: I'm ready to attack.
- Head down, tail between legs: I submit—you're the boss. However, even cowering dogs may bite out of simple fear.
- Front legs stretched out, back curved down: Let's play.
- Head up, tail wagging, barking: I'm not sure who you are and I think my boss ought to know you're here (but I still might bite you).

Handling a dog:

- Control the head—a dog attacks with his teeth. Hold large dogs on the side and back of the head to control it.
- To lift a dog, use one hand to control the head, then support the dog's weight by lifting from behind his rear legs or, for small dogs, under the abdomen.
- Use a muzzle. This disarms the dog's primary weapon—his teeth. TIP: An improvised muzzle can be made from surgical tubing or a tourniquet (this will not work on very blunt-nosed dogs). **WARNING**: *Do not muzzle a panting dog!* Dogs do not sweat. They vent excess heat by panting. A muzzled dog may die very quickly from overheating.
- Try to work from a distance: use nooses, catchpoles, and pole nets to hold the dog so that others can move around behind it to immobilize its head to get a muzzle on (see chapter 3)
- Take a leash with you. Many dogs respond positively to being leashed.
- Small dogs can be controlled by "scruffing" like cats.
- Transfer the dog into a pen or enclosure as soon as possible.

Fig 16.3 One way to control a dog. Trina Hudson holds the head in the crook of her arm, with her hand on the back of the dog's head. The other arm controls the dog's hind quarters. Note the heavy gloves.

Fig 16.4 Small dogs can be "scruffed." Grasp the loose skin behind the dog's head and pull tight to control the head. Support the dog's weight with the other hand.

Fig 16.6 Scruff a cat to keep it from biting or scratching you. Grab the loose skin behind the neck to control the head and front legs, and use the other hand to hold the back legs.

Fig 16.5 A rescuer climbs a ladder to rescue a cat trapped by flood waters in Cincinnati, OH.

Cats

Attack profile: scratches, then bites.
- Cat bites and scratches are prone to infection.
- House cats don't attack unless cornered, but see large cats below.
- Flattened ears are a sign of imminent attack.

Approach: lower profile, avoid eye contact, talk in a low, soothing voice. Approach slowly. Wear gloves and forearm protection!

Handling a cat
- TIP: A large towel or blanket often makes an effective tool for capturing a cat. Throw it over the cat and quickly wrap it up.
- Control a cat by "scruffing" it. Grasp it by the loose skin on the back of the neck and shoulders and pull tight. This controls the head and keeps the cat from biting you, as well as keeping his front claws away. Then grab the cat's back legs with your other hand. This immobilizes the cat and makes it impossible for him to attack you.
- Use a sack or large bag for temporary holding or transport. Put the cat in the bag butt-first and quickly secure the top. Make sure it has adequate ventilation. *Don't* use a pillow case—the cat will shred it in short order and escape.
- Use a cage or carrier for long-term holding or transporting a cat.

Fig 16.7 Large animals like horses present different problems for rescuers. Here a volunteer tries to round up a horse stranded during the January 1997 floods in Central California.

Horses

Attack profile: Runs away if possible, but will kick, strike with its hooves and bite if cornered or threatened.

Safety considerations

- Horses can kill. A horse is a large, powerful animal that can cause severe injury in a confined space like a stall, even if it means you no harm. Be careful!
- Beware of being kicked. The danger zone for kicking is 3-10' (1-3m) from the horses hind quarters. Either stay right next to the animal or move a safe (15-20', 5-7m) distance away. Watch the back legs—they will reach further than you think.

Body language

- If the horse's eyes widen, his muscles noticeably tense up, and he blows out through his nostrils, he is getting agitated. Exercise caution.
- Flattened ears are a sigh of an imminent attack.
- Most horses are gentle and naturally curious. They are used to being led and to being around people (this obviously does not apply to wild horses). Some are skittish, however, and you need to reassure the horse that you mean it no harm.
- Announce your presence by whistling or talking gently. Don't shout. Use a low, calming voice when talking. Don't face the horse directly.

- Use bait to exploit a horse's natural curiosity. Rattle pebbles in a can, crinkle paper in your pocket, or offer food.
- Horses have eyes on the sides of their heads, which gives them a wide field of vision to see predators. However, unlike humans, they do not see very well straight ahead. Try to approach a horse from the front quarter rather than from straight ahead to avoid startling it.
- Use slow, deliberate movements around horses, who sometimes spook easily. Don't wave or jump.
- Do not use the lights or sirens on emergency vehicles around horses.

Handling

- Horses are herd animals. They like to be with other horses and to follow an established leader. You can sometimes use this instinct to your advantage. If you can lead one horse where you want it to go, especially a dominant one, you can often get the others to follow.
- If possible, try to use the horse's owner, or someone experienced in handling horses, to help with the rescue. Handling horses in a disaster requires experience.
- The best place to stand is just behind the horse's left shoulder. Stroke or scratch the horse's shoulder and talk to him in a soothing voice. Be patient and win his confidence.
- Put on a halter. This gives you a way to control the horse's head. If you can do this you will control the rest of him. Walk next to the horse's left shoulder about an arm's length away.
- Lead the horse with a halter if possible. If you can't lead the horse, you may be able to herd it where you want it to go. Don't try to improvise a halter unless you know what you're doing. You can blindfold an agitated horse to calm it.
- Don't tie yourself to a horse, especially in the water. Don't tie anything to a horse unless you know what you're doing or have been directed to do so by an experienced horse handler. You can injure or kill the horse and yourself.
- Don't get into the water with a horse or try to ride a horse through water, especially moving water.
- Most boats are not suitable for carrying horses. Horses are tall and raise the center of gravity enough to make almost any boat unstable, not to mention that the animal's reactions may be unpredictable.

Fig 16.8 A rescuer leads a horse to safety during the 1997 California floods. The cattle have chosen the high ground.

Cows

Cattle are herd animals who like to stay together. Unlike horses, they are not used to being handled individually, but rather herded as a group.

Attack profile: Butts with its head. They can also kick with their hind legs.

Safety considerations

- Generally, cattle are less dangerous than horses (this does not apply to bulls).
- Some breeds have large horns that can cause serious injury.
- Some breeds weigh well over a thousand pounds are can cause serious injury just by their size, especially in confined areas like stalls.

Body language

- Stands its ground and lowers its head: attack likely
- Turns head and moves away: submission

Approach

- Approach slowly and deliberately. Talk in a slow, low, soothing tone.
- Keep a safe distance away.
- Allow yourself an escape route.

Handling

- Herd cattle where you want them to go rather than trying to capture them. Keep them together and lead or drive them into a pen or enclosure.
- Herd slowly. Don't wave your arms and shout.

Pigs

Attack profile: Charges and bites.

Safety considerations

- Farm pigs are domesticated but not used to being handled.
- A recent trend has been the keeping of pet pigs such as the Vietnamese Pot-Bellied at home. These pigs are more used to people and to being handled. However, they set up an bloody racket when picked up.
- Wild pigs have large tusks that can cause serious injury.
- Pigs have strong jaws and can cause serious wounds. Their bites are "dirty" and can easily become infected.
- Use adequate safety precautions when entering an enclosure with a large pig. Leave yourself an escape route.
- Approach sows with litters with caution.

Handling:

- Pigs are intelligent, stubborn and unpredictable. They are also strong and can move quickly. Even a small pig can be difficult to physically capture.
- Pigs have a limited cooling ability and overheat quickly. They do not handle stress well.
- Pigs are difficult to herd and often seem to do exactly the opposite of what you want them to do. They can also exhibit a "mob" reaction under stress; that is, if one pig squeals, the rest may attack.
- Best to deal with pigs separately rather than in a group. Use a large object like a board, stretcher or litter, to direct the pig(s) into a pen, cage, or enclosure. Pigs can be guided (sometimes) by tapping them on their sides or heads with a stick.
- Use food to induce the pig to go where you want him to go.

Sheep

Sheep are herd animals with a strong instinct to stay together. If separated from the flock sheep may panic and injure themselves.

Attack profile: Rams butt with their heads. Otherwise sheep don't kick or bite and are relatively easy to handle.

Body language: Lowered head is a sign of imminent attack.

Safety considerations

- Relatively low chance of serious injury.
- Sheep overheat easily.

Handling

- Sheep are best handled by experienced shepherds and sheep dogs.
- Herd or guide the flock with boards, panels, litters, etc.
- Approach slowly, reduce threat posture, and speak softly. Don't yell or wave.
- Don't grab a sheep by the wool if you need to handle it. This can damage the wool and injure the sheep. If you need to handle a single sheep, place one hand on the animal's chin or chest to stop it, then place the other hand on the animal's rear just under the tail (dock). Move the sheep where you want it to go.
- Because of sheep's tendency to panic if separated from the flock, it is usually better to capture a single sheep directly from the flock rather than cutting it out.

Goats

Goats are herd animals and like to stay together. They are more intelligent than sheep and can be pesky critters to deal with.

Attack profile: Males butt with their heads. Attacks are more likely than with sheep.

Body language

- Lowered head means attack imminent.
- Goats will put on a threat display by vocalizing and stamping their feet, but will not attack when captured.

Safety considerations

- Little danger of serious injury to handlers
- Goats can be easily injured by rough handling. Their bones are delicate and easily broken. They are prone to stress when captured and restrained.

Approach: Approach slowly, minimize threat profile, and speak softly.

Handling

- Goats are quite agile, are good jumpers, and can often quickly figure out how to escape restraint (opening gates, chewing ropes, etc.).
- Capture and guide like a sheep: one hand in front of chest and the other beneath the tail.
- You can grab a goat's horns to capture it, but they don't like to be led this way.
- Herd goats by directing them into a pen or enclosure.

EXOTICS

In some areas (e.g. south Florida), many people keep exotic pets, so you never know what you will encounter. In a recent flood in Mississippi, several hundred alligators floated out of a alligator farm in high water. These animals are not "domesticated" in the normal sense of the word.

Ostrich, Emu, etc.

These large, flightless birds have been kept in recent years as "stock."

Attack profile: Prefers to run, but if cornered will kick with feet.

General attack sequence is:

- Threat display: puffs up, stands erect
- Bluff attack
- Real attack

Safety Considerations

- Do not wear anything shiny (watches, glasses, jewelry), because like many birds they will peck at it.
- Their kick is extremely powerful and is capable of severely injuring or even killing an adult.
- These birds can kick straight forward only. Avoid attack by moving laterally back and forth in front of the bird so that he can't "square" on you. He can't kick at an angle.

Handling

- These birds are not terribly intelligent and are relatively easy to outsmart.
- An umbrella is a good tool. Open it to intimidate the birds by making yourself look larger.
- Use the crook of the umbrella or a similar tool to pull the bird's head down. He can't kick with his head down.
- To subdue, get the bird's head *down* and under control. Then put something opaque over it, like a hood or jacket. This will calm the bird. Leave the hood on when moving or transporting the bird.
- To move the bird, put one person on each side to control it.

Iguanas

Often kept as pets in tropical regions.

Attack profile: Prefers to run, but if cornered will attack by biting and clawing. The Iguana also has a long, sharp, tail that it uses like a whip. It can cut you.

Body language: Opens mouth and hisses when threatened or annoyed.

Safety considerations

- When handling an Iguana be sure to keep positive control of his head, feet and tail.
- Wear turnouts, gloves, or other protective clothing.

Handling

- Iguanas are good swimmers and good climbers. They generally try to escape by climbing. They are afraid of dogs.
- A good way to catch them is to run them into a high perch, then put a noose over their head.

Large cats (cougars, ocelots, etc.)

Some people keep large cats like Ocelots and cougars for pets.

Attack profile: Bites, claws. Larger cats can inflict serious injury. Unlike smaller cats, they may stand their ground on their own territory or even stalk humans.

Safety considerations:

- Minimum distance for attack/retreat is about 25'/7.6m.
- Big cats view size changes differently than most animals mentioned so far. They are programmed to stalk and attack smaller animals. Therefore you do not want to reduce your profile suddenly, as by kneeling, since this may encourage an attack. Keep small animals and children away, and do not approach by yourself.
- Don't make sudden moves or stalk them.
- Don't corner them if possible.
- Large objects that appear solid (blue tarp, cardboard, etc.) can be used to herd the cat where you want him to go.

Body language: crouched, ears laid back means unpredictable behavior.

Handling: Making a twirring sound and slowly blinking eyes has a calming effect.

Llamas and Alpacas

These have become popular in recent years as pack animals. They are herd animals; the Alpaca more so than the Llama and therefore easier to control.

Attack profile: Not particularly dangerous, but both species have "fighting teeth" that can cause injury.

Safety considerations: Beware of "berserk male syndrome:" a male sees who humans as competing males and will attack them.

Handling

- Llamas and Alpacas often lie down when threatened. This can be frustrating when you are trying to rescue them.

- Both Llamas and Alpacas have a nasty habit of spitting when annoyed. Those animals who have close human contact have often been trained not to spit.
- These animals dislike dark places. They will resist being loaded into an enclosed trailer.
- If they have a halter on, use it to lead the animal. Some owners leave the halters on all the time.
- They do not show pain in an obvious way. This can make it difficult to assess injuries.

Snakes

Snakes are often kept as pets, and may escape during floods. This can include very large ones like boa constrictors, and extremely poisonous ones. Unless you are very sure what you are dealing with it is usually better to let a herp expert do it. If you suspect non-native snakes are involved, do not assume you can identify which ones are dangerous and which ones are not.

Attack profile

- Poisonous snakes strike with their fangs.
- Constrictors wrap around their prey and suffocate them. Extremely large constrictors are dangerous to humans.

Safety considerations

- Beware of snakes during floods! Rising water runs snakes out of their holes and you have to be very careful where you step. They may end up on islands and on floating debris with human or other animal victims.
- Snakes generally have threat displays and warn before they strike.
- Many snake bites happen when someone inadvertently steps on a snake. If conditions permit wear heavy boots in snake-infested areas.

Handling

- Smaller snakes can be safely captured using a piece of PVC pipe with a bag on the end. The snake is looking for a small, dark place to hide in. It will enter pipe and can then be cinched down in the bag.
- If you must handle a snake, control the head. Even a non-poisonous snake can give you a painful bite.

Wolves, Coyotes, and Wolf-dog Hybrids

Wolf-dog hybrids have become fairly common as pets and may be encountered even in urban areas. Most attacks are by hybrids, whose behavior tends to be unpredictable. No rules apply to coyotes. Like other canines these are pack animals whose behavior is similar to other dogs.

Identification: large canine teeth, black spot on tail, short rounded ears with furry insides; dense shoulder ruff; coarse guard hairs on coat, long tail that does not curl; slanted eyes.

Attack profile: bites

Approach
- Don't point.
- Avoid staring.
- Keep your body profile low and small to reduce perceived threat.
- Keep menstruating females away.

Handling
- In general, handled like other canines. Expect the worst and exert positive control.
- A muzzle is *always* a goc ' ' '

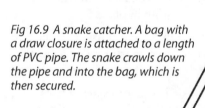

Fig 16.9 A snake catcher. A bag with a draw closure is attached to a length of PVC pipe. The snake crawls down the pipe and into the bag, which is then secured.

Handling and Transport of Animals

Disease And Other Hazards. Use normal protective measures against blood-borne pathogens: gloves, splash protection, eye and face protection, etc. It is a good idea to minimize contact with any animal whose medical history you are unsure of, or who you have just pulled from the water. Decontaminate both yourself and the animal after rescue, and make sure your immunizations are up to date. Some animal diseases are *zoonotic* that is, they can be transmitted from animals to humans. Some of the most common (but by no means all) are rabies, cat scratch fever, brucellosis and leptospirosis.

The best defense against disease is to be proactive. If bitten, wash the wound site immediately with soap and water, then consult with a physician or your medical control as soon as possible. Report all animal bites, even minor ones. Clean all scratches well and monitor for infection. Decontaminate immediately as outlined below.

Contamination
- Floods are, by definition, HAZMAT incidents. Assume that all the animals you are rescuing are contaminated and that they will contaminate you. Many animals have fur or hair that may become saturated with contaminants, and these will be transferred to you when you handle them.
- Wear normal medical protection: gloves, mask, and protective clothing when handling animals during or after rescue.
- Set up immediate post-rescue decontamination for animals as well as humans. Do this before the animals are transferred to permanent holding facilities.
- Wear protective clothing when on or near the water, as conditions permit. Protective clothing that excludes water (dry suits) protects better than that that does not (wet suits). Attempt to minimize animal contact to unprotected areas of your body (e.g. face and hands).
- Many animals, like dogs, instinctively shake themselves after getting wet and may spread contamination this way.
- Don't forget to decon equipment, too: trailers, cages, nets, poles, clothing, PPE, etc.
- Decontaminate personnel often. Track them throughout the event. Where were they today?

Ten Tips to Help You Avoid Developing A Serious Zoonotic Illness

Reprinted with permission from Animal Sheltering Magazine.

1. Stay current on your appropriate vaccinations, particularly tetanus and rabies.

2. Wash your hands frequently with antibacterial soap, especially after handling any animal and prior to eating and smoking.

3. Wear long pants and sturdy shoes or boots—no sandals or shorts.

4. Use gloves (preferably disposable) when changing litter pans, washing food and water dishes, or cleaning up feces, urine or vomit.

5. Disinfect scratches and bite wounds thoroughly.

6. Do not allow animals to lick your face or any wounds.

7. Learn safe and humane animal handling techniques and use proper equipment.

8. Seek assistance when handling animals whose dispositions are questionable.

9. Report any bites or injuries to your supervisor and to your physician.

10. Tell your physician that you work closely with animals, and visit her or him regularly.

Basic Animal First Aid

Because of their unfamiliarity with veterinary medicine and lack of specialized medical equipment, rescuers are limited in their abilities to give first aid to animals.

Safety considerations

● Rescuers should not put themselves at risk attempting to rescue or assist an injured animal. Many animals, especially larger ones, are capable of seriously injuring or killing a rescue worker.

● When approaching a "downed" large animal, stay clear of the animal's legs, even if it appears to be unresponsive. Instead, approach from the animal's back. Announce your approach to avoid startling it.

● Small animals are more likely to bite or scratch if they are injured. Wear heavy gloves when handling them and observe handling precautions already discussed (i.e. "scruffing').

Initial actions

Rescuers should locate, identify, and triage animals in the risk zone, and get this information back to veterinary professionals. Based on the circumstances, a decision will have to be made whether to attempt to evacuate the animal to a veterinary facility or to treat (or possibly euthanize) the animal in place.

LOCATE: injured animals

IDENTIFY: species, physical description (color, height, weight etc.), nature and severity of injuries.

MARK: location of animal

ASSESS: the animal's injuries and indicate whether injuries are minor, moderate, or severe.

EVACUATE: if the situation and the nature of the animal's injuries permit.

COMMUNICATE: the location and nature of the animal's injuries to veterinary professionals if evacuation is not possible.

Treatment

● Before beginning treatment insure that you are not placing yourself in danger i.e. either that the animal is docile and or that it is restrained.

● An excellent indication of the status of many animals is *Capillary Refill Time* (CRT). Check by pressing a finger against the animal's gums until the membrane blanches, then note the time that it takes for it to return to normal color. For most animals this will be 1-2 seconds.

Cuts, lacerations, and open bleeding wounds.
Handled more or less the same way as on humans:

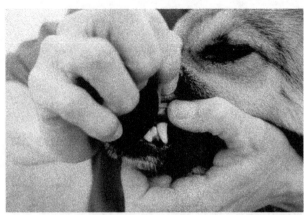

Fig 16.10 *Check Capillary Refill Time by pressing in on the animal's gum, then releasing pressure and seeing how long it takes to return to normal color.*

direct pressure until the bleeding stops, followed by bandaging. For large wounds on large animals, use duct tape to hold the dressing in place.

Impaled objects. Leave in place if possible. If the object projects you may need to trim it down closer to the body.

Broken limbs. These are best left to veterinarians. If the animal is docile you can splint it according to normal practice. Large animals with broken limbs are often destroyed, but this is a decision for a veterinarian.

Eye injuries. Should always be considered serious and left to a veterinarian. If the animal will permit it, cover with a clean, moist dressing.

Hypothermia

The most common injury in floods (Heath 1999). It is treated more or less the same way as in humans: the animal is evacuated to a warm, dry place, dried off and if necessary given supplemental heat from an outside source like a heater. Heath recommends using a "hot box;" a small enclosure heated by an electric fan heater to warm a hypothermic animal. More advanced protocols are available, but should be administered by a veterinarian. Watch animals for hypothermia after decontamination.

Evacuation

The larger an animal, the more effort required to evacuate it.

Dogs:
● Handle alert, ambulatory dogs as described previously.
● Dogs that are unconscious or unable to rise may have to be either moved to a safer location or evacuated.

Fig 16.11 Improvise a temporary muzzle by using a couple of wraps from the dog's leash.

Fig 16.12 A manufactured muzzle fits better and can be left on for a longer period of time.

- Small dogs can be lifted and placed in a carrier. Support the dog on its side and place it gently in the carrier. Restrain and muzzle it. Wear heavy gloves to avoid bites.
- Large dogs should be placed on a litter or stretcher. One can be improvised from a piece of plywood or something similar. Using a litter is safer for both dog and rescuers.
- Approach the dog from the back. Try to establish control of the head. Muzzle the dog if possible. Grasp the dog by the nape of the neck and loose skin on his back and slide him backward away from harm or onto the litter. Don't pull on the legs or get in front of the dog, since this makes it easier for him to bite you.
- If the surface is smooth and you don't need to lift the dog, slide the dog with a large towel or rug. This also allows you to avoid direct contact.

Horses and other large animals
- Under most circumstances, rescuers will not be able to physically move large non-ambulatory animals.
- Approach the animal from the back. Stay out of range of the horse's legs. Assess the animal's injuries and provide treatment as necessary.
- Ambulatory horses may be evacuated by trailer. However, you will need experienced horse handlers for this option.
- It is possible to sling-load large animals by helicopter. This is, as always, a high-risk option requiring experienced crews in the air and experienced horse handlers on the ground (see Appendix 6 for more information).

Animal rescue and handling equipment
Wear your river PPE around large animals, including helmets. This provides extra physical protection.

Leashes, halters, etc.
Use a species-specific leash or halter to lead an animal if one is available. A leash for dogs or an adjustable halter for horses works better than trying to improvise something on the spot. Many animals (e.g. dogs and horses), are accustomed to these and are more easily controlled with something familiar.

Muzzles. Essential for dogs, handy if available for cats. Remember not to muzzle a panting dog or cat. TIP: A muzzle can be improvised from surgical tubing.

Catch poles. This is a pole with a noose on the end. The noose is slipped over the animal's head and tightened to capture it. This allows the rescuer to stay at a safe distance and the pole prevents the animal from coming at you.

Nets
Throw net. A open mesh net that can be thrown over small animals for quick capture.

Capture net. A net on a pole (similar to a butterfly net but larger). A small animal is scooped up in the net and then transferred to a cage or pen.

Graspers. A padded set of jaws mounted on a handle. This allows a rescuer to humanely grasp the animal's neck and push it down to the ground, pinning the animal prior to capture.

Cages, pens, carriers, bags and sacks.
- Put small captured animals in a sack, bag, or carrier for evacuation to a holding facility. Make sure the animal can breathe and that the sack is strong enough to hold it.
- Transfer the animal as soon as possible to a holding facility with adequate food and water. Decontaminate the animal first.
- Choose a pen or enclosure large enough, if possible, to allow the animal to stand and groom itself.

Animal PFDs

Commercial PFDs are available for medium to large dogs. TIP: a dog PFD can be improvised from a human PFD. Slide it over the dog's front legs and zip it down his back.

Fig 16.14 Tools of the trade (left to right),: a capture net a catch pole, grasper, and gloves.

Fig 16.13 A catch pole provides positive control.

TIP: You can make an improvised catch pole from a firefighting pike pole. Drill two holes a few inches apart in the handle end. Then thread a piece of rope knotted at one end through the holes.

Conclusion

Animal rescue is an integral part of flood and swiftwater that can no longer be ignored or left to others. Evacuations run more smoothly and are more complete if animals are included. While participation of professional animal handlers and medical professionals like veterinarians is vital, much of the actual rescue will done by swiftwater professionals who will need basic animal rescue skills. Rescue of humans always comes first, but animal rescue must be integrated into the plan and implemented accordingly.

Fig 16.15 A grasper can be used to control small animals.

Swiftwater Training— a Suggested Program
Chapter 17

Fig. 17.0 Jim Segerstrom coaches a student on how to swim over a strainer. Realistic training is a must for swiftwater rescuers.

By now it should be obvious that any agency needs a swiftwater rescue *program*. Rescuers must train together with the equipment they will use. Training programs must also be ongoing to avoid skills decay. At the outset, however, it is often difficult to get a swiftwater rescue training program started. One of the biggest enemies is complacency. Commonly heard excuses are "we already have a water rescue program" or "another agency handles all water rescue calls" or "we only do body recoveries" or "we haven't had a call like that since I've been here." However, flash flooding is the number one weather-related fatality nationwide, and about one-third of the population of the U.S. live in flood-prone areas. Floods have also caused more property damage than any other type of natural disaster, including hurricanes. Yet, strangely, flood rescue gets little emphasis.

As we have seen, the dangers of swiftwater rescue are unique. Swiftwater rescue is quite different than other types of water rescue, and things happen in a

hurry. Because of the time factor, *any* public safety agency—fire, police, sheriff's deputies, EMS, and search and rescue, as well as ordinary citizens—can end up responding to a swiftwater rescue. This shows the need for *at least* awareness-level training for all public safety agencies.

Sometimes the need is not obvious. Multiyear droughts often precede flood years. This was the case in the San Marcos incident described at the beginning of this book, and in California in the 1980s. It may also seem strange to be teaching swiftwater rescue in the desert, but there have been major floods in cities like Albuquerque and Phoenix.

Then there is the "Silver Bullet" syndrome—an overreliance on equipment. If we can just get the right piece of equipment, so the thinking goes, our problems are solved. It is certainly easier to buy that new boat or hovercraft (and possibly even cheaper) than to initiate a lengthy training program and keep up with it. It can also be frustrating to spend training time and money on personnel who then leave the organization. However, as Jim Segerstrom noted, "the trend in emergency services is to rely on progressively more sophisticated equipment. Recent flood operations have shown boats, hovercraft, and other vessels suffering mechanical problems and being lost. Simple skills, such as paddle-boat handling and knowing how to set a 'ferry angle' could have made a difference in such cases." The reality is that training is more important than equipment. Trained, experienced rescuers can make do with very simple, even marginal, equipment, while rescuers untrained in swiftwater are likely to fail, perhaps even to die in the process, no matter what equipment they are using.

Pride can also be a factor. The phrase "we've got it under control" sometimes conceals a reluctance to learn from others, or to even admit that our own program is less than perfect. Swiftwater rescue is a relatively recent and dynamic field, and while the principles remain constant, new techniques and equipment are constantly being added.

Once the decision has been made to implement the program, the two inevitable questions arise: how much will it cost and how much training time will it take? In a very general way skill levels can be divided roughly into four levels:

- **Awareness/Familiarization.** The goal here is to help people recognize the special dangers of swiftwater and stay out of trouble, not to perform rescues. These broad-based educational programs are often quite effective and can apply to anyone—the general public (e.g., citizen awareness programs such as CERT), public officials, and those public service personnel (e.g., law enforcement, utilities personnel, etc.) whose primary duty is not water rescue but who may end up at a flood or swiftwater incident.
- **Operational/Basic.** Here, personnel can assist in rescues without being a liability, but cannot necessarily perform them. Most important, they have been trained in self-rescue. This corresponds to the first responder level, or the ASTM/NASAR basic water rescuer. This is the *minimum* level of training for rescue personnel who may be intentionally dispatched to a swiftwater incident.
- **Professional/Technician.** These are people who have had an intensive course of instruction on swiftwater rescue, and can perform basic swiftwater rescues competently and safely. However, they cannot necessarily lead them. This corresponds to Swiftwater Rescue Technician I (SRT 1).

- **Specialist/Expert.** Extensive training in swiftwater rescue, including incident command and related specialized fields like high ropes, scuba, etc.

When setting up a training program, agencies must consider the numbers of people to be trained and the levels to which they are to be trained. Obviously, everyone does not need to be trained to the specialist/expert level. By the same token, however, almost everyone needs *some* training if they are not to be an on-scene liability.

As a general rule, *everyone* should have awareness-level training. Anyone who might be expected to be on-scene on a swiftwater rescue needs to be at least a first responder. The people who plan to actually get into the water to make rescues should have at least technician level training, and incident commanders and technical specialists should be trained to specialist level. Agencies should avoid overdependence on a few individuals or agencies (i.e., "they" do all the swiftwater rescues). Recall the rescuer team structures in Chapter 12—there were a mix of firefighters and lifeguards. There were also separate search and rescue teams, as well as first responder teams.

Agencies must also consider skill decay. River skills, knot-tying, and the setup of complex technical systems must be practiced to be effective. This is even more true in the off-season or times of drought.

Training priorities must also be set. Hardly a week passes, it sometimes seems, without some new training responsibility being added to rescue agencies—confined space rescues, HAZMAT, and much, much more. It is impossible, especially for volunteers, to be prepared for everything. Therefore, agency planners consider a number of factors—historical record of calls, community growth and development trends, and the relative dangers of different types of rescue—in developing a mission essential task list, or METL. This is yet another idea borrowed from the military, specifically the reserves, who found that they did not have nearly enough time to train for all the missions they might have to perform. Instead, they analyzed their most likely missions and trained accordingly. For instance, if a unit was slotted for deployment to Europe and would most likely be engaged in a defensive battle against odds, it made more sense to concentrate training to that end. Similarly, a rural rescue team might have little need for training in building collapse, but have frequent calls for river rescues.

The METL must also consider the danger factor. While it is true that agencies must prioritize their time, equipment, and training, this is no reason to ignore swiftwater rescue, a very common problem. Experience has shown that while swiftwater calls may constitute only a small proportion of calls, they account for a disproportionately large number of rescuer deaths.

Any effective training program also carries with it a certain element of paradox. The more realistic the situation, the more its training value; however, the more dangerous it is for the students. Whereas agencies and the public are willing to accept injuries and even deaths as the price for a successful rescue operation, this willingness seldom applies to training situations, even when realistic training may save lives in the long run. We have already seen how this applies to helicopter training situations (*you can do it, but you can't train to do it*).

There are several ways to implement a swiftwater training program. An agency can send its personnel out for training, although this is expensive in terms of travel expenses and lost time. It is much more cost-effective to train an in-house agency instructor. Local whitewater outfitters may offer guide's schools, which can be excellent sources of swiftwater boat-handling training. Whitewater kayak schools and raft trips offer a fun way to learn about moving water. The American Canoe Association also offers a whitewater rescue course aimed primarily at recreational paddlers. Finally, there are private organizations that specialize in swiftwater rescue training for fire and emergency service personnel, some of which are listed in Appendix 4.

Conclusion

Who needs a swiftwater rescue program? You do.

Fig. 17.1 A flood rescue in Texas. Is your agency prepared for an incident like this?

Afterword
An Analysis of the San Marcos Incident

By now readers should have a better understanding of what went wrong at the San Marcos incident described in the Prologue. It was chosen because it cuts across many of the areas covered in this book. Here, we will make a detailed analysis. Our purpose is not to assign blame or second-guess the rescuers, but to learn from their mistakes.

The initial series of mistakes were made by the kayaker. He was a relatively inexperienced river runner, boating alone early in the season, with the river (which was admittedly easy) at flood stage. The American Whitewater Affiliation Safety Code recommends a minimum boating party of three persons. Further, because of a prolonged drought that limited paddling opportunities, his boating skills had probably decayed somewhat. His biggest single mistake was to run the dam. It is impossible now to know what he was thinking, but this is not the only case of a paddler who should have known better making a decision that seems, in retrospect, foolish.

Tom Goynes, who wrote the incident account in the Prolog, comments that "boaters should never boat floodwater alone. Too many surprises happen on familiar creeks that are high for the first time in a while. At least with another experienced boater along, the chance of a rescue by someone who understands rivers is possible."

The next major mistake, shared by the victim and most of the rescuers alike, was in assessing the exact nature of the river feature that held the kayaker. It was a breaking wave, *not* a hydraulic. This is a surface feature that, while it will hold a buoyant object, will not hold or recirculate a submerged one. Here is perhaps the ultimate irony of the whole incident: had the kayaker come out of his boat instead of staying in it, he would have washed under the wave and downstream. Thus, he could have rescued himself at any time. To many of the rescuers, who also assumed that this was a hydraulic, it seemed suicidal to send in a rescue boat, when in fact it was a relatively low-risk operation for trained rescuers, especially since they had hedged their bets by tethering the raft. Another irony of the situation was that while almost everyone on site (including the victim) had

some swiftwater rescue training, they almost all failed to recognize the true nature of the situation.

Goynes recommends that "swiftwater rescue training ... make more of an effort to explain hydraulic currents and to differentiate between hydraulics and other dam-related phenomena like stopper waves. Many firemen have seen so many scenes of motorboats being eaten by dams that they have perhaps too great a fear of dams in general. Boaters also need to understand these differences."

Unfortunately, this misunderstanding set the stage for what would happen next. Considering his own expertise and those with him, Goynes' plan for rescue—going up to the wave with a tethered raft—was quite reasonable. However, the incident chain of command was unclear, and the IC was not familiar with the capabilities of rescuers or the available rescue options. Nor did he understand what the rescuers were trying to do. "I assumed," said Goynes, "that once I arrived on the scene I would be allowed to do whatever needed to be done. In retrospect, I should have found the incident commander, reported in, and explained my plan to him."

This misunderstanding then precipitated the next mistake—the IC's decision to call a helicopter. This was a high-risk option, the more so because it was a military helicopter whose crew had not been trained in swiftwater rescue. The IC did not have a clear idea of what the helicopter could and couldn't do, and the helicopter crew did not understand the nature of the rescue they were about to make. The use of the jungle penetrator was also unfortunate, since it offered no way to cinch or secure the victim. When the kayaker suddenly grabbed the rescuer on the device, he rendered him helpless and unbalanced the aircraft at the same time.

Still, the situation was not hopeless. The best solution would have been to drag the kayaker free of the dam and toward shore, where he could have been picked up by rescuers on shore or in a raft, or to short-haul him to shore. Apparently, however, the aircraft's SOP called for all victims to be lifted to the fuselage.

"Hopefully," says Goynes, "people will realize that helicopters should be a last resort when it comes to certain river rescue situations. When they are used in whitewater situations, the crews should be trained in the mechanics thereof. The Texas Department of Public Safety now uses a ring that is either lowered to the victim from the helicopter or handed to the victim by the rescuer who is lowered from the aircraft. Once the ring is around the victim and the rope is tightened, a strap tightens around the victim, making it virtually impossible for him to fall (i.e., cinch collar or Res-Q-Ring™–SR). Furthermore, DPS policy is to never lift the victim any higher than is absolutely necessary, but to keep him close to the ground in the event that something should go wrong. Hopefully, the Army will learn to implement some of these techniques."

Like many accidents, this one depended on a series of miscalculations, each building on another. It points up the need for the things we have discussed in some detail, such as:

- Swiftwater education and experience, especially the need for retraining and re-education after a prolonged drought.
- A clear chain of incident command.
- The use of low- to high-risk rescue options.
- A clear understanding of the capabilities of helicopters, and appropriate training for both ground and air crews.

On the positive side, none of the rescuers were killed or injured, or had to be rescued.

Then there was the aftermath of the incident. Goynes describes his feelings:

"Probably no incident that I have been involved in has affected me like this one. I was not alone in the many hours of lost sleep and the agonizing over what went wrong and how I might have been able to do something differently. My heart goes out to the helicopter crew, especially the young man that was lowered into the river and actually had contact with Bruce right before he fell to his death. It was amazing how the entire community grieved the loss of this kayaker who very few locals had ever met. People who had not even been eyewitnesses reported losing sleep. It was in the middle of this grieving process when the Martindale Fire Department received a letter from Bruce's widow.

"Mrs. R, in an uncommon show of human compassion, especially considering what she had just gone through, had written a thank-you letter to the department. She talked of the times that Bruce had come home late at night from a failed rescue mission of his own or from a fire where someone had lost their home or perhaps their life, and how he had agonized over all

of the things that he might have been able to do differently. She knew our grief and she wrote to tell us that it wasn't our fault and that she appreciated our efforts. She sent along a check for $100 for the department to purchase much-needed equipment. It was in this one act of kindness that many of our members found healing. I pray that this account will prove to be a positive action as well, perhaps preventing another such accident along the way."

Goynes also took corrective action to avoid a repetition of the incident.

"Less than a week after this incident, I was speaking at a San Antonio canoe club meeting and a fireman present said that there wasn't a fire department raft in the entire city. While he said that current restrictions prohibit volunteers from helping firemen in rescue situations, he believed that an individual could certainly offer to help train firemen and to offer equipment to departments that have none. Since the incident, I have formed the Martindale Swiftwater Rescue Team, which will automatically be in charge of future river rescues in the area. Furthermore, I can have river people volunteer for the team even though they are not regular members of the firefighting team. We must realize that just because someone understands fires doesn't mean he understand rivers."

Common Search and Rescue Helicopters

(not to scale)

TYPE III

Bell B206 B-3 JetRanger. A very widely-used helicopter, although no longer in production, the Jet Ranger is economical, light, and easy to maintain. It can carry a respectable payload at most elevations and has a fair-sized baggage compartment. It is used for firefighting, recon, passenger transport, and medical evacuation.

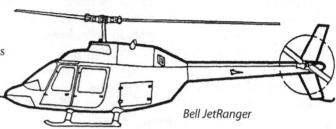

Bell JetRanger

Bell B206 L-4 LongRanger, 407. The LongRanger 407 is a larger variant of the JetRanger with more cabin space, a four-bladed rotor and a bigger engine, which gives it better performance at high elevations and allowing for a larger payload. This workhorse helicopter performs a wide variety of missions including firefighting, recon, rappelling, long-line, and transport of personnel and equipment. It is used by many air ambulance services and government agencies.

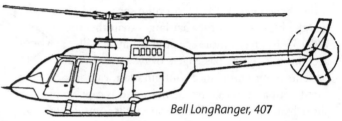

Bell LongRanger, 407

MD Helicopters MD-500E, 520N, 530F, 600N. A very common helicopter, the 500D is small and very maneuverable in confined areas. It has great visibility for long-line work, but its small size restricts its internal carrying capacity. There is no baggage compartment, so cargo must be either carried in baskets or strapped inside. Later models are NOTARs, in which the exhaust is vented through the tail boom to control yaw in lieu of a tail rotor. The 530F has more power and performs better at altitude. The 600N is a stretched version of the 520N with a 8-person capacity and a six-bladed rotor.

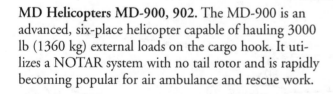

MD-500

MD Helicopters MD-900, 902. The MD-900 is an advanced, six-place helicopter capable of hauling 3000 lb (1360 kg) external loads on the cargo hook. It utilizes a NOTAR system with no tail rotor and is rapidly becoming popular for air ambulance and rescue work.

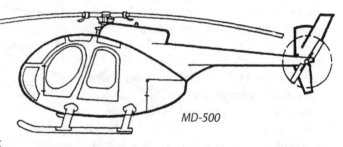

MD-900

Eurocopter AS-350 Series Ecureuil (Squirrel). These French designed helicopters are widely used by police, military, and air ambulance services as well as for search and rescue. The AS-355 is a twin-engined version. The Eurocopter 130 is a larger improved version with an ducted tail fan instead of a rotor.

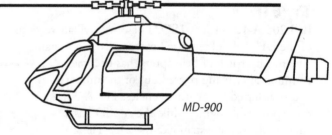

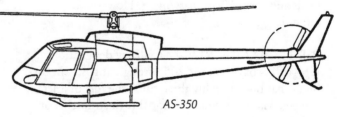

AS-350

Eurocopter EC-135. This twin-engine light helicopter has proven to be very popular for civilian, military, and police, air ambulance and rescue work. Capable of accommodating up seven passengers or two patients in air ambulance mode, it is reliable and economical. Featuring a ducted-fan Fenestron tail instead of a rotor, it may be fitted with a hoist for rescue work and dual cargo hooks for short hauls. External cargo capacity is 1360 kg/2998 lb.

EC-135

Eurocopter EC-145. A twin-engine light helicopter (developed from the BK-117) widely used by military, police, and rescue units, and recently adopted by the US Army and National Guard as the UH-72 "Lakota" light utility helicopter. With a fast cruise speed of 135 kts, the EC-145 is somewhat larger than the EC-135 and has a more spacious cabin. It will accommodate nine passengers or two stretcher patients in air ambulance configuration, can be fitted with an external rescue hoist and is capable of short-haul operations. Cargo capacity is some 1,713 kg (3,777 lbs), and rear clamshell doors offer excellent access to the interior. Noise level is exceptionally low.

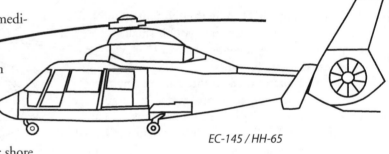

EC-145

Eurocopter AS-365 Dauphin 2. A twin-engine medium helicopter featuring retractable landing gear and a ducted-fan Fenestron tail, the AS-365 serves rescue, police, military and civilian markets. It has been adopted by the US Coast Guard as the HH-65 "Dolphin" and is that service's short range recovery (SRR) air-sea rescue helicopter. Equipped with an external rescue hoist, it may be deployed from either ship or shore and is often seen on inland rescue missions. It features advanced avionics, including an auto-hover capability.

EC-145 / HH-65

TYPE II

Bell 205 A-1, A-1S, A-1SD. These are civilian versions of the UH-1H "Huey" used extensively in Vietnam. The pickup truck of helicopters, has become common after being phased out of the military and transferred to civilian agencies. The A-2 has a more powerful engine, while the A-1SD has a later-model rotor system like the 212. The 205 is adept at most rescue tasks; it can lift heavy loads and carry 10–12 passengers. However, it is rather large.

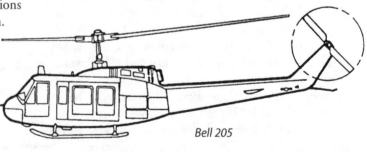

Bell 205

Bell 212, 412. The 212/412 (military designation UH-1N) is based on the 205 fuselage. It has twin turbines that burn roughly three times the fuel of a Jet Ranger. Many are equipped for IFR (instrument) opera-

(not to scale)

tions. The 412 differs in having a hingeless four-bladed rotor system. These aircraft are used for passenger and cargo transport, air ambulance and medical evacuation, long-line and bucket work, and rappelling. They have also been used for air deployment of rescue swimmers and personal watercraft.

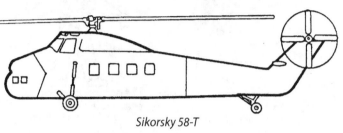

Bell 212 / 412

Sikorsky 58-T. The design of this venerable aircraft dates from the early 1960s, when it was powered with a piston engine. In the early 1970s, it was converted to a twin turbine design and has evolved into a solid, reliable aircraft for rescue work. Although no longer common, it has an excellent safety record as well as first-class high density altitude performance and lift capacity. Many of these aircraft are equipped with hoists, and they have been used for just about all helicopter rescue tasks. Although it is classified as a Type II helicopter, this large aircraft requires a 30' x 30' (9.1 x 9.1 m) landing pad.

Sikorsky 58-T

TYPE I

Boeing 234. This is the civilian version of the Army's CH-47 Chinook, also used extensively by the military. It has two counter-rotating turbine-powered rotors in tandem and no tail rotor, providing a very stable platform even in crosswinds or tailwinds. The 234 is very large and expensive to operate and maintain, but is capable of lifting very heavy loads (up to 27,000 lbs) or carrying large numbers of personnel.

Boeing 234

Sikorsky S-61N. The "Pelican" has been used extensively by the Coast Guard for offshore search and rescue but has been phased out by the newer Dolphin and "Jayhawk." In the civilian world they are used for helicopter logging and sometimes for rescue. Military models often have rescue hoists.

S-61N

Sikorsky UH-60 /S-70 Blackhawk.
Developed for the US military to replace the aging UH-1, The UH-60 (civilian designation S-70) has spawned a number of variants including the Firehawk (civilian firefighting), Pave Hawk (military CSAR), and Jayhawk (US Coast Guard SAR). Some National Guard Blackhawks have also been fitted with hoists and used for rescue. It is a large, complex, expensive aircraft with special maintenance requirements. The rotor wash is comparable to heavy-lift helicopters, and it requires a large helibase with plenty of separation for takeoffs and landings. However, it can carry 14+ passengers and lift 9,000 lbs/4000 kg externally on a cargo hook.

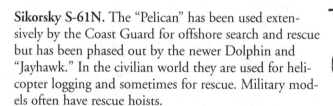

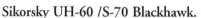

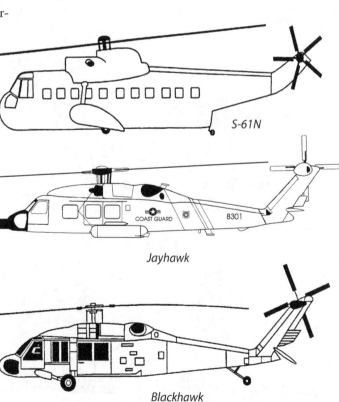

Jayhawk

Blackhawk

(not to scale)

CLEAR TO START ENGINE
make a circular motion above head with right arm.

HOLD ON GROUND
extend arms out at 45°. thumbs pointing down.

MOVE UPWARD
arms extended sweeping up.

MOVE DOWNWARD
arms extended sweeping down.

HOLD HOVER
arms extended with clenched fists.

CLEAR TO TAKE-OFF
extend both arms above head in direction of take-off.

LAND HERE, MY BACK IS INTO THE WIND
extend arms toward landing area with wind at your back.

MOVE FORWARD
extend arms forward and wave helicopter toward you.

MOVE REARWARD
arms extended downward using shoving motion.

MOVE LEFT
right arm horizontal, left arm sweeps over head.

MOVE RIGHT
left arm horizontal, right arm sweeps over head.

MOVE TAIL ROTOR
rotate body with one arm extended.

SHUT OFF ENGINE
cross neck with right hand, palm down.

FIXED TANK DOORS
open arms outward, close arms inward.

RELEASE SLING LOAD
contact left forearm with right hand.

WAVE OFF DO NOT LAND
wave arms from horizontal to crossed overhead.

Appendix 3
Death Investigation Checklist

901 REPORTING FORMAT:

● Assignment:
  Date & Time
  Location
  Synopsis

● Investigation:
  Date & Date Of Occurrence
  Location Of Occurrence
  Present Location Of Victim
  Description Of Scene
  Environmental Conditions
      Temperature
      Weather Conditions & Precipitation
      Sunrise/Sunset Time
  Names Of All Persons Present Including All EMS Personnel
      List All Agencies Investigating (FBI, FAA, NTSB, etc.)
  Identify Victim And Describe Method Of Identification
  Cause Of Death—If Known
      List Obvious Injuries
      Describe Any Body Fragmentation
  Attending Physician's Name Who Declared Victim Dead
  Date And Time Attending Physician Declared Victim Dead

● Medical History:
  Name Of Family Physician & Dentist, Including Phone #
  Medical History
  Determine Any History Of Drug/Alcohol Abuse
  Any Current Ailments
      Corrective Lenses
      Hearing Aid Or Other Disabilities
  List All Medications Prescribed (see evidence section)

● Interviews:
  Determine When Victim Arrived In Area
  Determine Victim's Plans While In Area
  Victim's Food & Beverage Intake:
      Where Victim Ate/Drank
      What Victim Ate/Drank
      How Much Victim Ate/Drank
      When Victim Ate/Drank
      How Much Of An Appetite Victim Had
  Victim's Sleep
      How Much Did Victim Sleep?
      When Did Victim Last Sleep?
      Did Victim Complain Of Sleeping Too Much Or Too Late?
  Did Victim Complain Of Any Recent Medical Ailments?
  Did Victim Complain Of Unusual Pains?
  Victim Right- Or Left-Handed?

● Identification:
  Height
  Weight
  What Types Of Identification Found On Victim
  List All Scars, Marks, Tattoos
  List All Scrapes, Cuts, Bruises
  Clothing & Footwear (wet/dry?)
      Color, Size, Fabric, Style, Manufacturer & Markings
  Body Temperature & Skin Condition
  Decomposition Observed

● Notification:
  Medical Examiner
      Date & Time Notified
      Which One Was Notified & By Whom
  Criminal Investigator
      Date & Time Notified
      Which One Was Notified & By Whom
  Supervisor & Agency Superiors
      Date & Time Notified
      Which One Was Notified & By Whom
  Family/Next Of Kin
      Date & Time Notified
      Who Was Notified & By Whom

● Evidence:
  Prescription Medication
      Drug Name (Trade & Generic Name)
      Prescription Number & Date
      Pharmacy & Phone Number
      Number Issued & Number Remaining
      Refills (Date & Number)
      Prescribing Physician
      Directions On Container
      Patient Name
      Location Found
  All Personal Property
      Inventory
      Chain Of Custody

● Photographs:
  Overall Or Aerial
  All Angles Far
  All Angles Close-Up
  Any Distinguishing Marks, Scars, Tattoos, Cuts, Bruises
  Consider Videotaping Scene If Available
  Use Photo Log And Fill-in Flash

- **Body Transport:**
    Date & Time
    Where It Was Transported & By Whom

- **Attachments:**
    Sketch
        Exact Body Position With Measurements
        Note Exact Location Of Evidence
        Include General Location Map
        Witness Statements

- **Disposition:**
        Whatever Applies

INVESTIGATIVE CONSIDERATIONS:

- **Background:**
    Criminal History
    Contact Local P.D. Of Victim
    Local Lodging & Back Country Records

- **Aircraft/Vehicle Accident:**
    Make, Model, Year, Color, Owner, Registration Number
    Flight Origin, Destination, Flight Number
    Location Of Subjects In Aircraft/Vehicle
    Seatbelts In Use
    Fuel In Tank(s)
    Insurance Company

- **Toxicology Results:**
    Blood Alcohol
    Carboxyhemoglobin
    Urine Drug Screen
    Brain Lactic Acid

- **Accidental Falls:**
    Handrails, Slope, Hazards At Location Of Fall
    Personal Items Left By Subject At Scene
    Distance To Resting Point From Site Of Fall
    Indication Of Tripping/Falling Accidentally ?

- **Drowning:**
    Describe Point Of Entry
    Distance From Point Of Entry To Recovery Location
    Swimming Ability Of Subject
    PFD, Helmet Or Wetsuit?

- **Firearm:**
    Manufacturer
    Caliber
    Type/Model
        Serial Number
    Owner
    Identification Of Cartridges
    Location Of Spent Casings

- **Hypothermia/Hyperthermia:**
    Maximum Temperature Past 24 Hours
    Minimum Temperature Past 24 Hours
    Wind Conditions & Wind-Chill Factor
    Humidity
    Duration Subject Was In Hot/Cold Environment

- **Suicide Note:**
    Location Found
        Identification Of Handwriting

AIDS TO DETERMINING TIME OF DEATH:

RIGOR MORTIS: Begins 24 hours postmortem, is complete in 12 hours. Begins developing in all muscles at the same time, but is perceived in smaller muscles. Test for rigor at the same time body temperature is recorded. Does position of body "fit" position found at scene?

LIVOR MORTIS: Begins forming immediately after death, perceptible in 12 hours, usually fully developed in 4 hours, fixed in 8–12 hours, always in dependent portion of body of deceased. (If position of body changed between 4–12 hours, livor mortis will move.) Cherry red color in monoxide and cyanide poisoning. Does discoloration blanch? Possibility body moved after death?

BODY TEMPERATURE: Usually no change from 99.6 degrees rectal or liver temperature during first hour after death. Rate of fall then 1½ degrees F. temperature/hour. *Take at least two temperature readings at least one hour apart. Record temperature, environmental temperature and time of reading!*

STATE OF DECOMPOSITION: If body has reached environmental temperature, body temperature will not decrease any further. Record of putrefaction is seen. Is body distended, discolored? Does body have a foul odor? Describe skin appearance and any body defects possibly made by animals at scene.

INSECTS: Flies lay eggs in body orifices. The maggots and their eggs and cocoons are useful to entomologists. Gather several samples of the insects, especially the larger ones and any cocoon-like structures. Also collect any beetles or other insects on the body. Label the container filled with 70% alcohol with the DATE, TIME and NAME OF DECEASED, and send to forensic entomologist for evaluation.

BOTANICAL SPECIMENS: Bodies found outdoors will possibly have been lying on grass, weeds or other botanical specimens. Collect these specimens in a brown paper bag from directly under the body and send to forensic botanist who can evaluate these specimens for arrest of growth pattern.

DROWNING FACT REFERENCE:

DROWNING is an asphyxia death in which the victim inhales large amounts of a liquid, which quickly increases blood volume, causing pulmonary edema and an overtaxed heart. Death can occur within 3–5 minutes of submersion.

NEAR-DROWNING or
FATAL POST-IMMERSION SYNDROME occurs when a drowning victim survives for as long as several days after submersion in a liquid due to immediate resuscitation, but later succumbs to either pneumonia or brain damage from cerebral anoxia.

AUTOPSY FINDINGS: Diagnosis of drowning is made on basis of *history, circumstances surrounding death, and complete autopsy and toxicological studies.* There is no morphological finding diagnostic of drowning. Autopsy findings include pulmonary congestion and edema, and oftentimes a froth-cone of the nose and/or mouth in which mucous, air and water mix during respiration. Froth found in the subject's airway indicates the subject was alive at the time of submersion. Blood-tinged foam is not indicative of chest trauma. It is caused by the tearing of lung tissue due to increased pulmonary pressure.

Be sure to check Dilantin and barbiturate levels in drowning victims, as this might be your only clue to their epileptic condition. Absence of those drugs, or very low levels in known epileptics, may give reason for drowning.

BLOOD ALCOHOL: In freshwater drowning, blood alcohol may be decreased due to increased blood volume. In saltwater drowning, there may be a slight increase in level. Putrid sample analysis for blood alcohol content is difficult to interpret.

SUBMERGED BODY CHANGES & INJURIES:
A body submerged after death can contain water and aquatic foreign materials in the stomach and lungs.

A body will sink to the bottom of the water, unless air is trapped in the clothing, and will float face down until bloating begins, at which time the body will surface. A body may reappear on the surface of the water 2–3 days after its entry during the summer months, and *weeks to months* later in winter months.

It is not possible to accurately estimate the length of submersion when the period exceeds four weeks.

Decomposition and putrefaction proceed at a rate determined by the water temperature. Cold and swiftly moving water preserves bodies; whereas heavy clothing and stagnant water hasten decomposition. Cold water also retards rigor mortis.

ADIPOCERE is a soap-like covering over a body, formed by the breakdown of the subcutaneous tissue due to prolonged submersion in water.

CUTISANSERINA (gooseflesh) is not an external diagnostic sign of drowning. "Washerwoman's Skin" is the wrinkling of the skin of the hands and feet of persons submerged for long periods of time. The "glove" of skin from a hand can be used to obtain fingerprints for identification purposes.

POSTMORTEM INJURIES can occur when a body scrapes bottom of water's baseground. The usual areas of injury are the prominent parts of the face, the chest and extremities. Fish and other aquatic animals may produce gnawing injuries to unclothed parts of the body. Boat propellers can cause multiple, parallel smooth-margined injuries to submerged bodies.

COLD is main threat to life during prolonged immersion in very cold water. The length of time immersed and the temperature of the water determine the survival time. A person wearing ordinary street clothes and life preserver can only survive 45 minutes in water less than 35° F.

Scuba-diving deaths can occur from air embolism, pneumothorax or pulmonary emphysema. Check equipment very carefully. The deaths are usually a result of either drowning or pressure changes associated with ascent or descent. The "bends" are rarely fatal. They are caused by air bubbles in a diver's blood produced by a body's response to depth of the water and length of time submerged.

Compiled by Ken Phillips, Search and Rescue Coordinator, Grand Canyon National Park (05-93)

Appendix 4
Swiftwater Rescue Training

Rescue 3 International (11084-A Jeff Brian Lane, PO Box 1050, Wilton, CA 95693, 800-45-RESCUE, www.rescue3.com, info@rescue3.com). Since 1979 Rescue 3 has trained more than 150,000 students worldwide, with an instructor network throughout the U.S. and 32 foreign countries. Today, it directs the Swiftwater Rescue Technician™ program, the most widely taught and recognized swiftwater rescue curriculum.

American Canoe Association (108 Hanover St., Fredricksburg, VA 22401, 540-907-4460, www.americancanoe.org, aca@americancanoe.org). The ACA, long known for its paddling instruction curriculum, also teaches a river rescue course aimed at recreational canoe and kayak paddlers.

Ohio Department of Natural Resources, Division of Watercraft (2045 Morse Road, Building A, Columbus, OH 43229-6693, 614-265-6480, www.ohiodnr.com/Watercraft). ODNR's pioneering swiftwater rescue course was the first widely taught river rescue course in the country. Their particular area of expertise is low-head dam rescue.

Indiana River Rescue School (1222 South Michigan St., South Bend, IN 46601, 574-235-7555, www.indianariverrescue.com, admin@indianariverrescue.com). The Indiana River Rescue School's course utilizes the East Race artificial canal at South Bend, which provides an excellent training site. Their curriculum is generally similar to ODNR.

Pennsylvania Fish & Boat Commission, Bureau of Boating and Access (PO Box 6700, Harrisburg, PA 17106-7000, 717-705-7834, www.fishandboat.com, ra-be@state.pa.us). Pennsylvania was also an early adopter of swiftwater rescue courses. Their course is widely recognized and includes power boat instruction.

Canyonlands Field Institute (Box 68, Moab, UT 84532, 800-860-5262, www.cfimoab.org, info@canyonlandsfieldinst.org). CFI specializes in teaching river rescue courses oriented toward commercial and recreational whitewater rafters. The emphasis is on in-water rescues on big rivers.

K38 Water Safety (14455 Sandbrook Drive, Tustin California 92780, 562-298-6916, K38Rescue@aol.com, www.shawnalladio.com). K38's area of expertise is the personal watercraft in all environments including swiftwater.

Other recommended rescue schools:

Whitewater Rescue Institute (210 Red Fox Rd., Lolo, MT 59847, 406-207-2027, cody@whitewaterrescue.com, www.whitewaterrescue.com).

Wave Trek Rescue (P.O. Box 236, Index, WA 98256, 360-793-1508, info@wavetrekrescue.com, www.wavetrekrescue.com).

Sierra Rescue (PO Box 63, Taylorsville, CA 95983, 800-208-2723, info@sierrarescue.com, www.sierrarescue.com)

Pet Owners—Be Prepared with a Disaster Plan!

The best way to protect your family from the effects of a disaster is to have a disaster plan. If you are a pet owner that plan must include your pets. Being prepared can save their lives.

Different disasters require different responses. But whether the disaster is a hurricane or a hazardous spill, you may have to evacuate your home.

In the event of a disaster, if you must evacuate, *the most important thing you can do to protect your pets is to evacuate them, too.* Animals left behind in a disaster can easily be injured, lost or killed. Animals left inside your home can escape through storm-damaged areas, such as broken windows. Animals turned loose to fend for themselves are likely to become victims of exposure, starvation, predators, contaminated food or water, or accidents. Leaving dogs tied or chained outside in a disaster is a death sentence.

1. Have a safe place to take your pets

Evacuation shelters generally do not accept pets except for service animals who assist people with disabilities. It may be difficult, if not impossible, to find shelter for your animals in the midst of a disaster, so plan ahead. *Do not wait until disaster strikes to do your research.*

- Contact hotels and motels outside your immediate area to check policies on accepting pets and restrictions on number, size, and species. Ask if "no pet" policies could be waived in an emergency Keep a list of "pet friendly" places, including phone numbers, with other disaster information and supplies. If you have notice of an impending disaster, call ahead for reservations.
- Ask friends, relatives, or others outside the affected area whether they could shelter your animals. If you have more than one pet, they may be more comfortable if kept together, but be prepared to house them separately.
- Prepare a list of boarding facilities and veterinarians who could shelter animals in an emergency; include 24-hour phone numbers.
- Make a list of boarding facilities and veterinary offices that might be able to shelter animals in disaster emergencies; include 24-hour telephone numbers. Ask your local animal shelter if it provides foster care or shelter for pets in an emergency. This should be your last resort, as shelters have limited resources and are likely to be stretched to their limits during an emergency.

2. Assemble a portable pet disaster supplies kit

Whether you are away from home for a day or a week, you'll need essential supplies. Keep items in an accessible place and store them in sturdy containers that can be carried easily (duffel bags, covered trash containers, etc.). Your pet disaster supplies kit should include:

- Food and water for at least five days for each pet, bowls and a manual can opener if you are packing canned pet food.
- Medications and medical records stored in a waterproof container and a first aid kit. A pet first aid book is also good to include.
- Cat litter box, litter, garbage bags to collect all pets' waste, and litter scoop.
- Sturdy leashes, harnesses, and carriers to transport pets safely and to ensure that your pets can't escape. Carriers should be large enough for the animal to stand comfortably, turn around and lie down. Your pet may have to stay in the carrier for hours at a time while you are away from home. Be sure to have a secure cage with no loose objects inside it to accommodate smaller pets. These may require blankets or towels for bedding and warmth, and other special items.
- Current photos and descriptions of your pets to help others identify them in case you and your pets become separated and to prove that they are yours.
- Pet beds and toys, if you can easily take them, to reduce stress.
- Information about your pets' feeding schedules, medical conditions, behavior problems, and the name and number of your veterinarian in case you have to board your pets or place them in foster care.
- Other useful items include newspapers, paper towels, plastic trash bags, grooming items and household bleach.

3. Know what to do as a disaster approaches

Often, warnings are issued hours, even days, in advance. At the first hint of disaster, act to protect your pet.

- Call ahead to confirm emergency shelter arrangements for you and your pets.

- Check to be sure your pet disaster supplies are ready to take at a moment's notice.

- Bring all pets into the house so that on won't have to search for them if you have to leave in a hurry.

- Your pet should be wearing up-to-date identification at all times. You can buy temporary tags or put adhesive tape on the back of your pet's ID tag, adding information with an indelible pen. This includes adding your current cell phone number to your pet's tag. It may also be a good idea to include the phone number of a friend or relative outside your immediate area—if your pet is lost, you'll want to provide a number on the tag that will be answered even if you're out of your home.

You may not be home when the evacuation order comes. Find out if a trusted neighbor would be willing to take your pets and meet you at a prearranged location. This person should be comfortable with your pets, know where your animals are likely to be, know where your pet disaster supplies kit is kept, and have a key to your home. If you use a petsitting service, they may be available to help, but discuss the possibility well in advance.

Planning and preparation will enable you to evacuate with your pets quickly and safely. But bear in mind that animals react differently under stress. Outside your home and in the car, keep dogs securely leashed. Transport cats in carriers. Don't leave animals unattended anywhere they can run off. The most trustworthy pets may panic, hide, try to escape, or even bite or scratch. And, when you return home, give your pets time to settle back into their routines. Consult your veterinarian if any behavior problems persist.

4 If you don't evacuate in the event of a disaster, shelter in place

If your family and pets must wait out a storm or other disaster at home, identify a safe area of your home where you can all stay together. Be sure to close your windows and doors, stay inside, and follow the instructions from your local emergency management office.

- Bring your pets indoors as soon as local authorities say there is an imminent problem. Keep pets under your direct control; if you have to evacuate, you will not have to spend time trying to find them. Keep dogs on leashes and cats in carriers, and make sure they are wearing identification.

- If you have a room you can designate as a "safe room," put your emergency supplies in that room in advance, including your pet's crate and supplies. Have any medications and a supply of pet food and water inside watertight containers, along with your other emergency supplies. If there is an open fireplace, vent, pet door, or similar opening in the house, close it off with plastic sheeting and strong tape.

- Listen to the radio periodically, and don't come out until you know it's safe.

5. After the storm

Planning and preparation will help you survive the disaster, but your home may be a very different place afterward, whether you have taken shelter at home or elsewhere.

- Don't allow your pets to roam loose. Familiar landmarks and smells might be gone, and your pet will probably be disoriented. Pets can easily get lost in such situations.

- While you assess the damage, keep dogs on leashes and keep cats in carriers inside the house. If your house is damaged, they could escape and become lost.

- Be patient with your pets after a disaster. Try to get them back into their normal routines as soon as possible, and be ready for behavioral problems that may result from the stress of the situation. If behavioral problems persist, or if your pet seems to be having any health problems, talk to your veterinarian.

For more information, contact The Humane Society of the United States, Disaster Services, 2100 L Street, N.W., Washington, DC 20037, 202-452-1100, www.humanesociety.org)

Adapted from "Disaster Preparedness for Pets," used with the permission of The Humane Society of the United States and the American Red Cross. For a free single copy of the brochure send a stamped, self-addressed business-size envelope to The HSUS at the above address.

Appendix 6
The Anderson Sling

Charles Anderson, a welder from Potter Valley, California, and founder of Care for Disabled Animals, designed a specialized equine sling that has been used successfully for helicopter short hauling of stranded or injured animals. He designed the sling after consultation with Dr. John Madigan and Richard Morgan of the UC Davis School of Veterinary Medicine.

The Anderson Sling was used to rescue a group of mules and a horse who were stranded in heavy snow near the Dodge Ridge Ski Resort in California in November 1991. This was a good test situation since the animals were not injured, the overhead access was clear and the weather conditions were mild. By the end of the day, all the animals had been successfully airlifted four miles to solid ground using the sling. Since then, it has been used to rescue a horse stranded on the rocky cliffs above the American River in and to rescue two foals and four adult horses in the flooded areas in Southern California in 1993.

What makes the sling unique is its support system. In the past, equine slings did not support the horse evenly, lifting mostly from the abdomen. Many horses were uncomfortable with this lack of support and would panic, often jumping and twisting themselves free. The Anderson Sling supports the horse from the skeletal system, distributing the horse's support and restraint system. Because the head and neck constitute a good portion of the horse's body weight, the sling helps steady the horse, preventing the jumping and twisting that occurred with earlier slings. The Anderson Sling also has considerable padding and leg supports to keep the sling in place and make it more comfortable for the animal. In addition to rescue, the sling has been used for long term rehabilitation of horses unable to stand. The Anderson Sling is available in various sizes to the public and to rescue organizations.

However, organizations should not attempt helicopter short hauls without adequate prior training, as well the support of experienced horse handlers. For more information on short hauls, see Chapter 10.

For more information on the Anderson Sling and related products, contact Care for Disabled Animals/ CDA Products, PO Box 53, Potter Valley, CA 95469, www.andersonsling.com, (707) 743-1300.

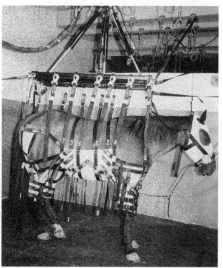

A Glossary of Swiftwater Terms

Boil Line: A white, frothy upwelling of water at the base of a river hydraulic. Water upstream of the boil line typically has a light color and moves back upstream, while the water downstream of the boil line is usually darker in color and moves downstream. See *Hydraulic*.

Bottom Load: One of three river loads. Objects in the bottom load are pushed along the river bottom by the current. Bottom loads become significant during flood conditions. See *Top Load* and *Suspended Load*.

Boulder Sieve: A collection of boulders in the river's current that acts as a strainer, i.e. they allow water to pass through but not solid objects like boats and people. Boulder sieves are often caused by flash flooding on side creeks. See *Strainer, Entrapment, Pinning*.

Chute: A clear tongue of water flowing between two obstacles. See *Tongue, Downstream V.*

Downstream V: Hydraulic effect in the shape of a "V" pointing downstream, caused by the convergence of downstream water flow in the channels of least resistance. The largest V pointing downstream indicates the main channel (which may not be in midstream). See *Tongue, Chute.*

Eddy: Horizontal reversal of water flow where the differential between the current's pressure on the upstream and downstream sides of an obstacle (such as a rock) causes the water behind the obstacle to flow upstream. See *Helical Flow, Laminar Flow.*

Eddy Line, Eddy Fence, Eddy Wall: An obvious line in the river where the current moves in opposite directions on either side. This current differential between the eddy and the downstream current ranges from a surface line (eddy line) to a wall of water dropping around the obstacle and recirculating horizontally (eddy fence).

Eddy Turn: The maneuver by which a boat leaves the main current and enters an eddy.

Entrapment: The process whereby an extremity or a person's whole body is forced into a crack, crevice, or undercut in an obstacle or the riverbed and held there by the force of the current or, when a person is held by the collapse of a watercraft such as a kayak caused by the force of the current. See *Pinning, Strainer, Boulder Sieve*.

Ferry: The process by which a boat moves across the current without going downstream.

Flood Channel: An artificial channel constructed for the purpose of moving floodwaters out of urban areas. Usually designed as part of a stormwater drainage system. The current speed in these channels may be up to twice that in a natural river.

Flush Drowning: Drowning by aspiration while swimming through big water. It is often associated with a gasping reflex caused by cold water, although lack of knowledge about when and where to breathe can also be a factor.

Gradient: The amount of elevation loss between two given points on a river. It is usually expressed as feet per mile or as a percentage (e.g. 5%). As such, it provides a rough guide to the expected difficulty of the river.

Helical Flow: The corkscrewing flow of the water between the banks and the main current. See *Laminar Flow, Eddy*.

High Side: Shifting the weight of a boat crew to the high (i.e., downstream) side of a boat to avoid flipping. This is done when the boat washes up against an obstacle, hits a large breaking wave, crosses an eddy line, or is caught in a hole.

Hole (Stopper, Breaking Hole): A river wave, usually caused by an underwater obstacle, that breaks back upstream. A hole is a surface phenomenon; it may flip or hold a buoyant object like a watercraft but it will not recirculate a swimmer.

Hydraulic (Pourover, Reversal, Sousehole, Keeper): A vertical reversal of water flow where the pressure of the current falling over a gradient (such as a dam) causes the water at the base of the gradient to be forced downward into a loop-style reversal and back to the surface. Part of the water continues downstream and part reverses back upstream to the base of the gradient. This reverse flow can cause an object to be recirculated in the hydraulic (hence the name "keeper"). The churning whitewater of a hole often consists of 40–60% air. See *Boil Line*.

Rule of Thumb: A "frowning" hydraulic tends to be a keeper by recirculating in on itself, while a "smiling" one tends to flush out due to the current at its sides.

Hydraulic Effect, Hydraulic: A movement of water caused by pressure.

Hypothermia: A lowering of the body's core temperature. It occurs when the body loses heat faster than it can generate it. Hypothermia can cause impaired judgement, debility, and eventually, death. *Immersion* hypothermia is caused when the body is exposed to cold water. See *Hypothermia-Induced Debility*.

Hypothermia-Induced Debility (HID): Loss of body strength and function caused by chilling of the extremities. With immersion hypothermia victims this may happen rapidly, rendering them unable to swim or assist with rescue efforts.

Laminar Flow: The layered, downstream flow of the river's main current. The layer in the center of the current just below the surface moves fastest, while the side and bottom layers are slowed somewhat by friction. See *Helical Flow, Eddy*.

Low-Head Dam, Weir: An artificial construction in the river for flood control, irrigation, or power generation. Water flowing over a low-head dam often causes a hydraulic, and they have been the site of numerous accidents.

Peel-Out: A maneuver in which a boat leaves an eddy and enters the main current.

Pillow: A reaction feature that forms upstream of a river obstacle such as a rock. The water pushes up into a rounded shape, sometimes considerably higher than the surrounding water, hence the name. A pillow forms a cushion on the obstacle, pushing objects like boats away from it.

Pinning: Caused when a watercraft is pushed against the upstream side of a river obstacle and held there. Recovering a pinned boat can be a difficult proposition. See *Entrapment, Strainer, Boulder Sieve*.

Rapids Rating System: An internationally-recognized system in which all river rapids are rated on a scale from I (easy) to VI (extreme danger). The system provides a rough but inexact guide.

River Left: The left bank of the river, looking downstream.

River Right: The right bank of the river, looking downstream.

Safe Swimming Position: A river swimming position where the swimmer's feet are kept near the surface and pointed downstream. This reduces the danger of foot entrapment and allows the swimmer to fend off rocks with his feet. It is commonly used in shallow, rocky rivers and by inexperienced swimmers. In deeper water, more conventional swimming techniques can be used. See *Entrapment*.

Scouting: Looking at a rapid before running it. This may be done from the bank or from an eddy.

Standing Wave (Haystack, Tail Wave): A wave or series of waves caused by the convergence of main channel currents, underwater obstacles or ledges, or an increasing river speed/gradient which converts the hydraulic energy of the river to a wave or series of waves, diminishing in size, that form downstream. Unlike a ocean wave, river waves remain fixed in place.

Strainer: Any river obstacle that allows water but not solid objects to pass through it. It is most often caused by trees, brush, or debris piles in the current. Strainers are extremely dangerous since a boat or swimmer may be trapped against them by the force of the current. See *Boulder Sieve, Entrapment, Pinning*.

Suspended Load: That part of the river's load that is suspended in the current. The amount of suspended load increases greatly during flood conditions (roughly as the cube root of the flow increase). As the river's flow decreases, it begins to drop the suspended load. See *Top Load, Bottom Load*.

Tongue: A roughly triangular-shaped flow of water, often raised above the level of the adjacent water, that marks the main current flowing downstream in an obstructed section. See *Downstream V, Chute*.

Top Load: That part of the river's load that floats on top of the water. See *Bottom Load, Suspended Load*.

Undercut: A rock or ledge in the current that is cut away under the surface. Usually identifiable by the lack of an upstream pillow. A boat or swimmer may be entrapped or pinned if forced into it by the current. See *Entrapment, Pinning, Pillow*.

Upstream V: A hydraulic effect in the form of a **V** pointing upstream. It is caused by downstream water flowing around an obstacle. Objects such as rocks submerged just below the surface form an upstream V.

Volume: The amount of water in a river, defined as the amount of water flowing by a given point in one second. Usually expressed as cubic feet per second (cfs) or cubic meters per second (cumecs).

This glossary was compiled with original material from Rescue 3.

BIBLIOGRAPHY

SWIFTWATER

Books

Bechdel, Les, and Slim Ray. *River Rescue*, 4th ed. Asheville, NC: CFS Press 2009

Bennett, Jeff *The Complete Whitewater Rafter* Camden, ME: Ragged Mountain Press, 1996

Brewster, B. Chris, ed. *The United States Lifesaving Association Manual of Open Water Rescue.* Englewood Cliffs, NJ: Prentice-Hall, 1995

Brooks, Dr. C. J. *Designed For Life: Lifejackets Through the Ages* Richmond, BC: Mustang Survival 1995

Ferrero, Franco, *White Water Safety and Rescue* Pesda Press 1998

Hunter, Jez *Swiftwater and Flood Rescue Field Guide* Sudbury, MA: Jones and Bartlett 2009

Munger, Julie and Abigail Polsby, *Swiftwater Rescue Workbook* Taylorsville, CA: Sierra Rescue 2011

Nealey, William. *Kayak* Birmingham, AL: Menasha Ridge Press 1986

Ray, Slim *The Canoe Handbook* Harrisburg, PA: Stackpole Books, 1992

River Rescue Columbus, Ohio: Ohio Department of Natural Resources, Division of Watercraft 1980

Reithmaier, Peter *Sicherheit im Wildwasser* Vienna, Austria: Naturfreunde Österreich, 1994

Rowe, Ray *The Chest Harness And Its Use in Whitewater Canoeing* Surry, UK: The British Canoe Union 1987

Smith, David S., and Sara J. Smith *Water Rescue: Basic Skills for Emergency Responders* St. Louis, MO: Mosby-Year Book, 1994

Technical Rescue Technology Assessment (FA-153 1/95). Washington, DC: U.S. Fire Administration and Federal Emergency Management Administration, 1995

Walbridge, Charles, and Wayne Sundmacher. *Whitewater Rescue Manual* Camden, ME: Ragged Mountain Press, 1995

Walbridge, Charlie ed. *River Safety Task Force Report 1996-1999* Springfield, VA: American Canoe Association 1999

Articles

Cross, Gary "K-38 Rescue Course Review" *Technical Rescue Magazine* #52 Watrlooville, Hampshire UK

O'Shea, Michael J. "Fluid flow, Newton's second law and river rescue" *Physics Education* Vol. 41 No. 2 2006 available online at http://iopscience.iop.org/0031-9120/41/2/003

Ray, Slim "Get Ready for Raging Waters: Equip your department for flood and swiftwater response." *FIRE-RESCUE* Jan. 2000

Rigg, Nancy "Water Rescue's New Wave?" *RESCUE Magazine* July/August 1995

Segerstrom, Jim "Swiftwater Crossing." *RESCUE Magazine* Jul/Aug 1994

_____ "Vehicle in the Water: Are You Ready?" *RESCUE Magazine* May/June 1995

_____ "Low-head Dams: a Fight for Life." *RESPONSE Magazine* Mar/Apr 1986

_____ "A Well-Kept Secret." *EMERGENCY Magazine* 1990

Smith, Gordon R. and Stephen R. Allen "Force on a Highline Caused by River Flow" International Technical Rescue Symposium 2009 available online at http://www.itrsonline.org/PapersFolder/2009/SmithGordon-Allen2009_ITRSPaper_Forces.pdf

"Water Rescue Craft" *Technical Rescue Magazine* #54 Watrlooville, Hampshire UK

Video

The Awesome Power. National Oceanographic and Atmospheric Agency 1988 available at http://www.archive.org/details/gov.ntis.ava17096vnb1

Car In The Water Public Education Series. Sacramento, CA: KXTV Channel 10

Flash Flood Safety Arizona Flood Plain Management Association, PO Box 18102, Phoenix, AZ 85005

Heads Up! River Rescue for River Runners Fredericksburg, VA: American Canoe Association 1993 DVD

No Way Out Los Angeles County Office of Education (RETAC), 9300 Imperial Highway, Downey, CA 90242

Staying Alive Wilton, CA: Rescue 3 International 2009 DVD/VHS Video

Swept Away: A Guide to Water Rescue Operations Chatham, NY: Alan Madison Productions 2005 DVD/VHS

Swiftwater Surviva. Public Education Series. Sacramento, CA: KXTV Channel 10

The Hidden Danger, Low Water Crossing NOAA/NWS available at http://www.weather.gov/om/water/XWATER/index.shtml

Water, The Timeless Compound Buzz Fawcett Productions 1980

RIGGING AND TECHNICAL ROPES

Books

Adkins, Jan. *Moving Heavy Things.* New York: Houghton Mifflin, 1980

Blandford, Percy W. *Knots and Splices.* New York: Arco Publishing, 1978

Brennan, Ken *Rope Rescue For Firefighting*, Saddle Brook, NJ: Fire Engineering Books and Videos 1998

Brown, Mike *Engineering Practical Rope Rescue Systems*, Delmar Cengage Learning 2000

Crockett, Ken. *Climbing Terms & Techniques*. Essex (UK): Fraser Stewart, 1993

Fleming, Steve. *Low Angle Rope Rescue*, 3rd ed. Ft. Collins, CO: Technical Rescue Systems, 1987

Frank, James A., ed. *CMC Rope Rescue Manual*, 4th ed. Santa Barbara, CA: CMC Rescue, 2010.

Frank, James A., ed. *CMC Rope Rescue Manual Field Guide*, 4th ed. Santa Barbara, CA: CMC Rescue, 2010

Frank, James A., and Donald E. Patterson. *CMC Rappel Manual*, rev. ed. Santa Barbara, CA: CMC Rescue, 1993

Guidelines for Rope Access Work Denver, CO: US Department of the Interior Technical Information Service 2004

Hudson, Steve, ed. *Manual of U.S. Cave Rescue Techniques*, 2nd ed. Huntsville, AL: National Speleological Society, 1988

Hudson, Steve and Tom Vines *High Angle Rescue Techniques*, 3rd ed. Mosby 2004

Long, John and Bob Gaines *Climbing Anchors*, 2nd ed., Guilford, CT: Falcon Guides 2006

Leubben, Craig *Knots for Climbers* Evergreen, CO: Chockstone Press 1993

Luebben, Craig *Rock Climbing Anchors: A Comprehensive Guide*, Seattle, WA: The Mountaineers Books 2007

March, Bill. *Modern Rope Techniques*, 3rd ed. Edison, NJ: Hunter Publishing, 1985.

May, W. G. *Mountain Search and Rescue Techniques*. Boulder, CO: Rocky Mountain Rescue Group, 1973.

MacInnes, Hamish *International Mountain Rescue Handbook* 4th ed., London: Frances Lincoln 2006

Mauthner, Kirk and Katie Mauthner, *Release Devices: A Comparative Analysis* 2nd ed.. Ouray, CO: Rigging for Rescue 2000

Mauthner, Kirk and Katie Mauthner *Gripping Ability on Rope in Motion* Ouray, CO: Rigging for Rescue 1994.

Merchant, D. F. *Life On A Line: A manual of modern cave rescue ropework techniques* issue 1.3 Published online at draftlight.net/lifeonaline 2002/2003

Owen, Peter. *The Book of Outdoor Knots*. New York: Lyons and Burford, 1993

Padgett, Allen, and Bruce Smith. *On Rope: North American Vertical Rope Techniques*. 2nd ed., Huntsville, AL: National Speleological Society, 1997

Setnicka, Tim. *Wilderness Search and Rescue*. Boston, MA: Appalachian Mountain Club, 1980

Thorne, Reed. *Rope Rescue Nomenclature*, 3rd ed. Lake Elizabeth, CA: Thorne Group, 1988.

Vines, Tom, and Steve Hudson. *High Angle Rescue Techniques*. Dubuque, Iowa: Kendall/Hunt Publishing, 1989

Walbridge, Charlie *Knots for Paddlers* Springfield, VA: American Canoe Association 1995

Weber, Richard *Technical Rope Rescue Training Manual* Rev. 7 Privately Published Muir Valley, KY 2010

Wheelock, Walt. *Ropes, Knots, and Slings for Climbers*, rev. ed. Glendale, CA: La Siesta Press, 1967.

Articles

Albuquerque & Ouray MR Team Members "Multi-Point Pre-Equalized Anchoring Systems" *Technical Rescue Magazine* #57 Watrlooville, Hampshire UK

Church, Greg "AutoLock Decenders" *Technical Rescue Magazine* #57 Watrlooville, Hampshire UK

Dill, John "Are You Really On Belay?" Pts. 1 &2 *Nylon Highway* June 1991

Eddie, Keith. "The Pulley Advantage." *RESCUE Magazine* Jan/Feb 1994

Fleming, Steve. "The Low-Down on Low Angle." *RESCUE Magazine* Jul/Aug 1994

Hunter, Jez "Over the Edge" *Technical Rescue Magazine* #51 Watrlooville, Hampshire UK

_____ "Evaluating Low Angle Rope Rescue Situations." *SearchLines Magazine* Nov/Dec 1993

King, Norma, and Richard Wright. "Does the Münter Hitch Measure Up?" *RESCUE Magazine* Sept/Oct 1994

O'Shea, Michael J. "Elasticity and mechanical advantage in cables and ropes" European Journal of Physics 28 (2007) 715–727

Owen, R. & S. Naguran "Self Equalising Anchors: A Myth?" *Technical Rescue Magazine* #37 Watrlooville, Hampshire UK

Pendley, Tom "Tying the 'True' Belay" *Fire Rescue Magazine* July 2010

Schafer, Keith T. "The Load Distributing Anchor System." *RESPONSE Magazine* Summer 1991

Thorne, Read "The Dreaded Disease of Redundatitus" *Technical Rescue Magazine* #59 Watrlooville, Hampshire UK

Valcourt, Greg. "Hang Time—Learning the Rescue Ropes" *RESCUE Magazine* Mar/Apr 1994

Wehbring, John. "The Two-Rope Raising/Lowering System" 1994 Conference Proceedings, National Association of Search and Rescue.

Video

Mauthner, Kirk and Katie Mauthner, "The What-If of Highline Failure...Is there a back-up?" Ouray, CO: Rigging for Rescue 1997. DVD/VHS

HELICOPTERS AND AVIATION
Books

Setnicka, Tim J. *Wilderness Search and Rescue*. Boston, MA: Appalachian Mountain Club, 1980.

Stoffel, Skip, and Patrick LaValla. *Personnel Safety in Helicopter Operations.* Olympia, WA: Emergency Response Institute, 1988.

Waters, John M. *Rescue At Sea* 2nd ed. Annapolis, MD: Naval Institute Press 1989

Interagency Helicopter Operations Guide Boise, ID: National Interagency Aviation Council June 2009

Interagency Helicopter Rappel Guide Boise, ID: National Interagency Helicopter Council 2011

Articles

Brockett, Bill. "How to Set Up a Landing Zone." *RESCUE Magazine* Nov/Dec 1994

Johnson, Sean "In-Water Using Helicopters" *Technical Rescue Magazine* #57 Watrlooville, Hampshire UK

_____ "Helos in Water Rescue pt 2" *Technical Rescue Magazine* #59 Watrlooville, Hampshire UK

Segerstrom, Jim. "Helo Ops: Planning and Training for Successful Helicopter Operations." *RESCUE Magazine* Nov/Dec 1994

"Gear Review: Bauman Screamer Suit" *Technical Rescue Magazine* #53 Watrlooville, Hampshire UK Watrlooville, Hampshire UK

INCIDENT COMMAND, MANAGEMENT, SEARCHES AND RECOVERIES

Associated Press. *The Flood of 1993—America's Greatest Natural Disaster.* New York: St. Martin's Press, 1993

Association of Flood Plain Managers *Coast to Coast: 20 Years of Progress* Boulder, CO The Natural Hazards Research and Information Center 1996

Alexander, David *Confronting Catastrophe; New Perspectives on Natural Disasters,* Oxford University Press 2000

Farabee Charles R. "Butch" Jr. *Death, Daring, & Disaster— Search and Rescue in the National Parks* Boulder, CO: Roberts, Rinehart Publishers, 1999

Steinberg, Ted *Acts of God: The Unnatural History of Natural Disaster In America* Oxford University Press 2000

Syrotuck, William G. *Analysis of Lost Person Behavior.* Rome, NY: Arner Publications, 1977

——— *An Introduction to Land Search Probabilities and Calculations.* 1975.

——— *Some Grid Search Techniques for Locating Lost Individuals in Wilderness Areas.* 1974

Articles

Collins, Sherrie "The Role of Peer Support Groups and Peer Counselors." *RESPONSE Magazine* Spring 1993

Edwards, Barry "Not So Gently Down the Stream." *RESCUE Magazine* Jul/Aug 1992

Fleming, Steve "Compassion Kills!" *SearchLines Magazine* Nov/Dec 1993

Gallagher, Mickey, and Steve Mosely "A History of the Los Angeles County Ocean Lifeguards and Swiftwater Rescue." *RESPONSE Magazine* Fall 1993

Kuhr, Steve. "The 4 Phases of Emergency Management: How to Prepare for Disasters and MCIs." *RESCUE Magazine* Jan/Feb 1995

Lawson, Norman W. "Legal Aspects of Search and Rescue" 1994 Conference Proceedings, National Association of Search and Rescue

McCurley, Loui "Preparing for the Big One: Qualifying Equipment & Personnel for FEMA/NIMS SAR Response" June 8, 2010 http://www.pmirope.com/news/blog6-8-10/

Rigg, Nancy "Before Rescue: How to Create an Effective Public Safety Education Program." 1994 Conference Proceedings, National Association of Search and Rescue

_____ "Survivor of Disaster—A Voice for Change." *RESPONSE Magazine* Fall 1993

Scott, Robert, Nancy Rigg, and Lane Contreras. "Beyond Rescue, the Psychodynamics of Complete Rescue." *RESPONSE Magazine* Summer 1994

Segerstrom, Jim "Raging Waters: Rescuers Tested by California Floods." *RESCUE Magazine* Mar/Apr 1995

Ray, Slim "A Flood is a Weapon of Mass Destruction" *Advanced Rescue Technology* Oct/Nov 2000

_____ "A New Flood Rescue Response Model" *Advanced Rescue Technology* Jan/Feb 2001

_____ "Flooded: A Night to Remember in Fort Collins" *9-1-1 Magazine* Mar/Apr 2000

_____ "It's Time to Overhaul North Carolina's Flood Response System" *Raleigh News & Observer* September 24, 2000

_____ "Notes From the Field: Lessons Learned on Flood Management" *Fire-Rescue* October 2001

_____ "Over Their Heads: Flood Response in Kansas City" *9-1-1 Magazine* Jan/Feb 1999

_____ "Overwhelmed: North Carolina's response to Hurricane Floyd" *9-1-1 Magazine* Mar/Apr 2000

_____ "Who You Gonna Call? Floodfighters! Floodproofing your department and community." *Fire-Rescue* Dec. 1999

MEDICAL

Books

American Red Cross CPR: Basic Life Support for the Professional Rescuer. American Red Cross, 1988.

Bierens Joost J. L. M. ed. *Handbook On Drowning* Springer-Verlag 2006

Emergency Care of the Sick and Injured, 4th ed. Chicago: The American Academy of Orthopedic Surgeons, 1987.

Giesbrecht, Gordon G. and James A. Wilkerson *Hypothermia Frostbite And Other Cold Injuries: Prevention, Recognition, Rescue, and Treatment,* Seattle, WA: The Mountaineers 2006

Issac, Jeff, and Peter Goth. *The Outward Bound Wilderness First-Aid Handbook.* New York: Lyons & Burford, 1991.

Mitchell, Jeffrey T. and George S., Jr. Everly *Critical Incident Stress Debriefing: An Operations Manual for CISD, Defusing and Other Group Crisis Intervention Services* 3rd ed., Ellicott City, MD: Chevron Publishing Corporation; 2001

Smith, David S., and Sara J. Smith. *Water Rescue: Basic Skills for Emergency Responders.* St. Louis, MO: Mosby-Year Book, 1994.

State of Alaska Cold Injuries Guidelines 2003 rev. 2005 http://www.umanitoba.ca/faculties/kinrec/research/media/AlaskaColdGuidelines05.pdf

Wilkerson, James A. ed. *Medicine for Mountaineering* 6th ed. Seattle, WA: The Mountaineers 2010

Articles

Edwards, Barry, and Jim Segerstrom. "Psychology 101: A Field Guide to Victim and Rescuer Behavior." *RESCUE Magazine* (Sept/Oct 1994)

Dickison, Anne E. "Drowning and Near-Drowning, Medical Problems Associated with Whitewater Activities." Unpublished manuscript

Giesbrecht, Gordon G. "Accidental Hypothermia" http://www.umanitoba.ca/faculties/physed/research/people/giesbrecht/Hypothermia.pdf

Greenhalgh, Tom, and Mike Bielmaier. "Spinal Immobilization in the Aquatic Environment." *RESCUE Magazine* (July/August 1995).

Johnson, Sean "Immersion Hypothermia" pt 1 *Technical Rescue Magazine* #53 Watrlooville, Hampshire UK

_____ "Immersion and Hypothermia" pt 2 *Technical Rescue Magazine* #54 Watrlooville, Hampshire UK

DVD Video

National Water Safety Congress *Cold Water Boot Camp* Mentor, OH 2008 DVD

_____ *Beyond Cold Water Boot Camp* Mentor, OH 2010 DVD

American Canoe Association *Cold, Wet and Alive* Fredericksburg, VA 1989 VHS, DVD

ANIMAL RESCUE

Irvine, Leslie *Filling the Ark: Animal Welfare in Disasters* Philadelphia: Temple University Press, 2009

Ray, Slim *Animal Rescue in Flood and Swiftwater Incidents* Asheville, NC: CFS Press 1999

Heath, Sebastian *Animal Management in Disasters* St. Louis: Mosby 1999

Articles

Ray, Slim "Reining Cats and Dogs" *Fire-Rescue Magazine* April 2000

INDEX

[Index created with **TExtract** / www.Texyz.com]

ACKNOWLEDGMENTS

Photographs:

AP/ Wide World Photos: 162, 192

Bettmann Archive: 11, 191

Jayson Mellom/The Tribune: 22

Paige L. Christie/NOC Retail: 37, 40

NRS: 43

Gary Fields/Asheville Citizen-Times: 44 (bottom)

Jonathan Ernst/Asheville Citizen-Times: 167

Paul Ratcliffe/Sierra Shutterbug: 27, 85, 86 (both), 88

Tony Wilson/NOC Photo: 33

John Porada: 44 (top), 45 (bottom)

J.E. Weinel: 47 (both), 48 (bottom), 52, 53 (both), 56, 71, 121

Gordon and Wendy Dalton: 111

Jeff Redding: 113

Bryan and Michelle Stewart: 125

Conterra, Inc.: 58

Travis County Star Flight: 147

Peter Linde/King County Sheriff's Dept. 154

Ken Phillips: 46, 92, 155 (both)

Rescue 3: 98, 147 (top), 161 (bottom), 213, 215, 233

Bill Wortell/Del Norte County Sheriff's SAR: 98, 139, 146 (right top & bottom)

Rex Myers/EMS Options: 94, 95

Linda Gheen: 211

Virgil Bodenhamer: 207

Bill Perry/Gannett News Service: 158, 170, 171 (bottom)

Karl Pittelkau/Warren Sentinel: 217

Nancy Rigg: 168, 174 (both), 175, 176 (all)

Annie Wells/Press Democrat: 129

Robert Gauthier/San Diego Union Tribune: 157

Nelvin Cepeda/San Diego Union Tribune: 171 (top)

Michael Patrick/Knoxville News-Sentinel: 163

Chris Moore: 22

HSUS/Dantzler: 219

John Davenport/San Antonio Express-News: 220

Todd Feeback/Kansas City Star: 222

Glenn Hartong/The Cincinnatti Enquirer: 223, 224 (top right)

Randy Pench/Sacramento Bee: 225

Erhard Krause/Sacramento Bee: 226

All photos not credited are by Slim Ray

Illustrations:

Pandra Williams/J.E. Weinel: 59, 80 (bottom), 81 (both), 209 (inset), 211 (both)

Horst Fürsattel/HF Kajaksport: 36 (top & bottom right), 105, 122

William Nealey: 7, 103

Jan AtLee: 89 (bottom)

PMI/ Petzl: 147

Ken Phillips: 208

The illustrations on pages 114, 125, 167, 168, and 175 are from sketches by Jim Segerstrom. The graphs and map on page 201 (the Ft. Collins flood) are based on data provided by John Weaver.

The Far Side cartoon by Gary Larson on page 137 is reprinted by permission of Chronicle Features, San Francisco, CA. All rights reserved.

All illustrations not credited are by Slim Ray and Jan Miles

About the Author

Slim Ray is an internationally-recognized authority on flood, swiftwater and white-water safety and rescue with over fifteen years experience in swiftwater rescue, including course development and instruction with Rescue 3, Canyonlands Field Institute, and the Nantahala Outdoor Center. He co-authored *RIVER RESCUE* with Les Bechdel and has written numerous articles on the subject. He is an active recreational kayaker and has worked as a raft guide and canoe and kayak instructor both in the US and abroad. He has been the American representative in several international river safety conferences and organized the International Safety Symposium (ISS '90) in the US.

How to Contact the Author

Slim Ray provides consulting services and instruction for agencies and individuals worldwide for flood and swiftwater rescue, team organization and training, and other matters relating to these subjects. He encourages readers to contact him regarding new ideas and techniques as well as any errors or omissions in this book. Anyone having newspaper clips, videos, or photographs of flood or swiftwater rescues is encouraged to send a copy to the author at:

CFS Press
8 Pelham Road
Asheville, NC 28803
www.cfspress.com
slimray@cfspress.com

Books by Slim Ray

RIVER RESCUE (with Les Bechdel)

THE CANOE HANDBOOK

ANIMAL RESCUE IN FLOOD AND SWIFTWATER INCIDENTS

SWIFTWATER RESCUE FIELD GUIDE